Soil Pollution: Impacts and Solutions

Soil Pollution: Impacts and Solutions

Editor: Jacey Murray

Callisto Reference
www.callistoreference.com

Callisto Reference,
118-35 Queens Blvd., Suite 400,
Forest Hills, NY 11375, USA

Visit us on the World Wide Web at:
www.callistoreference.com

ISBN: 978-1-64116-186-2 (Hardback)

The publisher's policy is to use permanent paper from mills that operate a sustainable forestry policy. Furthermore, the publisher ensures that the text paper and cover boards used have met acceptable environmental accreditation standards.

Printed in China.

Cataloging-in-Publication Data

Soil pollution : impacts and solutions / edited by Jacey Murray.
p. cm.
Includes bibliographical references and index.
ISBN 978-1-64116-186-2
1. Soil pollution. 2. Soil protection. 3. Soils--Environmental aspects.
I. Murray, Jacey.
TD878 .S65 2019
628.55--dc23

Table of Contents

Preface VII

Chapter 1 **Contamination of Soils and Substrates in Horticulture** 1
Wlodzimierz Breś and Barbara Politycka

Chapter 2 **Cyanobacterial Toxins Emerging Contaminants in Soils: A Review of Sources, Fate and Impacts on Ecosystems, Plants and Animal and Human Health** 20
Noureddine Bouaïcha and Sylvain Corbel

Chapter 3 **Copper Contamination in Mediterranean Agricultural Soils: Soil Quality Standards and Adequate Soil Management Practices for Horticultural Crops** 42
Daniel Sacristán and Ester Carbó

Chapter 4 **Treated Municipal Wastes: Are they Contaminating or Enriching the Soil?** 63
Ahmed Al-Busaidi and Mushtaque Ahmed

Chapter 5 **Edge of Field Technology to Eliminate Nutrient Transport from Croplands: Specific Focus on Denitrification Bioreactors** 83
Michael Aide, Indi Braden and Sven Svenson

Chapter 6 **The Molecular-Based Methods used for Studying Bacterial Diversity in Soils Contaminated with PAHs** 102
Anna Gałązka and Jarosław Grządziel

Chapter 7 **Simulation of Phosphorus Transport in Soil under Municipal Wastewater Application using Hydrus-1D** 122
Ali Erfani Agah, Patrick Meire and Eric de Deckere

Chapter 8 **Dual Soil Decontamination Procedures** 136
Yoshiharu Mitoma, Alina M. Simion and Cristian Simion

Chapter 9 **Soil Salinization and Mitigation Measures in Land Reclamation Regions** 155
Shiguo Xu, Yi Xu, Yanzhao Fu and Qi Wang

Chapter 10 **Radioactive Contamination of the Soil: Assessments of Pollutants Mobility with Implication to Remediation Strategies** 172
Ivana Smičiklas and Marija Šljivić-Ivanović

Chapter 11 **Mycoremediation of Atrazine in a Contaminated Clay-Loam Soil and its Adsorption-Desorption Kinetic Parameters**........ 196
Wilberth Chan Cupul and Refugio Rodríguez Vázquez

Chapter 12 ***Ex Situ* Surfactant-Enhanced Bioremediation of NAPL-Impacted Vadose Zone**........ 211
Roger Saint-Fort

Chapter 13 **The Electrokinetic Treatment of Polluted Soil by Hydrocarbon: From Laboratory to Field**........ 232
Maribel Pérez-Corona, Zsuzsanna Plank and Erika Bustos

Permissions

List of Contributors

Index

Preface

Every book is a source of knowledge and this one is no exception. The idea that led to the conceptualization of this book was the fact that the world is advancing rapidly; which makes it crucial to document the progress in every field. I am aware that a lot of data is already available, yet, there is a lot more to learn. Hence, I accepted the responsibility of editing this book and contributing my knowledge to the community.

An alteration in the natural soil environment, which is caused by the presence of xenobiotic chemicals or other factors, is known as soil pollution. The most common reasons for the contamination and degradation of soil are improper disposal of waste, use of agrochemicals, industrial activity, oil spills, intensive farming, acid rain, etc. Some of the chemicals responsible for the degradation of soil are polynuclear aromatic hydrocarbons, heavy metals, pesticides and petroleum hydrocarbons, besides others. The effects of soil contamination are not limited to animals but also affect the environment as a whole. Changes in plant metabolism, reduction in crop yields, soil erosion, alteration in the metabolism of microorganisms residing in a soil environment, etc. are some of the primary impacts of soil pollution. Bioremediation, mycoremediation, aeration and phytoremediation are some of the chief remediation strategies that can be used as clean-up interventions. This book provides significant information to help develop a good understanding of soil pollution and its impacts. It includes some of the vital pieces of work being conducted across the world, on various topics related to mitigation solutions and strategies. It will serve as a valuable source of reference for graduate and postgraduate students, as well as experts.

While editing this book, I had multiple visions for it. Then I finally narrowed down to make every chapter a sole standing text explaining a particular topic, so that they can be used independently. However, the umbrella subject sinews them into a common theme. This makes the book a unique platform of knowledge.

I would like to give the major credit of this book to the experts from every corner of the world, who took the time to share their expertise with us. Also, I owe the completion of this book to the never-ending support of my family, who supported me throughout the project.

Editor

1

Contamination of Soils and Substrates in Horticulture

Wlodzimierz Breś and Barbara Politycka

Additional information is available at the end of the chapter

Abstract

Contamination of the soil environment mostly is identified with industry, especially mining and road transport. Unfortunately, also in the commercial horticulture, there are numerous problems concerning the contamination of soils and substrates. Sources of contamination can be fertilizers and waste materials polluted by heavy metals, particularly by cadmium. In the greenhouses where traditional methods of cultivation are used, the soil pollution due to the application of excessively high doses of fertilizers constitutes an environmental hazard. Much faster similar effect occurs in greenhouses where an open system of fertigation is used. In addition to mineral impurities, organic compounds emitted by the plant or that are formed during decomposition of organic matter are the problem. This phenomenon is called allelopathy. In practice, it concerns the monoculture and perennial crops and especially is observed in nurseries, orchards, plantations of berries and asparagus. For this reason, in the later section, the soil sickness, replantation problem and toxicity of mulches in green areas are also discussed.

Keywords: overfertilization, heavy metals, allelochemicals, soil sickness, replantation

1. Introduction

Soil pollution is commonly associated with industry, mines and road transport. In the case of water contamination, the significant role of agriculture is primarily indicated. Horticulture can also adversely affect the soil environment. In this case, the basic problem is the use of very high doses of fertilizers and the heavy metals contained in some fertilizers. Contaminants can also occur in waste materials used to improve the properties of soil and horticultural substrates. Besides mineral impurities, toxic organic substances, which are metabolites of plants and micro-organisms or substances of anthropogenic origin used for the control of pests, pathogens and weeds can be released into soil and substrates. The ability to suppress other plants through the release of toxic substances from living parts or dead plant tissues during their decomposition is called allelopathy. Understanding of the causes and consequences of risks outlined above determines for rational decision-making in horticulture.

2. Heavy metals in fertilizers and waste materials

Horticulture is the branch of agriculture dealing mainly with the cultivation of vegetables, medicinal plants, ornamental plants and fruit trees and bushes. Horticultural plants are an important part of the human diet. For this reason, attention is paid to factors affecting the quantity and quality of yield.

Crop yield depends on many factors including variety, control of diseases and insects, and weather conditions. However, the essential role is played by the physical and chemical properties of the soil or growing medium. To improve soil quality, farmers use organic and mineral fertilizers. Unfortunately, fertilizers can be contaminated by substances that can potentially pose a risk to human and animal health and the environment. In the case of mineral fertilizers, this problem concerns mainly cadmium compounds.

The presence of cadmium in topsoil is a consequence of the use of phosphate fertilizers contaminated with this element [1–4]. Cadmium uptake by plants depends on pH of soil or growing medium. Under acidic conditions, cadmium solubility increases. In these conditions, the adsorption of cadmium by soil colloids, hydrous oxides, and organic matter is very low. However, zinc can reduce cadmium's availability to plants, by inhibiting calcium uptake and preventing it from moving from the roots to the shoots of the plants [5]. Lime fertilizers, as well as waste materials rich in calcium and/or magnesium, can also be contaminated by heavy metals [6–8]. Moreover, in this case, the list of potentially toxic elements is much longer: Cd, Cr, Co, Cu, Pb, Mo, Ni, Zn, As and Hg [8–11]. Essential and beneficial elements can become toxic only at high concentrations. In many countries, the use of fertilizers or waste materials contaminated with heavy metals is limited by the introduction of a maximum permissible content of these elements. However, the rules of individual countries are not unified [4]. For example, in Poland, the maximum permissible concentrations of heavy metals in fertilizers are as follows:

- in organic and organic mineral fertilizer (in 1 kg of dry weight of the fertilizer): Cr—100 mg, Cd—5 mg, Ni—60 mg, Pb—140 mg, Hg—2 mg,
- in agricultural lime (expressed per 1 kg CaO): Cd—8 mg, Pb—200 mg,
- in agricultural lime containing magnesium (calculated per 1 kg of the sum CaO + MgO): Cd —15 mg, Pb—600 mg, and
- in other mineral fertilizers (in 1 kg of dry weight of the fertilizer): As—50 mg, Cd—50 mg, Pb—140 mg, Hg—2 mg [12].

Heavy metals may be introduced into the soil and substrates also with soil improvers or mulches. 'Soil improver' is defined as a material added to soil in situ whose main function is to maintain or improve its physical and/or chemical and/or biological properties, with the exception of liming materials. 'Mulch' means a type of soil improver used as protective covering placed around plants on the topsoil whose specific functions are to prevent the loss of moisture, control weed growth and reduce soil erosion. According to a regulation of the European Union, the maximum content of heavy metals in the final product or constituent may not exceed the values shown in **Table 1**.

Element	Maximum content in the product (mg kg dw)
Cadmium (Cd)	1
Chromium total (Cr)	100
Copper (Cu)	100
Mercury (Hg)	1
Nickel (Ni)	50
Lead (Pb)	100
Zinc (Zn)	300

Table 1. Heavy metals limits for soil improvers, mulch and organic constituents of growing media [13].

In organic and mineral growing media, the content of heavy metals in the final product may not exceed the values shown in **Table 2**.

Element	Maximum content in the product (mg kg dw)
Cadmium (Cd)	3
Chromium total (Cr)	150
Copper (Cu)	100
Mercury (Hg)	1
Nickel (Ni)	90
Lead (Pb)	150
Zinc (Zn)	300

Table 2. Heavy metal limits for growing media, including mineral growing media [13].

The source of heavy metals may also be sewage sludge from municipal sewage treatment plants used to fertilize soil or compost from sewage sludge. The use of these materials in the EU is subject to a number of strict requirements. The most important are the Water Framework Directive 2000/60/EC on water protection, Directive 91/271/EEC on urban waste water treatment, Directive 96/61/EC concerning integrated pollution prevention and control, Directive 99/31/EC on the landfill of waste and Directive 86/278/EEC on the use of sludge in agriculture [13–17]. The limit values for heavy metals in sludge or in composts are defined in national regulations. The regulatory framework prevents harmful effects on soil, vegetation, animals and humans [18–20].

3. Effect of intensive fertilization on chemical composition of soil in greenhouse

3.1. Effect of long-term traditional fertilization

Traditional cultivation of plants in greenhouses or plastic tunnels is based on intensive organic and mineral fertilization of soil. Manure and compost are commonly used organic fertilizers. In temperate climate of central Europe, the cultivation of plants in greenhouses and tunnels is uneconomic due to short days and low light intensity as well as high heating costs from November to March. The gardening season begins in early spring and ends in late autumn. In this relatively short period, intensive fertilization is carried out. The doses of fertilizers used in greenhouses and plastic tunnels are much higher than the doses used in field crops. For example, for wheat, 230–360 kg NPK/ha is recommended, while for early varieties of cauliflower grown in the greenhouse, 450–580 kg NPK/ha is recommended. Moreover, due to the greenhouse effect, the average day and night temperatures in greenhouses and tunnels are significantly higher than the temperatures in the field. Plants grow faster and produce greater biomass. For this reason, the watering of plants is more intense, and therefore, the elution of components into the soil is stronger. A detailed documentation of this problem was presented by Breś and Roszyk [21, 22]. The authors selected five horticultural farms near Poznan (Poland) in which the plants were grown for 20–40 years. In the middle of the growing season, the authors took soil samples from the layers 0–20, 20–40, 40–60, 60–80, 80–100 and 100–120 cm. For the sake of comparison, the studies also included samples taken near the greenhouse from occasionally fertilized lawn. To evaluate the effect of long-term fertilization on the distribution of nutrients in the profile of soils, chemical analysis of samples was performed. For nutrient extraction, 0.03 M CH_3COOH was used. This method allows one to assess the amount of components readily available for plants. As an example, the content of N-NO_3, P, K, Ca, Mg, Cl and S–SO_4 in soil samples collected in two of the five test farms is given below. In **Table 3**, data refer to a greenhouse where vegetables and ornamental plants were grown for 40 years, while **Table 4** presents the results of analyses of soil samples from a greenhouse in which for 40 years only vegetables were cultivated. Most of the nitrogen, phosphorus, calcium, magnesium, chlorides and sulphates were found in a layer 0–40 cm deep. In extreme cases, the greenhouse in soil nitrogen content was 60 times, phosphorus 3 times and potassium 15 times higher compared to the soil next to the greenhouse. Greenhouse soils were very rich, even at a depth of 80 cm. The significant amount of sulphates in the soil in greenhouses is a result of more frequent use of potassium sulphate than potassium chloride. This practice is very common in horticulture. Based on the scale of pollution, it can be assumed that in these farms, the evaluation of fertilization requirements based on the chemical analysis of soil or substrate was conducted infrequently or not at all. The authors found that the range of changes in the chemical properties of the investigated soils depended most on the length of greenhouse utilization. Moreover, the soil of the farms where ornamental plants were grown exclusively contains more nutrients than the soil from farms specializing in the cultivation of vegetables. Soil texture had the least impact on the chemical composition of soils. Similar trends were observed by examining the content of micronutrients. The results of these studies clearly indicated strong leaching of nutrients and the threat of groundwater contamination. The soil contamination in the greenhouse reported in this study was so high that it became necessary to rapidly introduce new technologies friendly for the environment. As a method to reduce leaching of nutrients, wider use of slow-release, controlled-release and inhibitor-stabilized fertilizers was proposed. Another solution to the problem was soilless cultures and fertigation.

Layer of soil (cm)	$N–NO_3$	P	K	Ca	Mg	Cl	$S–SO_4$
	Content in the soil (mg/dm^3)						
Farm Ogrody—greenhouse							
0–20	314	248	491	4013	256	306	497
20–40	297	244	484	4600	230	346	342
40–60	77	261	376	1123	106	224	94
60–80	76	231	517	1322	12	215	69
80–100	65	152	676	752	136	93	55
100–120	89	138	586	556	108	151	93
Farm Ogrody—lawn							
0–20	5	77	29	2408	99	22	0
20–40	4	78	16	2624	111	21	0
40–60	5	89	12	2949	102	23	1
60–80	4	80	12	2793	94	20	0
80–100	4	71	16	2255	91	22	0
100–120	3	70	14	2140	80	20	0

Table 3. Effect of long-term fertilization on the distribution of nutrients in profile of greenhouse soil—farm Ogrody [21].

3.2. Soilless culture and fertigation

Soilless culture is the cultivation of plants in systems other than soil in situ, including hydroponics and another growing media or substrates. The main advantage of soilless culture is a pathogen-free root environment at the beginning of the crop cycle. Thanks to that fact, one can avoid costly and time-consuming soil replacement or sterilization [23]. An essential element of this technology is fertigation, that is the process in which fertilizers are applied with the irrigation. Fertigation can be carried out in an open or closed system. In the open system, an excess of the applied nutrient solution leaks into the soil. In the closed system, the

Layer of soil (cm)	$N–NO_3$	P	K	Ca	Mg	Cl	$S–SO_4$
	Content in the soil (mg/dm^3)						
Farm Marcelin—greenhouse							
0–20	159	255	309	2445	229	151	891
20–40	111	238	326	1379	160	126	779
40–60	90	160	431	728	95	56	284
60–80	69	99	541	1463	125	45	441
80–100	79	97	420	1491	91	33	296
100–120	49	66	476	2599	78	54	149
Farm Marcelin—lawn							
0–20	13	54	155	3101	77	124	84
20–40	9	71	103	1493	87	105	141
40–60	13	54	125	2340	72	80	43
60–80	14	50	110	1538	61	57	7
80–100	12	34	102	1103	49	49	40
100–120	11	28	119	1062	57	47	38

Table 4. Effect of long-term fertilization on the distribution of nutrients in profile of greenhouse soil—farm Marcelin [21].

excess of nutrient solution after disinfection returns to the fertigation system (recirculation of nutrient solution). In this system, drainage water does not contaminate the environment [24]. Fertigation would also provide less water and fertilizer utilization. In soilless cultures as a growing medium expanded clay aggregates, growstones, perlite, pumice, sand and wood fibre are used. However, the most commonly used substrates in soilless cultures are rockwool and coconut fibres. The described cultivation technology requires high-quality water and very good water-soluble fertilizers [25, 26]. According to the recommendations, in order to stabilize the concentration and the pH value of the solution in the root zone and in order to adjust the substrate moisture, the volume of nutrient solution must be higher than the nutritional requirements of plants [27]. For most soilless cultures, 30–50% overflow is recommended [28]. As an effect of open systems, the excess nutrient solution leaks from the growing medium and pollutes the soil. This process was documented by Breś [25]. The author measured the volume of leaking solution and analysed the chemical composition of leakage during the growth of cherry tomato in coconut fibre, as well as gerbera, rose, tomato and cucumber growing in rockwool. Concentrations of nutrients found in the drainage from soilless cultures were many times higher than the mean concentrations of components in the nutrient solution supplied to plants. This suggests that the basic cause of the increase in ion concentrations is a predominance of transpiration over nutrient uptake by plants [29]. The monthly deposition of elements transferred with drainage waters to the soil was also calculated. Some details from the publications of Breś [25] are given in **Table 5**. Notable is deposition of K (up to 413 kg/month/ha), $N–NO_3$ (up to 230 kg/month/ha), Ca (up to 220 kg/month/ha) and $S–SO_4$ (up to 101 kg/month/ha). Leaching of Na (up to 62 kg/month/ha) and Cl (up to 34 kg/month/

Nutrient	Tomato in rockwool	Cherry tomato in coconut fibres	Rose in rockwool
$N–NO_3$	30–230	23–177	13–83
P	7–54	2–18	3–16
K	53–413	36–282	17–106
Ca	23–178	28–220	9–54
Mg	7–57	5–38	3–21
Na	4–33	8–62	1–6
Cl	1–10	4–30	0.1–0.3
$S–SO_4$	13–101	12–90	4–24

Table 5. Ranges (kg/ha) of monthly losses of nutrients during plant cultivation in soilless culture with the application of open fertigation systems [25].

ha) was lower. A similar trend was found for *Anthurium* grown in expanded clay aggregates [29]. Some authors believe that the ratio of the uptake rates of NO_3, K and P, in comparison with the transpiration rate, decreased from May to September because the substrate temperature had a greater effect on nutrient uptake than on water absorption [30].

In research conducted by Uronen [31] during the cultivation of cucumbers grown in rockwool, phosphorus leakage was 35–47% while nitrate leakage amounted to 33–43% of the applied nutrients. Cultivation in organic substrates is characterized by a smaller run-off than in rockwool [25, 31]. Thus environmental pollution is reduced. The amount of nutrients leaking from 1 ha of agricultural field crops is distinctly smaller. For example, nitrogen seldom exceeds 140 kg N/ha/year [32, 33].

Besides the amount of fertilizers leaking from open fertigation systems, the vertical distribution of nutrients accumulating in the soil profile (mean content in subsequent soil layer), in relation to the duration of greenhouse operation, is also important. Such investigations were conducted in the years 2004–2011 in horticultural farms specializing in soilless plant cultivation [34]. The greenhouses were located in the Wielkopolska province (Poland). Every year, from February to November tomatoes were grown in rockwool. Before the first crop culture, soil samples were collected for chemical analyses at every 20 cm layer to the depth of 1 m. Successive samples were taken in autumn after the completion of 1, 2, 3 and 7 growing cycles. For nutrient extraction from soil, 0.03 M CH_3COOH was used. The amount of components readily available for plants was determined. Significant changes in the chemical properties of soils were detectable already after the first growth cycle of plants. **Figure 1** shows the dynamics of changes in electrical conductivity measured in soil layers. The degradation rate of the soil environment as a result of application of an open fertigation system depended primarily on the duration of greenhouse operation. The increase of nutrient contents in the soil profile during seven years of monitoring was very high: Ca 283%, Mg 325%, $N–NO_3$ 326%, K 666%, P 684% and $S–SO_4$ 2164%. Once again, it proved that the previously reported benefits of fertigation apply only for recirculating systems. Only in closed systems, it is possible to reduce water consumption by 15–35% and to limit losses of nutrient solution by 15–67% [35, 36].

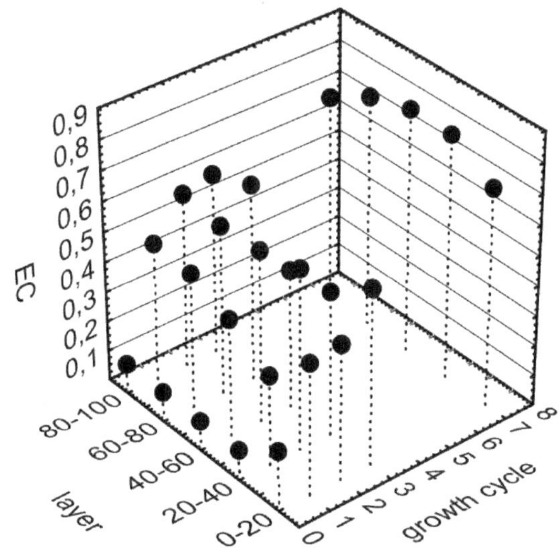

Figure 1. Relationship between duration of greenhouse operation (0—before first growth season and after 1, 2, 3...7 growth cycles), depth of soil sampling (cm) and electrical conductivity (EC mS/cm).

4. Organic contaminants released from plant residues

4.1. Post-harvest residues

There are many plant species that possess the ability to suppress other plants through the release of toxic substances from living parts or dead plant tissues. This phenomenon is called allelopathy. Allelopathy is a chemical interaction between plants defined as any direct or indirect, beneficial or harmful effects of one plant (donor plant) on another (recipient plant) through the production of chemical compounds that are released into the environment through root exudation, leaching, volatilization and decomposition of plant residues. A wide variety of phytotoxic substances exists in plant residues. Microbial decay of plant residues releases the toxic metabolites into the soil where they may adversely affect the growth and development of plants. In agro-ecosystems, decaying post-harvest residues are the main source of phytotoxic compounds, and they can provide a serious problem [37].

Allelopathic chemicals are generally secondary metabolites, and most of them have been identified as volatile terpenes and phenolic compounds [38]. Allelochemicals can be synthesized in every part of the plant. They can be found in seeds, flowers, fruits, pollen, leaves, stems and roots. Their content depends on the developmental stage of the plant or plant part. It was found that significantly larger amounts of them occur in young plants [39]. Different stress factors can enhance the production and release of allelochemicals by plants [40].

Some plant species with a high allelopathic potential release into the environment particularly high amounts of allelopathic compounds. These include crop plants from the families Fabaceae and Brassicaceae. Perennial crops and monocultures of these families are common in many parts of the world, and they cause a number of problems due to soil sickness, regeneration failure and replant problems. Allelochemicals from legumes are mainly polyphenols and propanoids [41]. Crops from the family Brassicaceae contain compounds called glucosinolates, which break down during the decomposition of post-harvest residues into powerful volatile allelochemicals—isothiocyanates, which can affect plant growth and microbial activity [42–44]. Also, plants belonging to the group of the world's worst weeds displaying great expansion and invasiveness properties such as quackgrass (*Agropyron repens*), Canada thistle (*Cirsium arvense*), field bindweed (*Convolvulus arvensis*), white pigweed (*Chenopodium album*) and Johnson grass (*Sorghum halepense*) exhibit high allelopathic potential [45, 46]. On the other hand, the weed suppressive ability of crop plants with allelopathic properties may also be considered as plant weed control in agricultural systems [47]. The use of allelopathic cover crops, inclusion of allelopathic plants in crop rotation and the use of their residues as mulches can be an economical and environmentally friendly form of weed control [48].

Allelopathic chemicals act in many ways. Some retard plant growth or inhibit seed germination by disrupting cell division. Some interfere with respiration and other physiological process. Many affect plant nutrition by reducing the water and nutrient uptake. Biological activity of phytotoxic substances depends on their chemical nature and concentration—at lower concentrations, they may exert stimulatory effects, whereas at higher concentrations, they may exert inhibitory effects [49].

The decomposition of crop residues is the result of complex microbial processes controlled by numerous environmental factors influencing the activity of microflora such as temperature, moisture, aeration, inorganic ions and pH [50, 51]. Allelochemicals released into the soil are also continuously removed from the soil solution by plant uptake, immobilized due to adsorption to soil particles and degraded by micro-organisms [52–55]. Moreover, allelopathic compounds are subjected to degradation by oxidation and photolysis as well as processes of removal by volatilization or leaching [53]. The type of soil is important in the accumulation of allelochemicals, for example, in poorly drained, clay soils, the allelochemicals are not leached easily. By contrast, in well-drained sandy soils, the allelochemicals have a tendency to leach. The difference between the speed of allelochemicals' release into the environment and the speed of their degradation will decide whether they will accumulate in the soil to a toxic level [49]. A low concentration of allelochemicals at a given point in time is not an argument against their allelopathic role or evidence of their activity at low concentrations, because the allelopathic effects depend on many factors interacting with them in the soil and may not be directly related to the actual concentrations. Soil factors and their interactions with microflora need to be considered in assessing the factors that determine the presence and stability of allelochemicals [56–58].

4.2. Soil sickness and replantation problem

The phenomenon of soil sickness is defined as a decrease in soil fertility as a result of the prolonged growth of the same plant species, in spite of its intensive cultivation and fertilization. Delayed development of plants and a significant reduction in yield are symptoms of soil sickness. It is widely assumed that soil sickness is a phenomenon caused by a complex combination of biotic and abiotic factors disturbing the biological balance in soil, that is deficiencies or imbalance of plant nutrients, degradation of soil properties, disproportionate development of various groups of micro-organisms in soil, increased infestation of pathogens, pests and weeds and accumulation of phytotoxic compounds [59]. The intensive modern agriculture with mechanization, indiscriminate use of fertilizers and pesticides and with an emphasis on reduced crop diversity has led to serious changes in the physical, chemical and biological properties of soil, which have adversely influenced plant development and crop yields. Soil sickness in modern agriculture is mainly due to specialized single crop based limited rotations. These systems do not follow the scientific principles of crop rotations. In horticulture, soil sickness concerns mainly monoculture and perennial crops with limited rotation, such as nurseries, orchards, plantations of berries and asparagus, lawns as well as greenhouse cultivations, where the same substrate is used many times [54, 60–62]. One of the main causes of soil sickness is the accumulation of phytotoxic compounds, that is plant and microbial phytotoxins, as well as remains of pesticides [59].

As a result of long-term growth of the same plant species, there occurs in the soil accumulation of homogeneous compounds secreted from plants and the products of microbial decomposition of plant post-harvest residues. The living plants can secrete allelochemicals and the decaying plant residues can release toxic metabolites into the soil. In soil sickness, the release of toxic substances from the dead plant tissues during their decomposition plays a greater role than their active secretion from the living plants. A specific kind of soil sickness is autotoxicity, which manifests when a plant species releases chemical substances that inhibit or delay the germination and growth of the same plant species. Many crop plants exhibit autotoxicity, i.e. self-destruction of a plant species through the production of metabolites that escape into the environment and directly inhibit the growth of that species [63]. Autotoxicity is a cause of soil sickness in the cropping of such vegetables as asparagus, carrot, cucumber, eggplant, pea and tomato [64–66]. This phenomenon is also observed in orchards and then is called the replantation problem. Cutting down an old, non-productive orchard and establishing a new one in the same place is associated with the replantation problem. It occurs most frequently in apple, peach, sour cherry and sweet cherry orchards. When an old orchard is removed, large amounts of root residues remain in the soil. They are a rich source of phytotoxic substances. For example, peach root bark contains two glycosides—amygdalin and prunasin—that under enzymatic hydrolysis in soil produce hydrogen cyanide, a powerful inhibitor of respiration [67]. The main cause of soil sickness in apple orchards is accumulation of the toxic dihydrochalcone—phlorizin, large amounts of which occur in the bark of apple roots. The release into the soil of these compounds from the decaying residues of tree roots after the liquidation of old trees prevents the normal growth of young trees in the replanted orchard [68].

Monoculture and perennial crops with limited rotation favour the proliferation of pathogenic fungi, which produce mycotoxins. *Aspergillus*, *Penicillium* and *Fusarium* are the major fungal genera producing secondary metabolites toxic not only to humans and animals but also to plants [54, 69]. The phytotoxic activity of mycotoxins manifests in their inhibitory effects on growth parameters and differs from their effects in plant diseases [69].

Pesticides are toxic chemicals used to control weeds, pests and pathogens in crops. It is normal practice to apply several different pesticides to a single crop in any given growing season. In intensive agriculture, the application of pesticides is frequently inappropriate or excessive. Although each pesticide is meant to kill a certain pest, pathogen or weed, a very large percentage of pesticides reach other destinations than their target. Instead, they enter the air, water and soil [70]. Some of these pesticides or their remains can act as toxins to plants when found in soil at sufficient concentrations. Accumulation refers to the build-up of pesticides resulting from repeated use. Excessive use of pesticides is one of the main factors causing soil pollution and can lead to several unintended, harmful effects on the environment, adversely affecting the soil micro-organisms and generally causing a decrease of soil fertility. The toxicity level of a pesticide depends on the kind of chemical, the dose, the length of exposure and the route of entry or absorption by the plant. The accumulation of pesticides in the soil can kill or reduce the populations of essential soil macro- and micro-organisms, including earthworms, insects, spiders, mites, fungi and bacteria, thus reducing or stopping important nutrient cycles [71, 72]. The fate of pesticides in soils varies greatly depending on their chemical nature, the type of soil, the climate conditions and the agricultural practices. In the soil, they are decomposed by soil micro-organisms, leached from the root zone, or they are adsorbed and accumulated by soil particles [73]. The amount of pesticide adsorbed to the soil varies with the type of pesticide, soil moisture, pH and texture. Pesticides are strongly adsorbed to soils that are rich in clay or organic matter, whereas they are not as strongly adsorbed to sandy soils. Pesticide degradation in soil generally results in a reduction in toxicity; however, breakdown products of some pesticides are sometimes more toxic than the substrate. Plant injury can be a problem resulting from adsorption of pesticides to soil particles. Injury can result when a pesticide used for one crop is later released from the soil particles in amounts great enough to cause injury to a sensitive rotational crop. It is also hard to predict the long-term effects of such changes in the soil microbial communities, which may lead to the occurrence of soil-borne pathogens [73].

4.3. Toxicity of mulches in green areas

Mulching is a popular form of soil care, especially in green areas. A mulch is a layer of material applied to the surface of soil. It limits weeding, improves soil moisture, stabilizes soil temperature, reduces soil compaction and increases soil nutrition, which indirectly contribute to better plant growth. For the preparation of mulches, various organic and inorganic materials are used. Natural materials such as bark, sawdust, straw, shredded or chipped wood, leaves, coniferous needles or dried grass clippings are used as organic mulches. Plant residues from a crop may also be used to form a mulch [43, 47]. However, most of these materials are not suitable in green belts because of poor aesthetic appeal [74].

Although mulches are multifunctional and in green areas, they are applied mainly for aesthetic purposes, mulching is one of the most effective methods for non-herbicide weed control [75]. Mulches can act only as a physical barrier that limits access of light to germinated weeds and reduces their ability to photosynthesis. Certain organic materials, especially shredded and chipped bark or wood, may control weeds chemically through the leaching of allelopathic compounds. Bark and wood mulches are often used for weed suppression in urban landscapes and gardens where herbicides are prohibited or unwanted [74]. Biological activity of phytotoxic substances depends on their chemical nature and the tree species from which they are derived. The results obtained by Rathinasabapathi and co-workers [76] showed the phytotoxic activity of wood chips from deciduous trees and conifers (*Acer rubrum*, *Quercus michauxii*, *Juniperus silicicola*, *Azadirachta indica* and *Magnolia grandiflora*).

Most commonly, the branches of various tree species are used as mulch material, fresh and without composting, because composting is a time- and cost-consuming process. Thus, the use of these wood wastes for the preparation of mulches is a simple way of recycling them. However, although the wood chips are easy to obtain and one of the cheapest organic materials for mulching, especially in green areas, their application may be associated with the release into the soil of phytotoxic substances. The use of wood chips for mulching the soil contributed to an increase in the content of phenolic compounds [77]. It was found that the strongly lignified wood wastes decomposed in the soil by micro-organisms are a rich source of phenolic compounds, even small amounts of which may adversely affect the growth and development of plants [77, 78]. According to Krasutsky [79], the bark of *Betula pendula* contains large amounts of polar triterpenes—betulin, betulinic acid and lupeol. Phytotoxicity of these compounds has been shown in numerous biological assays [80].

In recent years, interest has grown in mulches from a variety of wood wastes, which are crushed and coloured. Wood chips are durable and easy to use as an organic material for mulching. Their sources are sawmill wastes and wastes arising from logging or cutting trees and shrubs [81]. Sometimes processed wood is also used, for example manufactured product debris, discarded pallets and wood reclaimed from constructions and demolitions [82]. Depending on the source of the wood chips, they may contain toxic chemicals, which pollute soil and ground water. It has been found that some of the recycled waste wood used for making landscape mulch products is contaminated with various chemicals, such as creosote, chromium copper arsenate or lead-based paints used for wood preservation against fungi and insects [83–85].

Some problems can develop when hardwood bark is stored in overlarge or waterlogged piles, which creates anaerobic conditions. Then, anaerobic micro-organisms carry out fermentation and in the pile such products as acetic acid, methanol, ammonia and hydrogen sulphide accumulate. Application of such bark as mulch can cause direct plant injury. Damage symptoms including leaf scorch, bleached leaves and defoliation occur very quickly, and in the case of sensitive herbaceous plants, even plant death may occur [86].

Author details

Wlodzimierz Breś[1*] and Barbara Politycka[2]

*Address all correspondence to: wbnaw@up.poznan.pl

1 Department of Plant Nutrition, Poznań University of Life Sciences, Poznań, Poland

2 Department of Plant Physiology, Poznań University of Life Sciences, Poznań, Poland

References

[1] Mortvedt JJ.: Heavy metal contaminants in inorganic and organic fertilizers. Fertilizer Research. 1995;439(1):55–61.

[2] Lugon-Moulin N., Ryan L., Donini P., Rossi L.: Cadmium content of phosphate fertilizers used for tobacco production. Agronomy for Sustainable Development. 2006;26:151–155. doi:10.1051/agro:2006010

[3] Cheraghi M., Lorestani B., Merrikhpour H.: Investigation of the effects of phosphate fertilizer application on the heavy metal content in agricultural soils with different cultivation patterns. Biological Trace Element Research. 2012;145:87–92. doi:10.1007/s12011-011-9161-3

[4] Gonçalves Jr AC., Nacke H., Schwantes D., Coelho GF.: Heavy metal contamination in Brazilian agricultural soils due to application of fertilizers. In: Soriano M.C.H., editor. Environmental Risk Assessment of Soil Contamination. InTech; 2014. p. 105–135. doi:dx.doi.10.5772/57268

[5] Soil Contamination: Impacts on human health. Report produced for the European Commission DG Environment [Internet]. 2013. Available from: http://ec.europa.eu/science-environment-policy [Accessed: 2016-04-27]

[6] Jasiewicz C., Antonkiewicz J.: Selected properties of furnace waste and carbide lime considering their potential reclamation and environmental management. Problems of Landscape Ecology. 2009;XXIV:157–162. (in Polish)

[7] Davies BE., Paveley CF., Wixson BG.: Use of limestone wastes from metal mining as agricultural lime: Potential heavy metal limitations. Soil Use and Management. 1993;9: 47–52. doi:10.1111/j.1475-2743.1993.tb00928.x

[8] Conceição FT., Navarro GRB., Silva AM.: Anthropogenic influences on Cd, Cr,Cu, Ni, Pb and Zn concentrations in soils and sediments in a watershed with sugar cane crops at São Paulo State, Brazil. International Journal of Environmental Research. 2013;7(3): 551–560.

[9] Breś W., Golcz A., Komosa A., Kozik E., Tyksinski W.: Nutrition of Horticultural Plants. Poznan: Poznan University of Live Sciences; 2009. 192 p. (in Polish)

[10] Gabe U., Rodella AA.: Trace elements in Brazilian agricultural limestones and mineral fertilizers. Communications in Soil Science and Plant Analysis 1999;30(5–6):605–620. doi:10.1080/00103629909370231

[11] Nicholson FA., Smith SR., Alloway BJ., Carlton-Smith C., Chambers BJ.: An inventory of heavy metals inputs to agricultural soils in England and Wales. Total Environment. 2003;311:205–219. doi:10.1016/S0048-9697(03)00139-6

[12] Regulation of the Minister of Agriculture and Rural Development of 18 June 2008. On the implementation of certain provisions of the Act on fertilizers and fertilization.

Journal of Laws. 2008;No. 119–6515 – Pos. 765. [Internet]. 2008. Available from: http://isap.sejm.gov.pl/Details Servlet?id=WDU20081190765 [Accessed: 2016-07-09] (in Polish)

[13] Commission Decision (EU) 2015/2099 of 18 November 2015 establishing the ecological criteria for the award of the EU Ecolabel for growing media, soil improvers and mulch (notified under document C(2015) 7891). Official Journal of the European Union. L 303. [Internet]. 2015. Available from: http://eur-lex.europa.eu/legal-content/EL/TXT/PDF/?uri=CELEX:32015D2099&from=EN [Accessed: 2016-07-09]

[14] Council Directive 96/61/EC of 24 September 1996 concerning Integrated Pollution Prevention and Control (IPPC). Official Journal. L 257. [Internet]. 1996. Available from: http://eur-lex.europa.eu/legal-content/EN/TXT/PDF/?uri=CELEX:31996L0061&from=CS [Accessed: 2016-07-09]

[15] Water Framework Directive (Directive 2000/60/EC of the European Parliament and of the Council of 23 October 2000 establishing a framework for Community action in the field of water policy). [Internet]. 2000. Available from: http://eur-lex.europa.eu/resource.html?uri=cellar:5c835afb-2ec6-4577-bdf8-756d3d694eeb.0004.02/DOC_1&format=PDF [Accessed: 2016-07-09]

[16] Council Directive 91/271/EEC concerning urban wastewater treatment. 21 May 1991. [Internet]. 1991. Available from: http://eur-lex.europa.eu/legal-content/EN/TXT/PDF/?uri=CELEX:31991L0271&from=EN [Accessed: 2016-07-09]

[17] Council Directive 1999/31/EC on the landfill of waste. [Internet] 1999. Available from: http://eur-lex.europa.eu/legal-content/EN/TXT/PDF/?uri=CELEX:31991L0271&from=EN [Accessed: 2016-07-09]

[18] Council Directive 86/278/EEC of 12 June 1986 on the protection of the environment, and in particular of the soil, when sewage sludge is used in agriculture. [Internet]. 1986. Available from: http://eur-lex.europa.eu/legal-content/EN/TXT/PDF/?uri=CELEX:31986L0278&from=EN [Accessed: 2016-07-09]

[19] Regulation of the Minister of the Environment of 6 February 2015. On municipal sewage sludge. Official Journal Warsaw, 25 February 2015. Pos. 257 [Internet]. 2015. Available from: http://dziennikustaw. gov.pl/du/2015/257/1 [Accessed: 2016-07-09] (in Polish)

[20] Sewage Sludge Ordinance of 15 April 1992 (BGBl. I S. 912), last amended by Article 74 of the Regulation of 31 August 2015 (BGBl. I S. 1474) has been changed. [Internet]. 2015. Available from:. https:// www.gesetze-im-internet.de/bundesrecht/abfkl_rv_1992/gesamt.pdf [Accessed: 2016-07-09] (in German)

[21] Breś W., Roszyk J.: Effect of long-term fertilization on the content of nutritional elements in the profiles of greenhouse soils. Bulletin of Vegetable Crops Research. 1996;XVL:51–57. (in Polish)

[22] Breś W., Roszyk J.: Effect of long-term fertilization on the content of trace elements in the profiles of greenhouse soils. Bulletin of Vegetable Crops Research. 1997;XLVI:25–30. (in Polish)

[23] Gruda N., Prasad M., Maher M. J.: Soilless Culture. Encyclopedia of Soil Science. New York: Marcel Dekker, Inc.; 2004. doi:10.1081/E-ESS-120028194

[24] Breś W.: Influence of soilless cultures on soil environment. Ecological Chemistry and Engineering (A). 2010;17(9):1069–1076.

[25] Breś W.: Estimation of nutrient losses from open fertigation systems to soil during horticultural plants cultivation. Polish Journal of Environmental Studies. 2009;18(3):41–345.

[26] Breś W., Kleiber T., Trelka T.: Quality of water used for drip irrigation and fertigation of horticultural plants. Folia Horticulture. 2010;22(2):67–74.

[27] Sonneveld C., Voogt W.: Nutrient solutions for soilless cultures. In: Plant Nutrition of Greenhouse Crops. Springer Science+Business Media B.V.; 2009. p.257–275. doi: 10.1007/978-90-481-2532-6

[28] De Pascale S., Paradiso R.: Water and nutrient uptake of roses growing in two inert media. Acta Horticulturae. 2001;548:631–640.

[29] Kleiber T.: Pollution of the natural environment in intensive cultures under greenhouses. Archives of environmental protection. 2012;38(2):45–53.

[30] Duchein M., Baille M., Baille A.: Water use efficiency and nutrient consumption of a greenhouse rose crop grown in rockwool. Acta Horticulturae. 1995;408:129–135.

[31] Uronen KR.: Leaching of nutrients and yield of tomato in peat and rockwool with open and closed system. Acta Horticulturae. 1995;401:443–449.

[32] Gasser MO., Laverdiere MR., Lagace R., Caron J.: Impact of potato-cereal rotation and slurry applications on nitrate leaching and nitrogen balance in sandy soils. Canadian Journal of Plant Science. 2002;82(4):469–479. doi:10.4141/S01-050

[33] Bronswijk JJB., Hamminga W., Oostindie K.: Rapid nutrient leaching to groundwater and surface water in clay soil areas. European Journal of Agronomy. 1995;4(4): 431–439.

[34] Breś W., Trelka T.: Effect of fertigation on soil pollution during greenhouse plant cultivation. Archives of Environmental Protection. 2015;41(2):75–81. doi:10.1515/aep-2015-0021

[35] Pardossi A., Malorgio F., Incrocci L., Tognoni F.: Hydroponic technologies for greenhouse crops. In: Dris R., editor. Crops: Quality, Growth and Biotechnology. Helsinki: WFL Publisher; 2006. p. 360–378.

[36] Meric MK., Tüzel IH., Tüzel Y., Oztekin GB.: Effects of nutrition systems and irrigation programs on tomato in soilless culture. Agricultural Water Management. 2001;99:19–25.

[37] Rice EL.: Allelopathy. New York: Academic Press; 1984.422 p.

[38] Li ZH., Wang Q., Ruan X., Pan CD., Jiang DA.: Phenolics and plant allelopathy. Molecules 2010;15:8933–8952. doi:10.3390/molecules15128933

[39] Haig T.: Allelochemicals in plants. In: Zeng RS., Mallik AU., Luo SM., editors. Allelopathy in Sustainable Agriculture and Forestry. New York: Springer; 2013. p. 63–104.

[40] Cseke LJ., Kaufman PB.: Regulation of metabolite synthesis in plants. In: Cseke LJ., Kirakosyan A., Kaufman PB., Warber S., Duke JA., Brielman H., editors. Natural Products from Plants. Boca Raton: CRC Press; 2006. p. 101–141.

[41] Mondal MF., Asaduzzaman M., Asao T.: Adverse effects of allelopathy from legume crops and its possible avoidance. American Journal of Plant Sciences. 2015;6:804–810. doi:10.4236/ajps.2015.66086.

[42] Al-Khatib K., Libbey C., Boydston R.: Weed suppression with Brassica green manure crops in green pea. Weed Science. 1997;45:439–445.

[43] Boydston R., Hang A.: Rapeseed (*Brassica napus*) green manure crop suppresses weeds in potato (*Solanum tuberosum*). Weed Technology. 1995;9:669–675.

[44] Uremis I., Arslan M., Uludag A., Sangun MK.: Allelopathic potentials of residues of 6 Brassica species on Johnsongrass [*Sorghum halepense* (L.) Pers.]. African Journal of Biotechnology. 2009;8:3497–3501.

[45] Friebe A., Schulz M., Kuck P., Schnabl H.: Phytotoxins from shoot extracts and root exudates of *Agropyron repens* seedlings. Phytochemistry 1995;38:1157–1159.

[46] Golubinova I., Ilieva A.: Allelopathic effects of water extracts of *Sorghum halepense* (L.) Pers., *Convolvulus arvensis* L. and *Cirsium arvense* Scop. on early seedling growth of some leguminous crops. Pesticides and Phytomedicine (Belgrade) 2014;29:35–43. doi: 10.2298/PIF1401035G

[47] Kunz Ch., Sturm DJ., Varnholt D., Walker DF., Gerhards R.: Allelopathic ability and weed suppressive ability of cover crops. Plant Soil Environment. 2016;62:60–66. doi: 10.17221/612/2015-PSE

[48] Jabran K., Mahajan G., Sardana V., Chauhan BS.: Allelopathy for weed control in agricultural systems. Crop Protection 2015;72:57–65. doi:10.1016/j.cropro.2015.03.004

[49] Seigler, DS.: Chemistry and mechanisms of allelopathic interactions. Agronomy Journal. 1996;88:876–885.

[50] Politycka B., Adamska D.: Release of phenolic compounds from apple residues decomposing in soil and the influence of temperature on their degradation. Polish Journal of Environmental Studies. 2003;12:95–98.

[51] Magdoff F., Weil RR.: Soil Organic Matter in Sustainable Agriculture. Boca Raton, USA: CRC; 2004. 398 p.

[52] Cheng HH.: Characterization of the mechanisms of allelopathy: Modeling and experimental approaches. In: Inderjit, KMMD., Einhellig FA., editors. Allelopathy: Organisms, Processes and Applications. ACS Symposium Series 585. Washington, D.C.: American Chemical Society; 1995. p. 132–141.

[53] Vidal RA., Bauman TT.: Fate allelochemicals in the soil. Ciencia Rural. 1997;27(2):351–357.

[54] Grodzinsky AM.: Allelopathy in Soil Sickness. Jodhpur, India: Scientific Publishers; 2006. 258 p.

[55] Kobayashi K.: Factors affecting phytotoxic activity of allelochemicals in soil. Weed Biology and Management. 2004;4:1–7.

[56] Blum U.: Allelopathic interactions involving phenolic acids. Journal of Nematology. 1996;28:259–267.

[57] Blum U., Gerig TM., Worsham AD., King LD.: Modification of allelopathic effects of p-coumaric acid on morning-glory seedling biomass by glucose, methionine and nitrate. Journal of Chemical Ecology. 1993;19:2791–2811.

[58] Weidenhamer JD., Hartnett DC., Romeo JT.: Density-dependent phytotoxicity: Distinguishing resource competition and allelopathic interference in plants. Journal of Applied Ecology. 1989;26:613–624.

[59] Politycka B.: Soil sickness: An overview. In: Politycka B., Wu F., Sampietro DA., editors. Research Methods in Plant Sciences, vol 3. Soil Sickness. Houston, USA: Studium Press; 2011. p. 7–22.

[60] Narwal SS.: Allelopathy in Crop Production. Jodhpur, India: Scientific Publishers; 1994. 303 p.

[61] Politycka B., Wójcik-Wojtkowiak D., Pudelski T.: Phenolic compounds as a cause of phytotoxicity in greenhouse substrates repeatedly used in cucumber growing. Acta Horticulturae. 1984;156:89–94.

[62] Utkhede RS., Smith EM.: Biotic and abiotic causes of replant problems of fruit trees. Acta Horticulturae. 1994;363:25–32.

[63] Kumari A., Kohli RK., Autotoxicity of ragweed parthenium (*Parthenium hysterophorus*). Weed Science. 1987;35:629-632.

[64] Singh HP., Batish DR., Kohli RK.: Autotoxicity: Concept, organisms and ecological significance. Critical Reviews in Plant Sciences. 1999;18:757–772.

[65] Politycka B., Golcz A.: Soil sickness syndrome in organic substrates repeatedly used in eggplant (*Solanum melongena* L.) culture. Allelopathy Journal. 2006;18(1):39–46.

[66] Huang L-F., Song L-X., Xia X-J., Mao W-H., Shi K., Zhou Y-H.: Plant-soil feedbacks and soil sickness: From mechanisms to application in agriculture. Journal of Chemical Ecology. 2013;39:232–242. doi:10.1007/s10886-013-0244-9.

[67] Gur A., Cohen Y.: Causes of soil sickness in replanted peaches. I. The role of cyanogenesis in peach soil sickness. Acta Horticulturae. 1988;233:25–32.

[68] Utkhede RS.: Influence of cultural practices on the growth and yield of young apple trees planted in replant disease soil. Acta Horticulturae. 1998;477:27–38.

[69] Ismaiel AA, Papenbrock J.: Mycotoxins: Producing fungi and mechanisms of phytotoxicity. Agriculture 2015;5:492–537. doi:10.3390/agriculture5030492

[70] Pimentel D.: Amounts of pesticides reaching target pests: Environmental impacts and ethics. Journal of Agricultural and Environmental Ethics. 1995;8:17–29.

[71] Elmholt S., Frisvad JC., Trnac U.: The influence of fungicides on soil mycoflora with special attention to tests of fungicide effects on soil borne pathogens. In: Altman J., editor. Pesticide Interactions in Crop Production: Beneficial and Deleterious Effects. Boca Raton, FL: CRC Press. 1991. p. 227–243.

[72] Hart MR, Brookes PC. Soil microbial biomass and mineralization of soil organic matter after 19 years of cumulative field applications of pesticides. Soil Biology and Biochemistry. 1996;28:1641–1649.

[73] Aktar MW., Sengupta D., Chowdhury A.: Impact of pesticides use in agriculture: Their benefits and hazards. Interdisciplinary Toxicology. 2009;2(1):1–12. doi:10.2478/v10102-009-0001-7

[74] Chalker-Scott L.: Impact mulches on landscape plants and the environment—a review. Journal of Environmental Horticulture. 2007;25:239–249.

[75] Marble SC.: Herbicide and mulch interactions: A review in the literature and implications for the landscape maintenance industry. Weed Technology. 2015;29:341–349. doi: 10.1614/WT-D-14-00165.1

[76] Rathinasabapathi B., Ferguson J., Gal M.: Evaluation of allelopathic potential of wood chips for weed suppression in horticultural production system. HortScience. 2005;40:711–713.

[77] Bantle A., Broken W., Ellerbrock RH., Schulze ED., Weisser WW., Matzner E.: Quantity and quality of dissolved organic carbon released from coarse woody debris of different tree species in the early phase of decomposition. Forest Ecology and Management. 2014;329:287–294.

[78] Reigosa MJ., Pedrol M., editors. Allelopathy: From Molecules to Ecosystems. Enfield, New Hampshire: Science Publishers; 2002. 316 p.

[79] Krasutsky PA.: Birch bark research and development. Natural Product Reports. 2006;23:919–942.

[80] Macías FA., Simonet AM., Esteban MD.: Potential allelopathic lupane triterpenes from bioactive fractions of *Melilotus messanensis*. Phytochemistry. 1994;36:1369–1379.

[81] Duryea M.: Landscape mulches: What are the choices in Florida. Electronic Data Information Source of University of Florida. 2000. FOR 80: http://edis.ifas.ufl.edu

[82] Herms DA., Lloyd JE., Stinner BR.: Effect of organic mulches and fertilization on soil microbial activity nutrient availability and growth of river birch. In: Michel FC Jr., Rynk R., Hoitink HAJ., editors. Proceedings of International Symposium: Composting and Compost utilization. Emmaus, PA: The JG Press, Inc. 2002.

[83] Ghaly AE., Dave D., Zhang B.: Augmenting composting microbial community with thermophilic cellulolytic organisms for enhanced degradation of phenolic compounds in creosote treated wood waste. Journal of Bioremediation and Biodegradation. 2012;3:139. doi:10.4172/2155-6199.1000139

[84] Townsend TG., Solo-Gabriele H., Tolaymat T., Stook K.: Impact of chromated copper arsenate (CCA) in wood mulch. Science of the Total Environment. 2003;309:173–185. doi:10.1016/S0048-9697(03)00047-0

[85] Bebee E., England J.: Lead concentrations in processed C&D wood. Biocycle. 1998;39:32–33.

[86] Lerner R.: Sour mulch burns tender plants. In: Purdue Gardening News. West Lafayette: University of Pardue. 2004.2p. http://www.hort.purdue.edu/ext/sourmulch.html

2

Cyanobacterial Toxins Emerging Contaminants in Soils: A Review of Sources, Fate and Impacts on Ecosystems, Plants and Animal and Human Health

Noureddine Bouaïcha and Sylvain Corbel

Additional information is available at the end of the chapter

Abstract

In the last few decades, there has been a growing interest in the occurrence of cyanotoxins and their potential toxicity in the aquatic environment. However, the used of dried toxic cyanobacteria cells as fertilizer or the used of surface water contaminated with cyanotoxins for agricultural crops irrigation can be source of soil contamination. In addition, surface waters presenting dense toxic blooms of cyanobacteria and used for agricultural practices are not controlled and are often used without prior treatment. Once in soil, cyanotoxins may be transported again to water bodies by leaching, runoff and drainage processes or can be accumulated in soils and, therefore, may cause contamination of vegetation by absorption from soils or by surface pollution of plants. In addition to possible effects on human health, elevated levels of cyanotoxins in soils can negatively affect plant vigour, animal health, microbial processes and overall soil health. Consequently, the focus of this chapter of soil contamination is cyanotoxins as contaminants of emerging concern in the soil, identifying sources of contamination, determining their fate and effects in the soil, and understanding their bioaccumulation in agricultural plants used for feed and food and consequences on animal and human health.

Keywords: cyanotoxins, microcystins, soil, fate, phytotoxicity, plant, bioaccumulation

1. Introduction

The occurrence of toxic cyanobacterial blooms has become increasingly frequent throughout freshwater bodies in the world. To date, factors identified as contributing towards their global expansion included increased nutrient inputs, transport of cells or cysts via anthropogenic activities and/or migratory birds, increased aquaculture production and/or overfishing, altering food webs and permitting harmful species to dominate algal communities [1, 2]. It has also been shown that an increase in surface water temperatures and CO_2 concentrations due to changing global climate could play a role in the proliferation of cyanobacterial blooms [3–6] and may also affect the strain composition within a cyanobacterial community and consequently change the concentration of cyanotoxins, such as microcystins [7, 8]. The problems associated with cyanobacterial blooms in fresh waters are diverse, from the environment asphyxiation due to excessive consumption of oxygen to purely aesthetic problems in recreational areas when the blooms are a colourful and often smelly scum on the surface of the water [9]. To these problems possibly affecting the economic development of specific areas, productions of cyanotoxins as secondary metabolites can represent a human and animal health threat [10]. Humans can be indeed exposed to cyanotoxins through both direct routes, including contamination of drinking and recreational waters, and indirect routes, including food supplements made from cyanobacteria or through consumption of contaminated food after toxin accumulation in fish, shellfish and other aquatic organisms, as well as in vegetables after using contaminated water for irrigation [11]. In the case of use of surface waters contaminated by cyanotoxins for the supply of drinking water, the potential health risks are managed at the level of the treatment station. In general, a strengthening of clusters of treatment and a complete operation and correct of this station would avoid any risk of contamination of the drinking water [12–20]. By contrast, the raw water used in irrigation often comes from a natural water body or an artificial pond for agricultural

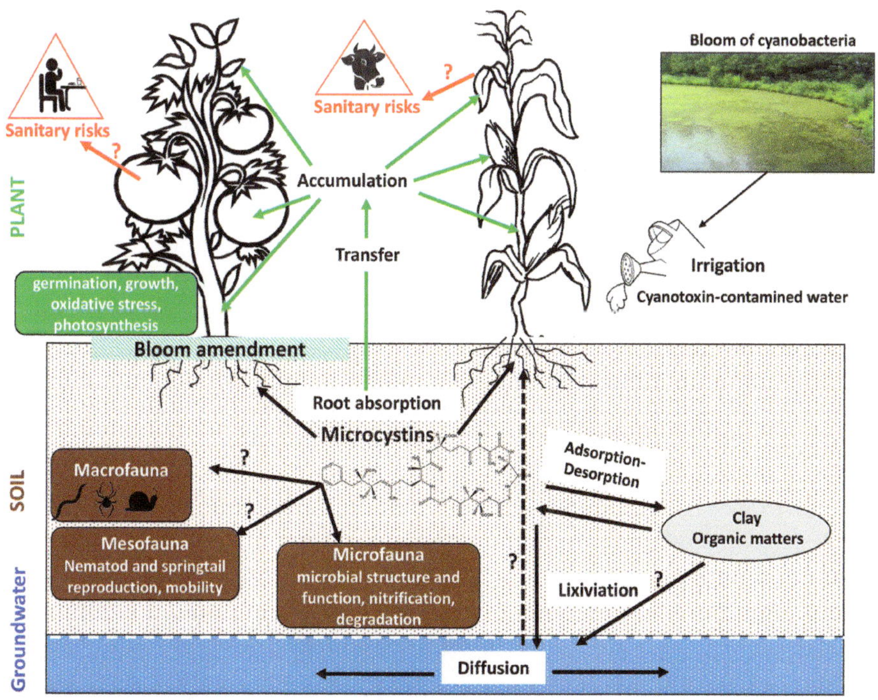

Figure 1. Schematic microcystins fate process in soil-plant systems and their impacts on human and animal health.

purposes and is not subject to any control or supervision. Consequently, the presence of cyanotoxins in irrigation water may cause toxic effects in the biological activity of the soil and in edible plants presenting therefore, a threat to animals and humans health (**Figure 1**). In fact, many studies have shown that the presence of microcystins in the irrigation water can have a considerable impact on the germination, growth and development of cultivated plants (reviewed in Ref. [10]). However, the fate of these toxins in the soil and their effects on the microfauna (protozoa, nematodes) and the microflora (bacteria, fungi and algae) of cultivated soils are scarce. This chapter aims to provide a current description of knowledge of cyanotoxins present in the irrigation water and their effects on soil and consequences on animals and public health.

2. Cyanotoxins and their producers

Some cyanobacteria species belong essentially to the genera *Microcystis, Anabaena, Aphanizomenon, Planktothrix, Oscillatoria, Cylindrospermopsis* and less often *Gomphosphaeria, Coelosphaerium, Gloeotrichia, Nodularia* and *Nostoc* are known to biosynthesize a diversity of alkaloid and peptide cyanotoxins that have been suggested to pose threats to human and environmental health worldwide [9, 21–24]. These cyanotoxins are essentially endotoxins that can be released in the environment following a cellular lyse during the senescence phase [25] or following treatment of cyanobacterial blooms with algaecides [26]. They can be classified into four families according to the organs on which they act: hepatotoxins (liver), neurotoxins (nervous system), cytotoxins (liver and kidneys) and dermatotoxins (irritant toxins). Hepatotoxins are divided into two groups: microcystins, cyclic heptapeptide hepatotoxins (MW 900–1200), that are regarded as the most frequently occurring and widespread of the cyanotoxins with more than 100 MC variants already reported [27, 28] and nodularins (MW 800–900) composed of five amino acids with only nine different natural analogues have been characterized [29–32]. Both microcystins and nodularins are water-soluble molecules and their cyclic structure provides them a high chemical stability [22]. Their toxicity resulted on a potent and specific inhibition of serine/threonine protein phosphatases [33]. They have also known to induce oxidative stress [34]. Cyanobacterial neurotoxins (for review, see Ref. [35]) are divided into three groups: anatoxins that are neuromuscular junction blocking agents [35], saxitoxins that block nerve cell voltage-gated sodium channels [36] and the unusual non-protein neurotoxic amino acid L-beta-*N*-methylamino-L-alanine (BMAA) that has been associated to the neurological disorder amyotrophic lateral sclerosis/Parkinsonium dementia complex (ALS/PDC) among the indigenous Chamorro people of Guam and other Marianas islands [37]. Its neurotoxicity may be mediated via glutamate regulation [38]. Anatoxins and the BMAA are specific of cyanobacteria, while saxitoxins are also synthesized by some marine dinoflagellates and associated with the human disease paralytic shellfish poisoning or PSP [39]. Cytotoxins are represented by the hydrophilic alkaloid cytotoxin, cylindrospermopsin (MW 415), that has been first isolated from the filamentous *Cylindrospermopsis raciborskii* [21], and further from other species *Aphanizomenon ovalisporum* [40, 41], *Anabaena bergii* [42], *Umezakia natans* [43] and *Raphidiopsis curvata* [44]. It inhibits the synthesis of protein, resulting in a wide spread necrosis of the tissues of many organs such as liver and kidneys [45–47]. Two structural variants of cylindrospermopsin (7-epicylindrospermopsin and deoxycylindrospermopsin) have been characterized so far from bloom samples and isolated strains of cyanobacteria [41, 44, 48]. The dermatotoxins, irritant toxins such as lipopolysaccharides (LPS) commonly known endotoxins, are major components of the cell wall in most Gram-negative bacteria including cyanobacteria. They can elicit irritant and allergenic responses in human and animal tissues with contact [49–51].

3. Sources and occurrence of cyanotoxins in the soil

The main source of contamination of soils by cyanotoxins is by using cyanotoxin-contamined water for agricultural purposes. Among the cyanobacterial toxins, microcystins are the most widespread group with microcystin-LR (MC-LR) the more toxic and the main congener detected in freshwaters [10]. Recently, concerns are also focused in the increasing occurrence of the cytotoxic cylindrospermopsin in temperate areas [52]. However, cyanobacterial neurotoxins are less reported in the literature and studies regarding their effects on organisms of soils and plants are relatively scarce. The concentrations of microcystins in the surface water are generally comprised between 1 and 100 $\mu g\ L^{-1}$ [10] and the use of this microcystin-contaminated water for agricultural purposes has already been reported in several countries such as Morocco [53], Finland [54], Spain [55], New Zealand [56], Algeria [57], Australia [58], Tunisia [59], Turkey [60], Saudi Arabia [61], India [62], China [63] and Guatemala [64]. In addition to the contamination of soils by dissolved cyanotoxins and with the strong occurrence of cyanobacterial blooms worldwide, a strong quantity of cyanobacterial biomass (from thousand to million tons) is removed from water and discharged directly into croplands and forest land without another treatment [65]. This alternative represents a possible source of soil pollution with cyanotoxins. Another source of soil contamination by cyanotoxins consists of direct application of cyanobacterial biomass as an organic fertilizer as in China [66, 67]. In fact, since 1970s, the cyanobacteria were known for their interest in rice culture, as a biofertilizer. In wetland rice and wheat crops, free living cyanobacteria allowed nitrogen fixation to supplement soil nitrogen [68–70]. Cyanobacterial and rhizobacterial associations are used with the objective to increase soil fertility and crop yields, but the cyanobacteria and their secondary metabolites represent also interesting properties and can be involved as natural biocide or biocontrol agents (see review in Ref. [71]). In a recent study Han et al. [72], they related the use of algae waste as an organic fertilizer after composting. This process can allow the degradation of 90–95% of the total microcystins containing in cyanobacteria between 1 and 35 days [73, 74]. The microbial degradation of cyanotoxins, during composting, may be due to the diversity of microorganisms present, the conditions of composting and the type of cyanotoxins present in the bloom, as observed for microcystins by Dawson [75] and Kormas and Lymperopoulou [76]. In addition, several studies reported that the presence of cyanotoxins in *biological soil crust* (biocrust) samples in arid soils can be considered as another source of cyanotoxin-contaminated soils [77–79].

4. Fate and transport of cyanotoxins in agricultural systems

4.1. Persistence in the soil and adsorption in particles

The most abundant cyanotoxins, microcystins, have a cyclic structure that provides a high chemical stability in the environment. Once these toxins are present in the soil, they can be removed according to various processes such as photochemical degradation by UV and biodegradation by some bacteria species [10, 80–84]. The photochemical degradation of microcystins can last from 2 to 6 weeks [85, 86] in freshwater. But in the soil, this process was not studied, however as observed in water it depends on the adsorption on soil particles that is more important than in water. In fact, numerous studies on sediments and soil particles showed that the adsorption induced a diminution of photochemical degradation of microcystins [83]. The time of total degradation of MC-RR in cropland should be about 6 days according Bibo et al. [87], whereas Chen et al. [67] founded a relatively long time of microcystins persistence with a half-life ranging between 6 and 18 days. Another study, where the scums of *Microcystis aeruginosa* were dried on the shores of lakes revealed the persistence of high concentrations of microcystins for several months [88]. The results obtained by Miller et al. [89], on five soils with different physicochemical properties, showed the role of clay and organic carbon contents for microcystin-LR (MC-LR) and nodularin (NOD) adsorption. In fact, Miller and Fallowfield [90] found in batch experiments that the soil with the highest concentrations of organic carbon and clay content 2.9 and 16.1%, respectively, was the most effective at removing these toxins in comparison to the sandy soil. These results were supported by Morris et al. [91] works, who reported that sandy soil (98% sand) was incapable of removing microcystins; however, they confirmed the role of clay content and the clay quality for their adsorption. Consequently, the adsorption in soil particles depends on soil properties and the quantities of cyanotoxins brought. A laboratory study on cropland soil showed an adsorption of MC-RR from 3750 to 30,000 μg kg^{-1} [87]. In pound used to stock cyanobacterial bloom, the concentrations of adsorbed microcystins attained 65–200 μg kg^{-1} DW, whereas in China crop fields, the concentration after amendment was 6 μg kg^{-1} DW [65]. In addition, Chen et al. [67] reported that the adsorption mechanism of microcystins in soil is also due to chemical binding with the metal ions on the surface of particles. Therefore, with the possible adsorption onto soil particles, microcystins could be accumulated in soil for long times. Indeed, Corbel et al. [92] detected microcystins in soil in concentrations ranging from 1.3 to 3.9 μg MC-LR equivalent kg^{-1} (dry weight) after 90 days of silty-soil irrigation with water containing dissolved cyanobacterial extract containing 100 μg equivalent MC-LR L^{-1}. These results corroborate with an earlier study done by the same research team where they reported that the half-life of ^{14}C-MC-LR exceeded 60 days in the same agricultural soil [93]. However, in this last study the authors reported that only less than 14% of ^{14}C-MC-LR were adsorbed in soil particles, suggesting that a part of this toxin could be biodegraded [93]. In fact, it seems that the major dissipation process of microcystins in the soil is mainly via microbial degradation [67, 90, 94, 95]. Cylindrospermopsin can persist in the water for long periods because it has a very low photodegradation rate under natural conditions [81]. However, all the studies performed with cylindrospermopsin were carried out in a soil-free cultivation system, and therefore the persistence of this toxin in agricultural soils was not considered.

4.2. Transport and uptake into biota and infiltration in groundwater

As described above regarding the microcystins adsorption in cropland soils, it is suggested that the adsorption of these toxins is generally low, which can therefore potentially result in their higher bioavailability for plants and the groundwater contamination due to infiltration into the soil. Consequently, Eynard et al. [96] suggested that the soil was unable to protect groundwater contamination by microcystins. Chen et al. [67] reported that microcystins can migrate from the surface to deeper layers of the soil following precipitation, leading to possible groundwater contamination. In a recent study, Corbel et al. [93] showed that when the radiolabeled ^{14}C-MC-LR was introduced in a column of silty-sand agricultural soil, it underwent a weak microbial mineralization under aerobic conditions and therefore the large amounts of the toxin remained in soil aqueous extracts. In addition, the authors reported that the lixiviation of this toxin by $CaCl_2$ was even stronger than soil application was recent. These results were confirmed by other environmental measures, where microcystins were found in groundwater [61, 65]. For example, Chen et al. [65] found a concentration of 2.5 μg L^{-1} in lixiviate water that was higher than the WHO recommendation in drinking water (1 μg L^{-1}). The risk associated with the underground stock in water is the long-time persistence of toxins, in result of low microbial degrading activity. In fact, Holst et al. [97] did not detect any degradation of microcystins in groundwater maintained under oxic and anoxic conditions after a 100-day period. The toxins present in the soil solution are also available for soil organisms' uptake such as plants. For example, Pflugmacher et al. [98] demonstrated a rapid uptake of ^{14}C-MC-LR by aquatic plant (*Phragmites australis*). A recent study established the transfer of MC-LR from agricultural soil contaminated with radiolabeled MC-LR (18 mg ^{14}C-MC-LR kg^{-1}) to tomato seedling, with a final concentration of 6 μg MC-LR g^{-1} FW [93]. Several other studies reported an uptake of microcystins by plant roots and a presence of these toxins in shoots and leaves after culture on sand or agricultural soils [61, 65, 92, 99–102]. Concerning the other cyanotoxins, less detected in the surface waters, the soil-plant transfer data are scarce. However, Prieto et al. [103] reported the uptake of cylindrospermopsin by the roots of *Oryza sativa* plants. In a recent study, Contardo-Jara et al. [104] reported the transfer of the neurotoxin β-*N*-methylamino-L-alanine by *Triticum aestivum* in roots and shoots after irrigation with contaminated water at 100 μg L^{-1}.

5. Impacts of cyanotoxins on soil organisms

5.1. Microorganisms

Secondary, metabolites produced by cyanobacteria seem to have several activities as antiviral, antifungal and antibacterial [71]. In aquatic environments, several studies revealed an inhibition of bacterial growth after 8 days of exposure to cyanobacterial extract containing microcystins or pure microcystin standards [105]. In the same way, Giaramida et al. [106]

reported that the exposure to cyanobacterial extract containing microcystins induced changes in structure and physiology of bacterial communities. The measure of arylsulfatase, phosphatase, urease and β-D-glucosidase activities in the soil, after irrigation with cyanobacterial extract of *M. aeruginosa* (PCC7820) diluted between 5 and 100 μg equivalent MC-LR L^{-1} during 14 or 90 days, revealed an absence of the alteration of the activity of these enzymes linked to sulphur, phosphorus and nitrogen mineralization and cellulose degradation, respectively [107, 108]. In contrast, these studies revealed a stimulation of the potential of nitrification that was positively correlated to an increase in the abundance of ammonia-oxidizing bacteria, whereas the ammonia-oxidizing archaea were not impacted. In a recently study, El Khalloufi et al. [109] highlighted the effects of cyanobacterial extract containing microcystins on soil microorganisms from the rhizosphere of *Medicago sativa*. The authors exposed *M. sativa* to 100 μg equivalent MC-LR L^{-1} during 30 days at three times a week and a pyrosequencing analysis was further established to characterize the bacterial community of the rhizosphre. The results revealed fluctuations with an increase in *Betaproteobacteria* and a decrease in *Gammaproteobacteria* proportion. Furthermore, cyanobacterial extract containing microcystins used for irrigation seemed to be toxic towards *Actinobacteria, Gemmatimonas, Deltaproteobacteria* and *Gammaproteobacteria,* however other groups as *Clostridia, Opitutae* and bacteria related with *Betaproteobacteria,* were stimulated [109]. However, Lahrouni et al. [110, 111] reported that rhizobia-*Vicia faba* symbiosis was not impacted by microcystins. Nevertheless, several studies revealed the presence of heterotrophic bacteria in the soil containing a microcystin-gene cluster, *mlr*A, B, C and D essential for degradation of microcystins [89, 94, 95, 112]. For example, some species of the proteobacteria belonging to the genera *Sphingomonas, Methylobacillus* and *Paucibacter* are known to degrade microcystins [10, 76]. Additionally, Jia et al. [113] showed that a fungus, *Trichaptum abietinum,* was able to degrade microcystins. However, no studies have yet examined the effects of cylindrospermopsin and neurotoxins in soil microorganisms.

5.2. Invertebrates

The impact of cyanotoxins on aquatic invertebrates was well documented (for review, see Ref. [114]). However, the effects of these toxins on soil invertebrates are scarce. The effects of microcystins on soil nematods *Ceanorhabditis elegans* were studied by Li et al. [115, 116] and Holajjer et al. [117]. After exposure to 1 μg MC-LR L^{-1}, a reduction in lifespan, a delay of development, an increase in generation time, a decrease in brood size, a suppression of locomotion behaviour and a decrease in hsp-16-2-gfp expression were observed [115]. In addition, the neurotoxicity of MC-LR was demonstrated in *C. elegans* with significant severe defects of chemotaxis to NaCl and diacetyl, and thermotaxis [116]. Therefore, the application of toxic cyanobacteria in soil may reduce nematode infestation and finally increase plant yield (see review in Ref. [117]). Concerning the macrofauna, and to the best of our knowledge, only one study was reported in the literature on the survival and reproduction of the springtail *Folmiosa candida* after application to the soil of a cyanobacterial biomass containing different concentrations of microcystins from 21 to 3662 μg g^{-1} DW [118]. The results showed no adverse effects on survival and reproduction when the ratio cyanobacterial biomass/soil attained 4 g kg^{-1} DW soil.

5.3. Plants

The phytotoxicity of cyanotoxins was observed on aquatic plants but in the last years several studies investigated this field for terrestrial plants. As described in the review of Corbel et al. [10], the phytotoxicity of neurotoxins and cytotoxic alkaloids is less studied in comparison to microcystins. In laboratory conditions, several studies reported that the rate of germination of several plants decreased with an EC50 of 11 mg eq. MC-LR L^{-1} for *Triticum durum* [107, 119] and an EC50 comprised between 16 and 20 mg eq. MC-LR L^{-1} for tomatoes [120]. In these conditions, generally, the germination was impacted by microcystins for concentrations upper than 1 mg eq. MC-LR L^{-1} and responses differed according the sensitivity of plants. Indeed, Corbel et al. [107] highlighted the higher sensitivity of wheat in comparison with tomato and lettuce seeds. Chen et al. [121] reported that the rice seed were more resistant than the rape ones. In addition, Corbel et al. [107] reported that a crude extract of cyanobacteria containing microcystins induced a significant decrease in the radicle lengths of MicroTom and Saint-Pierre tomatoes plants for concentrations higher than 5 and 20 mg eq. MC-LR L^{-1}. Similar results, showing an inhibition of 44% of root growth, were obtained after exposition of *Triticum aetivum* exposed to 0.5 μg MC-LR L^{-1} [122]. Chen et al. [121] reported that high concentrations of MC-LR (>2 mg L^{-1}) inhibited root elongation, crown roots formation and lateral root formation from primordia for rice plants. In an earlier study, Gehringer et al. [123] observed a decrease in root and leaf biomasses of *M. sativa* with 5 and 10 μg eq. MC-LR L^{-1}. By contrast, other studies demonstrated that pure MC-RR at environmental concentrations (<10 μg L^{-1}) accelerated the rape growth of some plants [87]. In a recent study based on tomato irrigation for 14 days by cyanobacterial extract containing concentrations from 5 to 100 μg eq. MC-LR L^{-1}, Corbel et al. [107] showed similar results with an enhancement of aerial biomasses, whereas the root biomasses were not impacted by these treatments. In the same way, a chronic exposure with an experiment of duration 90 days revealed a stimulation of tomato growth during the first 40 days post-germination [124]. In addition to the toxicity of microcystins linked to the specific inhibition of serine/threonine protein phosphatases [33], the increase in antioxidant defences induced by these toxins suggests that oxidative stress is also a major mechanism contributing to their phytotoxicity (reviewed in Ref. [10]). Yin et al. [125] reported that the exposure of *Arabidopsis thaliana* cells to MC-LR at 5 mg L^{-1} induced a lipid peroxidation, a decrease in glutathione GSH content and increase in superoxide dismutase (SOD) and catalase (CAT) activities. Stüven and Pflugmacher [126] reported also that microcystins induced oxidative stress response in *Lepidium sativum* with an elevation of alpha- and beta-tocopherol concentrations and an increase in the activity of antioxidative enzymes (glutathione peroxidase, glutathione S-transferase and glutathione reductase). Peuthert et al. [99] observed lipid peroxidation in both the roots and shoots of several agricultural plants (*Pisum sativum, Cicer arietinum, Vigna radiate, Phaseolus vulgaris, Glycine max, M. sativa, Lens culinaris, T. aestivum* and *Zea mays)* that were exposed to MC-LR, either purified or in crude extract. Finally, the presence of microcystins in irrigation waters can imply modifications in the plant metabolism, notably on the photosynthesis. A study of Saqrane et al. [127] showed a decrease in chlorophyll concentrations in *Z. mays* and *L. esculenta* leaves after chronic exposure by irrigation for 30-day period to 4.2 and 2.1 mg eq. MC-LR L^{-1}, respectively. Consequently, the photosynthesis activity was disrupted as indicated by chlorophyll fluorescence. Similar results and conclusions were obtained by El Khalloufi et al. [120] when they exposed tomato plants with 22 mg eq. MC-LR L^{-1}. In contrast, in a recent study Corbel et al. [124] reported that chronic irrigation of tomato plants for a period of 90 days with lower concentrations (from 5 to 100 μg eq. MC-LR L^{-1}) did not induced a modification of chlorophyll-*a* and *b* concentrations or disturbed the photosynthesis metabolism. However, in a study performed by Gutiérrez-Praena et al. [102] in which tomato plants were exposed to MC-LR at 100 μg $^{L-1}$, changes were detected in the function of various proteins related to ATP synthesis, carbon fixation, photosynthesis and carbohydrate metabolism that appear to be linked with the observed decrease in photosynthetic efficiency. A decrease in the expression of some proteins involved in photosynthesis was also observed by Azevedo et al. [128] in rice plants exposed to 13 μg MC-LR L^{-1}. The contradictory results obtained in these different studies may be associated with differences in the microcystin concentrations used in each study and the nature of the toxin pure or present in a cyanobacterial crude extract. Furthermore, some studies have demonstrated that MC-LR can be responsible for changes in the mineral content of plants; in which the macro-mineral content of the roots is increased after exposing the plants to MC-LR in a concentration-dependent manner [120, 127, 129]. However, Freitas et al. [130] and Lahrouni et al. [129] reported that the exposure of *Lactuca sativa* and *V. faba*, respectively, to purified MC-LR and MC-LR contained in a cyanobacterial crude extract produced a decrease in the mineral content of the leaves. Compared to the different effects of microcystins on plants described above, the effects due to the alkaloid cylindrospermopsin (CYN) exposure is poorly documented. This cyanotoxin seems like microcystins to induce oxidative stress in plants [103, 131, 132]. For example, in a recent study Freitas et al. [130] reported that CYN induced in time- and concentration-dependent manner an increase in the GST activity in the roots of lettuce plants. However, the glutathione peroxidase (GPx) activity was significantly decreased in both the roots and the leaves of the same plant exposed to 100 μg CYN L^{-1} for 5 days. In the same study Freitas et al. [130] reported also that the exposure of lettuce to purified CYN, in contrast to MC-LR, produced an enhancement in leaf micro (Fe, Mn, Cu, Zn, Mo) and macro (Ca, Mg, P, K, Na) mineral content. In addition, in another study [133] reported a significant increase in the abundance of proteins involved in photosynthesis in lettuce plants exposed to CYN.

6. Bioaccumulation of cyanotoxins in agricultural plants and consequences on human and animal health

Humans were exposed to cyanobacteria toxins through many routes, including drinking water, recreational contact and health food products made from cyanobacteria, and food chain. While some of these routes are well enough informed the others are them less, notably that corresponding to the consumption of crop plants. Although, no case of poisoning by these products has been reported worldwide, this eventuality must not be ignored. Indeed, a recent epidemiological study showed that the excessive incidence of amyotrophic lateral sclerosis in the population of the islands of Guam in the Pacific was linked to a consumption of the seeds of cycas contaminated by a neurotoxin, β-methylamino-L-alanine (BMAA), produced by a species of cyanobacteria of the genus *Nostoc* living in symbiosis in the roots of this plant [134].

This last cited fact is gaining importance since plants could in a direct or indirect manner contribute to food chain cyanotoxin's transfer, and by the way constitute a potent health risk source. Therefore, the accumulation of cyanotoxins in cultivated plants could transform them into vectors of exposure as much for the herbivorous animals that for humans. However, it's important to notify that most of the published results on cyanotoxin's transfer on plants have been performed in hydroponic conditions, which can overestimated the availability of toxins to the root system. In addition, and as indicated previously the soil particles can adsorb microcystins, reducing therefore, their bioavailability for the plants' uptake. For example, recently, Kanzo et al. [135] reported that in hydroponic conditions, microcystins were able to accumulate in the roots, stems and leaves of *Brassica rapa* after exposure to 100 and 1000 μg MC-LR L^{-1}. However, in the same plant when cultivated in a soil system no accumulation was detected after exposure to the same concentrations of MC-LR.

Nevertheless, the ability of microcystins and cylindrospermopsin to accumulate in the tissues of different agricultural plants has been reported in the literature, and it was recently reviewed by Corbel et al. [10]. Microcystins have been detected in tissues of terrestrial plants [92, 93, 104, 122, 136, 137], indicating that they can be absorbed and transported in plants although their transport mechanism is unclear yet. However, the ability of absorbing microcystins and their accumulation in different tissues was variable among different plant species and depends on toxins' concentrations [99, 107, 127]. For example, Järvenpää et al. [138] reported that microcystins were detected on roots (a non-edible plant tissue for human but can be for animal) but not detected in leaves of mustard and broccoli. Furthermore, numerous studies concerning accumulation of cyanotoxins in agronomic plants growing in the soil were reported in radish roots, leaves of arugula and dill [61], in rice grains [65], in leaves of lettuce and cabbage [61, 139], in leaves and stems of water spinach [139] and in fruits and seeds of tomato and pepper [64]. However, a recent study based on the use of ^{14}C-labelled MC-LR showed that tomato fruits did not accumulate the toxin [92].

7. Conclusion and future directions

The occurrence of toxic cyanobacterial blooms, in surface waters that can be used without treatment for irrigation in agricultural purposes, has become increasingly frequent worldwide. With this increased awareness, research has been recently focused towards the fate of cyanotoxins in soils and health risk due to their potential transfer and accumulation in plants. Although there is much basic information on the concentrations of cyanotoxins found in freshwaters, there are very significant gaps in our knowledge of their effects on the biological activity of the soil and their bioaccumulation, and the role of detoxication and covalent binding in the agricultural plants irrigated with cyanotoxin-contaminated water. The great majority of the studies published recently were performed in hydroponic conditions and focused on microcystins (MCs) and specifically on a single MC variant (MC-LR) out of the almost more than 100 variants known and with high no relevant environmental concentrations. To protect consumers from the adverse effects of MCs, the WHO proposed a provisional upper limit in drinking water of 1 μg/L for the most toxic congener MC-LR and a tolerable daily intake (TDI) of 0.04 μg/kg body weight (bw). The available data on the phytotoxicity of microcystins indicate that their concentrations in edible tissues of various agricultural plants can exceed the WHO-TDI guideline. Consequently, more information on this aspect is urgently needed for risk assessment purposes such as

- The fate of cyanotoxins in agricultural soils and the biochemical, physiological and ecological processes that control their trophic transfer in different plants remain to be clarified.

- Furthermore, even the provisional guidelines that exist for MCs in water are only recommendations, and policy will not only need to clarify acceptable levels but also address to monitor and enforce these guidelines. As such, improvements, validation and standardization of methods for chemical analysis of MCs—towards effective monitoring and enforcement in agriculture food webs—will be crucial.

- Acceptable levels for foodborne cyanotoxins are based entirely on data from waterborne toxins and are not likely to be accurate in terms of exposure through agriculture foods; therefore, reliable exposure scenario and more good quality data should be collected before robust conclusions on the health risks.

Author details

Noureddine Bouaïcha* and Sylvain Corbel

*Address all correspondence to: noureddine.bouaicha@u-psud.fr

Ecology, Systematic and Evolution, University of Paris-Sud, CNRS, AgroParisTech, University Paris-Saclay, Orsay, France

References

[1] Heisler J, Glibert PM, Burkholder JM, Anderson DM, Cochlan W, Dennison WC, et al. Eutrophication and harmful algal blooms: a scientific consensus. Harmful Algae. 2008;8(1):3–13.

[2] Herry-Allani SE, Bouaïcha N. Cyanobacterial blooms in dams: environmental factors, toxins, public health and remedial measures. Dams: Structure, Performance and Safety Management. Nova Science Publisher, Hauppauge, NY, USA; 2013.

[3] Peperzak L. Climate change and harmful algal blooms in the North Sea. Acta Oecologica. 2003;24:S139–S144.

[4] Paerl HW, Huisman J. Climate change: a catalyst for global expansion of harmful cyanobacterial blooms. Environ Microbiol Rep. 2009;1(1):27–37.

[5] Paerl HW, Huisman J, et al. Blooms like it hot. Science. 2008;320(5872):57.

[6] Manganelli M, Scardala S, Stefanelli M, Palazzo F, Funari E, Vichi S, et al. Emerging health issues of cyanobacterial blooms. Ann DellIstituto Super Sanità. 2012;48(4):415–428.

[7] Van de Waal DB, Verspagen JM, Finke JF, Vournazou V, Immers AK, Kardinaal WEA, et al. Reversal in competitive dominance of a toxic versus non-toxic cyanobacterium in response to rising CO_2. ISME J. 2011;5(9):1438–1450.

[8] Yu G, Zhu M, Chen Y, Pan Q, Chai W, Li R. Polyphasic characterization of four species of Pseudanabaena (Oscillatoriales, Cyanobacteria) from China and insights into polyphyletic divergence within the Pseudanabaena genus. Phytotaxa. 2015;192(1):1–12.

[9] Ettoumi A, El Khalloufi F, El Ghazali I, Oudra B, Amrani A, Nasri H, et al. Bioaccumulation of cyanobacterial toxins in aquatic organisms and its consequences for public health. In: Kattel, G., editor. Zooplankton and Phytoplankton: Types, Characteristics and Ecology. Nova Science Publishers, New York; 201:1–34.

[10] Corbel S, Mougin C, Bouaïcha N. Cyanobacterial toxins: modes of actions, fate in aquatic and soil ecosystems, phytotoxicity and bioaccumulation in agricultural crops. Chemosphere. 2014;96:1–15.

[11] Testai E, Buratti FM, Funari E, Manganelli M, Vichi S, Arnich N, et al. Review and analysis of occurrence, exposure and toxicity of cyanobacteria toxins in food | Autorité européenne de sécurité des aliments. EFSA Support Publ. 2016;2016:EN(998): 309 p.

[12] Keijola AM, Himberg K, Esala AL, Sivonen K, Hiis-Virta L. Removal of cyanobacterial toxins in water treatment processes: laboratory and pilot-scale experiments. Toxic Assess. 1988;3(5):643–656.

[13] Himberg K, Keijola AM, Hiisvirta L, Pyysalo H, Sivonen K. The effect of water treatment processes on the removal of hepatotoxins from Microcystis and Oscillatoria cyanobacteria: a laboratory study. Water Res. 1989;23(8):979–984.

[14] Rositano J, Nicholson BC. Water treatment techniques for the removal of cyanobacterial toxins from water. Australian Centre for Water Quality Research. Salisbury, South Australia. 1994. 1–55.

[15] Bruchet A, Bernazeau F, Baudin I, Pieronne P. Algal toxins in surface waters: analysis and treatment. Water Supply. 1998;16(1):619–623.

[16] Hart J, Fawell JK, Croll B. The fate of both intra- and extracellular toxins during drinking water treatment. Water Supply. 1998;16(1):611–616.

[17] Rositano J, Newcombe G, Nicholson B, Sztajnbok P. Ozonation of NOM and algal toxins in four treated waters. Water Res. 2001;35(1):23–32.

[18] Newcombe G, Nicholson B. Treatment options for the saxitoxin class of cyanotoxins. Water Sci Technol Water Supply. 2002;2(5–6):271–275.

[19] Brooke S, Newcombe G, Nicholson B, Klass G. Decrease in toxicity of microcystins LA and LR in drinking water by ozonation. Toxicon. 2006;48(8):1054–1059.

[20] Rodríguez E, Majado ME, Meriluoto J, Acero JL. Oxidation of microcystins by permanganate: reaction kinetics and implications for water treatment. Water Res. 2007;41(1): 102–110.

[21] Hawkins PR, Runnegar MT, Jackson AR, Falconer IR. Severe hepatotoxicity caused by the tropical cyanobacterium (blue-green alga) *Cylindrospermopsis raciborskii* (Woloszynska) Seenaya and Subba Raju isolated from a domestic water supply reservoir. Appl Environ Microbiol. 1985;50(5):1292–1295.

[22] Sivonen K, Jones G. Toxic cyanobacteria in water: a guide to their public health consequences, monitoring and management. In: Chorus I. and Bartram, editors. Taylor & Francis; 1999. 41–111.

[23] Kuiper-Goodman IF, Fitzgerald J. Human Health Aspects. 1999 [cited 2016 Jul 12]; Available from: http://cdrwww.who.int/entity/water_sanitation_health/resourcesquality/toxcyanchap4.pdf

[24] Hitzfeld BC, Höger SJ, Dietrich DR. Cyanobacterial toxins: removal during drinking water treatment and human risk assessment. Environ Health Perspect. 2000;108(Suppl 1):113.

[25] Codd GA, Bell SG, Brooks WP. Cyanobacterial toxins in water. Water Sci Technol. 1989;21(3):1–13.

[26] Kenefick SL, Hrudey SE, Peterson HG, Prepas EE. Toxin release from *Microcystis aeruginosa* after chemical treatment. Water Sci Technol. 1993;27(3–4):433–40.

[27] del Campo FF, Ouahid Y. Identification of microcystins from three collection strains of *Microcystis aeruginosa*. Environ Pollut. 2010 Sep;158(9):2906–2914.

[28] Puddick J, Prinsep MR, Wood SA, Kaufononga SAF, Cary SC, Hamilton DP. High levels of structural diversity observed in microcystins from Microcystis CAWBG11 and characterization of six new microcystin congeners. Mar Drugs. 2014;12(11):5372–5395.

[29] Codd GA, Lindsay J, Young FM, Morrison LF, Metcalf JS. Harmful Cyanobacteria. In: Huisman J, Matthijs HCP, Visser PM, editors. Harmful Cyanobacteria [Internet]. Berlin/Heidelberg: Springer-Verlag; 2005 [cited 2011 Oct 24]. pp. 1–23. Available from: http://www.springerlink.com.gate1.inist.fr/content/g09v1w33400u2937/

[30] de Silva ED, Williams DE, Andersen RJ, Klix H, Holmes CFB, Allen TM. Motuporin a potent protein phosphatase inhibitor isolated from the Papua New Guinea Sponge Theonella swinhoei Gray. Tetrahedron Lett. 1992;33(12):1561–1564.

[31] Namikoshi M, Choi BW, Sun F, Rinehart KL, Evans WR, Carmichael WW. Chemical characterization and toxicity of dihydro derivatives of nodularin and microcystin-LR,

potent cyanobacterial cyclic peptide hepatotoxins. Chem Res Toxicol. 1993;6(2):151–158.

[32] Rinehart KL, Namikoshi M, Choi BW. Structure and biosynthesis of toxins from blue-green algae (cyanobacteria). J Appl Phycol. 1994;6(2):159–176.

[33] MacKintosh C, Beattie KA, Klumpp S, Cohen P, Codd GA. Cyanobacterial microcystin-LR is a potent and specific inhibitor of protein phosphatases 1 and 2A from both mammals and higher plants. FEBS Lett. 1990 May 21;264(2):187–192.

[34] Bouaïcha N, Maatouk I, et al. Microcystin-LR and nodularin induce intracellular glutathione alteration, reactive oxygen species production and lipid peroxidation in primary cultured rat hepatocytes. Toxicol Lett. 2004;148(1):53–63.

[35] Aráoz R, Molgó J, De Marsac NT. Neurotoxic cyanobacterial toxins. Toxicon. 2010;56(5): 813–828.

[36] Ikawa M, Wegener K, Foxall TL, Sasner JJ. Comparison of the toxins of the blue-green alga Aphanizomenon flos-aquae with the Gonyaulax toxins. Toxicon. 1982;20(4):747–752.

[37] Monson CS, Banack SA, Cox PA. Conservation implications of Chamorro consumption of flying foxes as a possible cause of amyotrophic lateral sclerosis—parkinsonism dementia complex in Guam. Conserv Biol. 2003;17(3):678–686.

[38] Papapetropoulos S. Is there a role for naturally occurring cyanobacterial toxins in neurodegeneration? The beta-N-methylamino-L-alanine (BMAA) paradigm. Neurochem Int. 2007;50(7):998–1003.

[39] Kaebernick M, Neilan BA. Ecological and molecular investigations of cyanotoxin production. FEMS Microbiol Ecol. 2001;35(1):1–9.

[40] Banker R, Carmeli S, Hadas O, Teltsch B, Porat R, Sukenik A. Identification of cylindrospermopsin in *Aphanizomenon ovalisporum* (Cyanophyceae) isolated from Lake Kinneret, Israel. J Phycol. 1997;33(4):613–616.

[41] Banker R, Teltsch B, Sukenik A, Carmeli S. 7-Epicylindrospermopsin, a toxic minor metabolite of the cyanobacterium Aphanizomenon o valisporum from Lake Kinneret, Israel. J Nat Prod. 2000;63(3):387–389.

[42] Schembri MA, Neilan BA, Saint CP. Identification of genes implicated in toxin production in the cyanobacterium *Cylindrospermopsis raciborskii*. Environ Toxicol. 2001;16(5): 413–421.

[43] Terao K, Ohmori S, Igarashi K, Ohtani I, Watanabe MF, Harada KI, et al. Electron microscopic studies on experimental poisoning in mice induced by cylindrospermopsin isolated from blue-green alga *Umezakia natans*. Toxicon. 1994;32(7):833–843.

[44] Li R, Carmichael WW, Brittain S, Eaglesham GK, Shaw GR, Liu Y, et al. First report of the cyanotoxins cylindrospermopsin and deoxycylindrospermopsin from *Raphidiopsis curvata* (Cyanobacteria). J Phycol. 2001;37(6):1121–1126.

[45] Ohtani I, Moore RE, Runnegar MT. Cylindrospermopsin: a potent hepatotoxin from the blue-green alga *Cylindrospermopsis raciborskii*. J Am Chem Soc. 1992;114(20):7941–7942.

[46] Falconer IR. An overview of problems caused by toxic blue-green algae (cyanobacteria) in drinking and recreational water. Environ Toxicol. 1999;14(1):5–12.

[47] Runnegar MT, Xie C, Snider BB, Wallace GA, Weinreb SM, Kuhlenkamp J. In vitro hepatotoxicity of the cyanobacterial alkaloid cylindrospermopsin and related synthetic analogues. Toxicol Sci. 2002;67(1):81–87.

[48] Norris RL, Eaglesham GK, Pierens G, Shaw GR, Smith MJ, Chiswell RK, et al. Deoxy-cylindrospermopsin, an analog of cylindrospermopsin from *Cylindrospermopsis raciborskii*. Environ Toxicol. 1999;14(1):163–165.

[49] Kaya K. Chapter 8: Toxicology of microcystins. In: Mariyo F. Watanabe, Ken-ichi Harada, Wayne W. Carmichael, Hirota Fujiki, editors. Toxic Microcystis, CRC Press. 1996. 175–202.

[50] Mayer H, Weckesser J. Unusual'lipid A's: structures, taxonomical relevance and potential value for endotoxin research. Handb Endotoxin. 1984;1:221–247.

[51] Stewart I, Webb PM, Schluter PJ, Shaw GR. Recreational and occupational field exposure to freshwater cyanobacteria—a review of anecdotal and case reports, epidemiological studies and the challenges for epidemiologic assessment. Environ Health. 2006;5(1):6.

[52] de la Cruz AA, Hiskia A, Kaloudis T, Chernoff N, Hill D, Antoniou MG, et al. A review on cylindrospermopsin: the global occurrence, detection, toxicity and degradation of a potent cyanotoxin. Environ Sci Process Impacts. 2013;15(11): 1979–2003.

[53] Oudra B, Loudiki M, Sbiyyaa B, Martins R, Vasconcelos V, Namikoshi N. Isolation, characterization and quantification of microcystins (heptapeptides hepatotoxins) in *Microcystis aeruginosa* dominated bloom of Lalla Takerkoust lake—reservoir (Morocco). Toxicon. 2001;39(9):1375–1381.

[54] Spoof L, Vesterkvist P, Lindholm T, Meriluoto J. Screening for cyanobacterial hepatotoxins, microcystins and nodularin in environmental water samples by reversed-phase liquid chromatography—electrospray ionisation mass spectrometry. J Chromatogr A. 2003;1020(1):105–119.

[55] Aboal M, Puig MÁ. Intracellular and dissolved microcystin in reservoirs of the river Segura basin, Murcia, SE Spain. Toxicon. 2005;45(4):509–518.

[56] Wood SA, Briggs LR, Sprosen J, Ruck JG, Wear RG, Holland PT, et al. Changes in concentrations of microcystins in rainbow trout, freshwater mussels and cyanobacteria in Lakes Rotoiti and Rotoehu. Environ Toxicol. 2006;21(3):205–222.

[57] Nasri H, El Herry S, Bouaïcha N. First reported case of turtle deaths during a toxic Microcystis spp. bloom in Lake Oubeira, Algeria. Ecotoxicol Environ Saf. 2008;71(2): 535–544.

[58] Everson S, Fabbro L, Kinnear S, Wright P. Extreme differences in akinete, heterocyte and cylindrospermopsin concentrations with depth in a successive bloom involving *Aphanizomenon ovalisporum* (Forti) and *Cylindrospermopsis raciborskii* (Woloszynska) Seenaya and Subba Raju [Internet]. 2011 [cited 2016 Jul 4]. Available from: http://www.sciencedirect.com/science/article/pii/S1568988310001204

[59] El Herry S, Fathalli A, Rejeb AJB, Bouaïcha N. Seasonal occurrence and toxicity of Microcystis spp. and Oscillatoria tenuis in the Lebna Dam, Tunisia. Water Res. 2008;42(4):1263–1273.

[60] Gurbuz F, Metcalf JS, Karahan AG, Codd GA. Analysis of dissolved microcystins in surface water samples from Kovada Lake, Turkey. Sci Total Environ. 2009;407(13):4038–4046.

[61] Mohamed ZA, Al Shehri AM. Microcystins in groundwater wells and their accumulation in vegetable plants irrigated with contaminated waters in Saudi Arabia. J Hazard Mater. 2009;172(1):310–315.

[62] Prakash S, Lawton LA, Edwards C. Stability of toxigenic Microcystis blooms. Harmful Algae. 2009;8(3):377–384.

[63] Liu Y, Chen W, Li D, Huang Z, Shen Y, Liu Y. Cyanobacteria-/cyanotoxin-contaminations and eutrophication status before Wuxi drinking water crisis in Lake Taihu, China. J Environ Sci. 2011;23(4):575–581.

[64] Romero-Oliva CS, Contardo-Jara V, Block T, Pflugmacher S. Accumulation of microcystin congeners in different aquatic plants and crops—a case study from lake Amatitlán, Guatemala. Ecotoxicol Environ Saf. 2014;102:121–128.

[65] Chen W, Jia Y, Li E, Zhao S, Zhou Q, Liu L, et al. Soil-based treatments of mechanically collected cyanobacterial blooms from Lake Taihu: efficiencies and potential risks. Environ Sci Technol. 2012;46(24):13370–13376.

[66] Chen W, Li L, Gan N, Song L. Optimization of an effective extraction procedure for the analysis of microcystins in soils and lake sediments. Environ Pollut. 2006;143(2):241–246.

[67] Chen W, Song L, Gan N, Li L. Sorption, degradation and mobility of microcystins in Chinese agriculture soils: risk assessment for groundwater protection. Environ Pollut. 2006;144(3):752–758.

[68] Mishra U, Pabbi S. Cyanobacteria:a potential biofertilizer for rice. Resonance. 2004;9(6): 6–10.

[69] Prasanna R, Jaiswal P, Nayak S, Sood A, Kaushik BD. Cyanobacterial diversity in the rhizosphere of rice and its ecological significance. Indian J Microbiol. 2009;49(1):89–97.

[70] Nain L, Rana A, Joshi M, Jadhav SD, Kumar D, Shivay YS, et al. Evaluation of synergistic effects of bacterial and cyanobacterial strains as biofertilizers for wheat. Plant Soil. 2010;331(1–2):217–230.

[71] Gupta V, Ratha SK, Sood A, Chaudhary V, Prasanna R. New insights into the biodiversity and applications of cyanobacteria (blue-green algae)—prospects and challenges. Algal Res. 2013;2(2):79–97.

[72] Han W, Clarke W, Pratt S. Composting of waste algae: a review. Waste Manag. 2014;34(7):1148–1155.

[73] Wang L, Xie L, Yang G, Yan Q, Ruan W. Impact of different bulking agents and compound microbial inoculant on blue algae composting. Chin J Env Eng. 2009;12:2261–2265.

[74] JIANG J, Du J, Chang Z, Jin H. Changes of contents of nutrients and microcystins during composting of blue algae. Jiangsu J Agric Sci. 2012;28(2):314–319.

[75] Dawson R. The toxicology of microcystins. Toxicon. 1998;36(7):953–962.

[76] Kormas KA, Lymperopoulou DS. Cyanobacterial toxin degrading bacteria: who are they? BioMed Res Int. 2013;2013:e463894.

[77] Metcalf JS, Richer R, Cox PA, Codd GA. Cyanotoxins in desert environments may present a risk to human health. Sci Total Environ. 2012;421:118–123.

[78] Richer R, Banack SA, Metcalf JS, Cox PA. The persistence of cyanobacterial toxins in desert soils. J Arid Environ. 2015;112:134–139.

[79] Dulić T, Meriluoto J, Palanački Malešević T, Gajić V, Važić T, Tokodi N, et al. Cyanobacterial diversity and toxicity of biocrusts from the Caspian Lowland loess deposits, North Iran. Quat Int [Internet]. [cited 2016 Apr 25]; Available from: http://www.sciencedirect.com/science/article/pii/S1040618216001725

[80] Zhang W, Fang T, Xu X. Study on photodegradation of cyanobacterial toxin in blooms of Dianchi Lake. Zhongguo Huanjing Kexue. 2000;21(1):1–3.

[81] Wörmer L, Huerta-Fontela M, Cirés S, Carrasco D, Quesada A. Natural photodegradation of the cyanobacterial toxins microcystin and cylindrospermopsin. Environ Sci Technol. 2010;44(8):3002–3007.

[82] Klitzke S, Beusch C, Fastner J. Sorption of the cyanobacterial toxins cylindrospermopsin and anatoxin-a to sediments. Water Res. 2011;45(3):1338–1346.

[83] Thirumavalavan M, Hu YL, Lee JF. Effects of humic acid and suspended soils on adsorption and photo-degradation of microcystin-LR onto samples from Taiwan reservoirs and rivers. J Hazard Mater. 2012;217–218:323–329.

[84] Maghsoudi E, Prévost M, Vo Duy S, Sauvé S, Dorner S. Adsorption characteristics of multiple microcystins and cylindrospermopsin on sediment: implications for toxin monitoring and drinking water treatment. Toxicon. 2015 Sep 1;103:48–54.

[85] Tsuji K, Naito S, Kondo F, Ishikawa N, Watanabe MF, Suzuki M, et al. Stability of microcystins from cyanobacteria: effect of light on decomposition and isomerization. Environ Sci Technol. 1994;28(1):173–177.

[86] Welker M, Steinberg C. Rates of humic substance photosensitized degradation of microcystin-LR in natural waters. Environ Sci Technol. 2000;34(16):3415–3419.

[87] Bibo L, Yan G, Bangding X, Jiantong L, Yongding L. A laboratory study on risk assessment of microcystin-RR in cropland. J Environ Manage. 2008 Feb;86(3):566–574.

[88] Jones GJ, Falconer IR, Wilkins RM. Persistence of cyclic peptide toxins in dried *Microcystis aeruginosa* crusts from lake Mokoan, Australia. Environ Toxicol Water Qual. 1995 Feb 1;10(1):19–24.

[89] Miller MJ, Critchley MM, Hutson J, Fallowfield HJ. The adsorption of cyanobacterial hepatotoxins from water onto soil during batch experiments. Water Res. 2001 Apr;35(6): 1461–1468.

[90] Miller MJ, Fallowfield HJ. Degradation of cyanobacterial hepatotoxins in batch experiments. Water Sci Technol. 2001 Jun 1;43(12):229–232.

[91] Morris RJ, Williams DE, Luu HA, Holmes CFB, Andersen RJ, Calvert SE. The adsorption of microcystin-LR by natural clay particles. Toxicon. 2000;38(2):303–308.

[92] Corbel S, Mougin C, Nélieu S, Delarue G, Bouaïcha N. Evaluation of the transfer and the accumulation of microcystins in tomato (Solanum lycopersicum cultivar Micro-Tom) tissues using a cyanobacterial extract containing microcystins and the radiolabeled microcystin-LR (14C-MC-LR). Sci Total Environ. 2016;541:1052–1058.

[93] Corbel S, Bouaïcha N, Mougin C. Dynamics of the toxic cyanobacterial microcystin-leucine-arginine peptide in agricultural soil. Environ Chem Lett. 2014;12(4):535–541.

[94] Bourne DG, Riddles P, Jones GJ, Smith W, Blakeley RL. Characterisation of a gene cluster involved in bacterial degradation of the cyanobacterial toxin microcystin LR. Environ Toxicol. 2001;16(6):523–534.

[95] Manage PM, Edwards C, Singh BK, Lawton LA. Isolation and identification of novel microcystin-degrading bacteria. Appl Environ Microbiol. 2009 Nov;75(21):6924–6928.

[96] Eynard F, Mez K, Walther JL. Risk of cyanobacterial toxins in Riga waters (Latvia). Water Res. 2000;34(11):2979–2988.

[97] Holst T, Jørgensen NOG, Jørgensen C, Johansen A. Degradation of microcystin in sediments at oxic and anoxic, denitrifying conditions. Water Res. 2003;37(19):4748–4760.

[98] Pflugmacher S, Ame V, Wiegand C, Steinberg C. Cyanobacterial toxins and endotoxins: their origin and their ecophysiological effects in aquatic organisms. Wasser Boden. 2001;53(4):15–20.

[99] Peuthert A, Chakrabarti S, Pflugmacher S. Uptake of microcystins-LR and -LF (cyanobacterial toxins) in seedlings of several important agricultural plant species and the correlation with cellular damage (lipid peroxidation). Environ Toxicol. 2007 Aug 1;22(4):436–442.

[100] Crush JR, Briggs LR, Sprosen JM, Nichols SN. Effect of irrigation with lake water containing microcystins on microcystin content and growth of ryegrass, clover, rape and lettuce. Environ Toxicol. 2008;23(2):246–252.

[101] Chen J, Dai J, Zhang H, Wang C, Zhou G, Han Z, et al. Bioaccumulation of microcystin and its oxidative stress in the apple (*Malus pumila*). Ecotoxicology. 2010;19(4):796–803.

[102] Gutiérrez-Praena D, Campos A, Azevedo J, Neves J, Freitas M, Guzmán-Guillén R, et al. Exposure of Lycopersicon esculentum to microcystin-LR: effects in the leaf proteome and toxin translocation from water to leaves and fruits. Toxins. 2014;6(6):1837–1854.

[103] Prieto A, Campos A, Cameán A, Vasconcelos V. Effects on growth and oxidative stress status of rice plants (*Oryza sativa*) exposed to two extracts of toxin-producing cyanobacteria (*Aphanizomenon ovalisporum* and *Microcystis aeruginosa*). Ecotoxicol Environ Saf. 2011;74(7):1973–1980.

[104] Contardo-Jara V, Schwanemann T, Pflugmacher S. Uptake of a cyanotoxin, β-*N*-methylamino-l-alanine, by wheat (*Triticum aestivum*). Ecotoxicol Environ Saf. 2014;104:127–131.

[105] Valdor R, Aboal M. Effects of living cyanobacteria, cyanobacterial extracts and pure microcystins on growth and ultrastructure of microalgae and bacteria. Toxicon. 2007;49(6):769–779.

[106] Giaramida L, Manage PM, Edwards C, Singh BK, Lawton LA. Bacterial communities' response to microcystins exposure and nutrient availability: linking degradation capacity to community structure. Int Biodeterior Biodegrad. 2013 Oct;84:111–117.

[107] Corbel S, Mougin C, Martin-Laurent F, Crouzet O, Bru D, Nélieu S, et al. Evaluation of phytotoxicity and ecotoxicity potentials of a cyanobacterial extract containing microcystins under realistic environmental concentrations and in a soil—plant system. Chemosphere. 2015;128:332–340.

[108] Corbel S, Bouaïcha N, Martin-Laurent F, Crouzet O, Mougin C. Soil irrigation with toxic cyanobacterial microcystins increases soil nitrification potential. Environ Chem Lett. 2015;13(2):459–463.

[109] El Khalloufi F, Oufdou K, Bertrand M, Lahrouni M, Oudra B, Ortet P, et al. Microbiote shift in the *Medicago sativa* rhizosphere in response to cyanotoxins extract exposure. Sci Total Environ. 2016;539:135–142.

[110] Lahrouni M, Oufdou K, Khalloufi FE, Benidire L, Albert S, Göttfert M, et al. Microcystin-tolerant Rhizobium protects plants and improves nitrogen assimilation in *Vicia faba* irrigated with microcystin-containing waters. Environ Sci Pollut Res. 2016;23:1–13.

[111] Lahrouni M, Oufdou K, Faghire M, Peix A, El Khalloufi F, Vasconcelos V, et al. Cyanobacterial extracts containing microcystins affect the growth, nodulation process and nitrogen uptake of faba bean (*Vicia faba* L., Fabaceae). Ecotoxicology. 2012;21(3):681–687.

[112] Imanishi S, Kato H, Mizuno M, Tsuji K, Harada K. Bacterial degradation of microcystins and nodularin. Chem Res Toxicol. 2005;18(3):591–598.

[113] Jia Y, Du J, Song F, Zhao G, Tian X. A fungus capable of degrading microcystin-LR in the algal culture of *Microcystis aeruginosa* PCC7806. Appl Biochem Biotechnol. 2012;166(4):987–996.

[114] Ferrão-Filho A da S, Kozlowsky-Suzuki B. Cyanotoxins: bioaccumulation and effects on aquatic animals. Mar Drugs. 2011;9(12):2729–2772.

[115] Li Y, Wang Y, Yin L, Pu Y, Wang D. Using the nematode *Caenorhabditis elegans* as a model animal for assessing the toxicity induced by microcystin-LR. J Environ Sci. 2009;21(3):395–401.

[116] Li Y, Ye H, Du M, Zhang Y, Ye B, Pu Y, et al. Induction of chemotaxis to sodium chloride and diacetyl and thermotaxis defects by microcystin-LR exposure in nematode *Caenorhabditis elegans*. J Environ Sci. 2009;21(7):971–979.

[117] Holajjer P, Kamra A, Gaur HS, Manjunath M. Potential of cyanobacteria for biorational management of plant parasitic nematodes: a review. Crop Prot. 2013;53:147–151.

[118] Lána J, Hofman J, Bláha L. Can cyanobacterial biomass applied to soil affect survival and reproduction of springtail Folsomia candida? Ecotoxicol Environ Saf. 2011;74(4):840–843.

[119] Saqrane S, Ghazali IE, Oudra B, Bouarab L, Vasconcelos V. Effects of cyanobacteria producing microcystins on seed germination and seedling growth of several agricultural plants. J Environ Sci Health Part B. 2008;43(5):443–451.

[120] El Khalloufi F, El Ghazali I, Saqrane S, Oufdou K, Vasconcelos V, Oudra B. Phytotoxic effects of a natural bloom extract containing microcystins on *Lycopersicon esculentum*. Ecotoxicol Environ Saf. 2012;79:199–205.

[121] Chen J, Song L, Dai J, Gan N, Liu Z. Effects of microcystins on the growth and the activity of superoxide dismutase and peroxidase of rape (*Brassica napus* L.) and rice (*Oryza sativa* L.). Toxicon. 2004;43(4):393–400.

[122] Pflugmacher S, Hofmann J, Hübner B. Effects on growth and physiological parameters in wheat (*Triticum aestivum* L.) grown in soil and irrigated with cyanobacterial toxin contaminated water. Environ Toxicol Chem. 2007;26(12):2710–2716.

[123] Gehringer MM, Kewada V, Coates N, Downing TG. The use of *Lepidium sativum* in a plant bioassay system for the detection of microcystin-LR. Toxicon. 2003;41(7): 871–876.

[124] Corbel S, Bouaïcha N, Nélieu S, Mougin C. Soil irrigation with water and toxic cyanobacterial microcystins accelerates tomato development. Environ Chem Lett. 2015;13:1–6.

[125] Yin L, Huang J, Huang W, Li D, Liu Y. Responses of antioxidant system in *Arabidopsis thaliana* suspension cells to the toxicity of microcystin-RR. Toxicon. 2005;46(8):859–864.

[126] Stüven J, Pflugmacher S. Antioxidative stress response of *Lepidium sativum* due to exposure to cyanobacterial secondary metabolites. Toxicon. 2007;50(1):85–93.

[127] Saqrane S, Ouahid Y, El Ghazali I, Oudra B, Bouarab L, del Campo FF. Physiological changes in *Triticum durum*, *Zea mays*, *Pisum sativum* and *Lens esculenta* cultivars, caused by irrigation with water contaminated with microcystins: a laboratory experimental approach. Toxicon. 2009;53(7–8):786–796.

[128] Azevedo CC, Azevedo J, Osório H, Vasconcelos V, Campos A. Early physiological and biochemical responses of rice seedlings to low concentration of microcystin-LR. Ecotoxicology. 2014;23(2):107–121.

[129] Lahrouni M, Oufdou K, El Khalloufi F, Baz M, Lafuente A, Dary M, et al. Physiological and biochemical defense reactions of *Vicia faba* L.—Rhizobium symbiosis face to chronic exposure to cyanobacterial bloom extract containing microcystins. Environ Sci Pollut Res. 2013;20(8):5405–5415.

[130] Freitas M, Azevedo J, Pinto E, Neves J, Campos A, Vasconcelos V. Effects of microcystin-LR, cylindrospermopsin and a microcystin-LR/cylindrospermopsin mixture on growth, oxidative stress and mineral content in lettuce plants (*Lactuca sativa* L.). Ecotoxicol Environ Saf. 2015;116:59–67.

[131] Beyer D, Suranyi G, Vasas G, Roszik J, Erdodi F, Márta M, et al. Cylindrospermopsin induces alterations of root histology and microtubule organization in common reed (*Phragmites australis*) plantlets cultured in vitro. Toxicon. 2009;54(4):440–449.

[132] Máthé C, Vasas G, Borbély G, Erdodi F, Beyer D, Kiss A, et al. Histological, cytological and biochemical alterations induced by microcystin-LR and cylindrospermopsin in white mustard (*Sinapis alba* L.) seedlings. Acta Biol Hung. 2013;64(1):71–85.

[133] Freitas M, Campos A, Azevedo J, Barreiro A, Planchon S, Renaut J, et al. Lettuce (*Lactuca sativa* L.) leaf-proteome profiles after exposure to cylindrospermopsin and a microcystin-LR/cylindrospermopsin mixture: a concentration-dependent response. Phytochemistry. 2015;110:91–103.

[134] Banack SA, Murch SJ, Cox PA. Neurotoxic flying foxes as dietary items for the Chamorro people, Marianas Islands. J Ethnopharmacol. 2006;106(1):97–104.

[135] Kanzo Y, Ruike K, Inamori R, Suzuki R, Qin Xu K, Inamori Y. Analysis of absorption and accumulation of blue-green algae toxin Microcystin in the Komatsuna cultivation. J Bioind Sci JBS. 2013;2(1):12.

[136] Monferran MV, Wunderlin DA, Nimptsch J, Pflugmacher S. Biotransformation and antioxidant response in Ceratophyllum demersum experimentally exposed to 1,2-and 1,4-dichlorobenzene. Chemosphere. 2007;68(11):2073–2079.

[137] Pereira S, Saker ML, Vale M, Vasconcelos VM. Comparison of sensitivity of grasses (*Lolium perenne* L. and *Festuca rubra* L.) and lettuce (*Lactuca sativa* L.) exposed to water contaminated with microcystins. Bull Environ Contam Toxicol. 2009;83(1):81–84.

[138] Järvenpää S, Lundberg-Niinistö C, Spoof L, Sjövall O, Tyystjärvi E, Meriluoto J. Effects of microcystins on broccoli and mustard and analysis of accumulated toxin by liquid chromatography—mass spectrometry. Toxicon. 2007;49(6):865–874.

[139] Li YW, Zhan XJ, Xiang L, Deng ZS, Huang BH, Wen HF, et al. Analysis of trace microcystins in vegetables using solid-phase extraction followed by high performance liquid chromatography triple-quadrupole mass spectrometry. J Agric Food Chem. 2014;62(49):11831–11839.

3

Copper Contamination in Mediterranean Agricultural Soils: Soil Quality Standards and Adequate Soil Management Practices for Horticultural Crops

Daniel Sacristán and Ester Carbó

Additional information is available at the end of the chapter

Abstract

This chapter increases the knowledge on the management of Cu-contaminated Mediterranean agricultural soils, by analysing the current soil quality standards for different Mediterranean regions and proposing new criteria for their establishment based on the influence of soil properties and type of crop. We evaluate the effect of Cu and its interaction with soil properties on biomass production of lettuce (*Lactuca sativa* L.) and tomato (*Solanum lycopersicum* L.), by establishing the effective concentrations EC_{50} and EC_{10} (effective concentrations of Cu in soil that reduces biomass production by 50 and 10%, respectively), and its absorption, translocation and accumulation in the different parts of the plant. Two different biomass assays were carried out in seven types of Mediterranean agricultural soils (four from Europe and three from Australia) contaminated with different Cu concentrations. When lettuce was grown, similar toxic effects and accumulation values were obtained for both of the agricultural areas under analysis. In both cases, the maximum threshold value was obtained for the soil having the highest pH and clay content, independently of the soil type. When comparing both crops in the European Mediterranean soils, toxicity values calculated for tomato were higher, and translocation of Cu to the fruit was constantly low, independently of the Cu dose. Moreover, tomato showed an important phytoremediation potential, extracting Cu from not only low–medium but also from highly (>1700 mg/kg) Cu-contaminated basic agricultural soils, and having low translocation rates to fruits. The analysis of the influence of soil properties on the effect of Cu on plant biomass production led to similar conclusions in both assays. SOM, clay content and CEC are the most relevant properties affecting the dynamic of Cu in soil. Considering this, for the type of crops and soils considered, the effect of Cu on plant biomass production was the most relevant of those analysed, and pH, clay content, SOM and CEC the most relevant soil properties. Therefore, these aspects should be considered when establishing adequate soil quality standards and proposing adequate soil management practices.

Keywords: agricultural soils, Cu toxicity, soil quality standards, European Mediterranean region, Australian Mediterranean region

1. Introduction

The contamination of soils, especially agricultural ones, with heavy metals is an extended soil degradation process that affects vast areas of the planet [1–7].

In a world with a productive model based on extensive areas with intensive inputs, some of which are sometimes hazardous and destructive, direct (solid waste disposals, mine residues, etc.) and indirect (inadequate agricultural practices) soil contamination processes are very likely to continue happening, especially in agricultural areas. These can lead to serious environmental problems, linked to soil degradation processes due to excessive accumulation of these toxic substances and can affect different ecosystems. Furthermore, this excessive accumulation of heavy metals in agricultural soils may not only result in environmental contamination but can also cause an increase on the heavy metal uptake by crops, affecting this way food quality and safety. According to [8], soil plays a central role in food safety as it determines the possible composition of food and feed at the root of the food chain.

The heavy metal contamination of soil is one of the most pressing concerns in the debate about food security and food safety in Europe [9] and globally [10]. However, the quality of the resource soil, defining this as the potential impact on human health derived from the propagation of harmful elements through the food chain, has not been properly studied in Europe due to the lack of adequate data, in terms of detail and reliability.

Of these harmful elements, those heavy metals considered micronutrients are particularly relevant, since plants tend to behave differently towards them, being more tolerant, and enhancing their absorption and accumulation in different plant tissues. Of special concern is Cu, since this heavy metal is extensively used as a fungicide; it is the main component of different chemical fertilisers and is present at high concentrations in sewage sludge and pig manure. Komárek et al. [11] carried out an extensive bibliographical research on the use of Cu as fungicide around the world and determined concentrations of Cu in agricultural soils of up to 3216 mg_{Cu}/kg.

In order to characterise contaminated soils, commonly, two different approaches have been developed: (i) establishment of soil quality standards and (ii) risk assessment [12]. The approaches based on soil quality standards have a great advantage, as the characterisation can be quick and cheap in many cases. However, difficulties arise if one considers the complexity of soils [13]. On the other hand, the approaches based on direct risk assessment are undoubtedly more realistic, but they require a degree of soil information that is not always available. Moreover, the costs associated with the application of these latter can be hardly undertaken in many cases [14].

Concerning the establishment of the soil quality standards, it is well known that different soil properties affect the dynamics of heavy metals in soils [15] and that different plants/crops behave differently in relation to toxicity problems and accumulation limits of heavy metals. However, these two aspects are not usually considered in the establishment of these values. Furthermore, high concentrations of elements such as Cu in soil can lead to toxicity problems to plants and the consequent reduction in plant biomass production [16] and/or to potential animal and human health risk because of the accumulation of Cu in vegetables, since, as commented previously, plant uptake from soil is the main way for Cu to enter the food chain [8, 17]. According to [18], some vegetables can accumulate relatively high levels of Cu from soil without any toxic effect. Therefore, both aspects (plant biomass production and Cu accumulation in the plant) are relevant when analysing Cu contamination of agricultural soils and toxicity in crops and necessary to establish/define adequate soil conservation and management strategies.

Regarding the accumulation of Cu in the edible part of the plant, some national and international legislations (e.g. [19, 20]) clearly establish the maximum Cu content in the edible part of the plant, which is 10 mg/kg in fresh weight basis. However, this is not so for the effect on biomass production.

Considering all the above, it arises the need to carry out better and more detailed analysis in order to define adequate soils quality standards taking into account these two factors. The consequences of not considering these two factors are that soil quality standards are commonly too indulgent, not reflecting the complexity of agricultural ecosystems and jeopardising the health of both ecosystems and humans.

The definition of adequate soils quality standards for different climatic areas, such as the Mediterranean region, and for different crops, such as the horticultural ones, will enable to suggest adequate agricultural practices to manage and preserve the resource soil under Cu contamination problems in the Mediterranean agricultural soils.

2. Study area and objectives

The study area selected was the Mediterranean Region. This area includes different parts of the world and covers all the countries with Mediterranean climate, in all or some part of it (**Figure 1**). This region is of special concern since it is said or considered to include the "orchards of the world" [21].

Within this region, one of the areas studied was the European Mediterranean region. The representative soils of this region were sampled from the Valencian Region, an area located in the south-east of Spain. This area can be considered as representative since climatic conditions and soil properties of this area are typical of the European Mediterranean Region. Furthermore, this area has undergone, over the recent decades, the same land use pattern changes as the one occurred in most of the European Mediterranean Region, where there has been an intensification of agricultural development, characterised by high consumption of agrochemicals, and an expansion of industrial-urban uses [23–25].

The other Mediterranean region considered was the Mediterranean area of Australia. Climatic conditions are similar to the ones describe previously, although the properties of the soils present in this area differ slightly, and include, for example, soils with lower pH values. Adequate representative soils were sampled from the Commonwealth Scientific and Industrial Research Organisation (CSIRO) Ginninderra Experiment Station (Australian Capital Territory —ACT), sampling those representative of the Mediterranean region [26] and that were dedicated to agriculture.

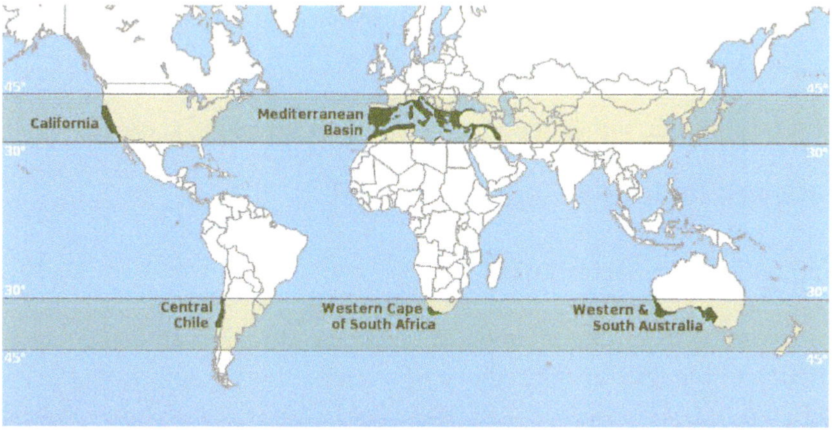

Figure 1. Distribution of the Mediterranean climate in the world [22].

Regarding the protection and conservation of soils, it is important to consider that many Mediterranean countries, including Spain (representative of the European Mediterranean) and Australia, use soil quality standards to characterise contaminated soils.

More specifically, in Spain, and according to the Spanish Royal Decree 9/2005 [27], any soil must be considered as potentially contaminated (or contaminated) when concentrations (or concentrations 100 times) above the corresponding baseline value are determined in them. In agricultural soils, the baseline value for the different elements is established taking into account the upper limit of the normal range of concentrations, which covers the natural variability of the metal in soil associated with background levels at regional level. This normal range of concentrations considers diffuse or nonpoint pollution (e.g. fertilisation and atmospheric deposition) but does not include point pollution due to local human activities (e.g. industries) [17, 28–30]. These values are useful to identify the current contents of heavy metals and to assess the degree of contamination by human activities [30]. Regarding the establishment of these values, Micó et al. [30] and Sánchez et al. [31] established the baseline values for different heavy metals in agricultural soils under vegetable crops of the Valencian Mediterranean region. The baseline for Cu was 65.9 mg/kg, and it is similar to those established in other Spanish Mediterranean regions [32, 33] and in other European Mediterranean regions [34, 35].

On the other hand, Australian guidelines for metal contaminant concentrations in soil and soil amendments are established at a state level (e.g. [36–38]) and are based on European regulations and research [20], which do not reflect the influence of both the soil and the climate of Australia.

Therefore, taking into account all of the above, the objective of the chapter is to analyse and discuss the results obtained by [39–41] concerning the definition of adequate soils quality standards for the Mediterranean region and the approach made to define adequate soil management practices, after considering the different soil properties of different representative soils of the Mediterranean region (European and Australian) and two horticultural crops representative of two different accumulation strategies: accumulator and non-accumulator. This will enable to suggest adequate agricultural practices to manage and preserve the resource soil under Cu contamination problems.

3. Materials and methods

3.1. Sampling and soil characterisation

Four agricultural plots from the Spanish Mediterranean region and three agricultural plots from the Australian Mediterranean region, all having different soil properties, were selected and sampled.

On one hand, the selection of the Spanish soils was performed considering the information and databases of previous studies [42, 43]. These classify them as representative of the European Mediterranean agricultural area. More specifically, the types of soils represented were two Calcaric Fluvisols with different soil properties (Sollana and Peníscola), a Gleyic Fluvisol (Nules) and a Salic Fluvisol (Rojales), according to the World Reference Base for Soil Resources [44]. The soils selected covered a wide range of the different types of soils devoted to vegetable crops in the European Mediterranean region [45].

On the other hand, the selection of the Australian soils was carried out considering the information of the Commonwealth Scientific and Industrial Research Organisation (CSIRO). The types of soils represented were a Chromic Luvisol (Soil 1), an Eutric Planosol (Soil 2) and a Pellic Vertisol (Soil 3), according to the World Reference Base for Soil Resources [44].

Soil properties were determined according to the official laboratory methods of the Spanish Ministry of Agriculture, Fishery and Food [46] for the soils of the Spanish Mediterranean Region, and to the official soil chemical methods for Australasia [47] for the soils of the Australian Mediterranean Region.

3.2. Experimental design

Three different sets of experiments were carried out and compared, each one including two different ecotoxicological assays (described later): one set of experiments with European Mediterranean soils and lettuce; another set with Australian Mediterranean soils and lettuce; the last set with European Mediterranean soils and tomato.

The sampled agricultural soils indicated previously were spiked with a Cu contaminant solution to achieve six different total Cu concentrations, the control (no Cu addition) and five different doses (65.9, 659.0, 1977.0, 3295.0 and 6590.0 mg_{Cu}/kg). These ranges of doses were selected and established after considering previous studies also carried out in Mediterranean agricultural soils [39, 48, 49].

Two different ecotoxicological assays were conducted in the contaminated soils: one to evaluate the effect of Cu over biomass production (28 days); and the other to analyse the absorption and accumulation of Cu in roots and stem and leaves for lettuce, or in roots, stem and leaves and fruit for tomato (3 months).

For the first assay, biomass production was assessed following the OECD test 208 [50], where 300 g of contaminated soils was placed in pots (10 cm in diameter) and ten lettuce or five tomato seeds were then seeded to 1 cm soil depth. Each treatment was replicated three times (three pots per Cu dose and three per control), and all pots were placed in a glasshouse. Experimental conditions were controlled and maintained according to the requirements specified in the biomass assay procedure [50].

For the accumulation assay, 1.2 kg of contaminated soils were placed in 25 cm diameter pots and ten lettuce or five tomato seeds were seeded to 1 cm soil depth, although only one of the germinated seeds was selected to grow until maturity. As for the biomass assay, each treatment was replicated three times (three pots per Cu dose and three per control) and all pots were placed in a glasshouse. Again, experimental conditions were controlled and maintained according to the requirements specified in the biomass assay procedure [50].

3.3. Biomass data analysis

Weight values obtained in the biomass assay were used to establish the EC_{50} and EC_{10} effective concentrations. Previous to this, homogeneity of variance and normality of weight data was checked using the Kolmogorov-Smirnov test and these were log-transformed when appropriate in order to stabilise variances. Dose-response data were fitted to a log-logistic curve according to Eq. (1) [51] for each of the soils tested in order to establish the EC_{50} and EC_{10}. TRAP© version 1.22 (Toxicity Relationship Analysis Program, United States Environmental Protection Agency) was used for this purpose [52–54].

$$y = \frac{y_0}{1 + e^{b(x-M)}} \tag{1}$$

where y = biomass (lettuce/tomato shoot weight of plants) produced (mg), x = log10(added Cu) (mg/kg), y_0 = biomass produced with non-added Cu (control) (mg), and M and b are parameters to be fitted, where M = log10(EC_{50}) and b is a slope parameter that indicates the inhibition rate. The concentration of Cu considered in the control dose was the initial Cu content of the soil assayed. The distribution of residuals, relationship between these and the fitted values and the adjusted coefficient of determination (R^2adj.) were examined in order to determine the model's adequacy. The EC_{10} was also calculated as described above.

3.4. Cu content in soils and plants

Stem and leaves, and root samples of the accumulation assay were grounded and 0.5 sieved prior to their analysis. Total Cu concentration in soils, stem and leaves, and roots was determined using the USEPA 3052 method [55]. Copper content in soils and plants was analysed by a Microwave Plasma Atomic Emission Spectrometer (MP-AES). The precision and the accuracy of the analysis were evaluated calculating the relative standard deviation (RSD) and the recovery of metal of external standards provided by the commercial house (Agilent) and different Certified Reference Materials (CRM). RSD values (from 4 to 9%) were smaller than 10% and were considered satisfactory [56]. Recoveries ranged from 83 to 111% and were within 80–120% interval proposed as satisfactory by [56].

In order to compare the Cu concentrations obtained in stem and leaves with the maximum Cu content in foodstuffs (10 mg/kg in fresh weight basis for lettuce) established by the identified legislation [57], different conversion factors were applied. These were calculated by assessing their moisture content through a gravimetric method [47]. Furthermore, and considering this maximum value, the critical limit that refers to the concentration of Cu in soil that results in the maximum concentration allowed in vegetable crops was defined when possible.

Moreover, to assess the accumulation and distribution of Cu in lettuce and tomato plants, and therefore their phytoremediation potential, three different concentration factors (CFs) were calculated. In this case study, the ratio between the heavy metal concentrations in root (mg/kg dry weight) and in soil; the ratio between the concentrations in stem and leaves and in root; and the ratio between the concentrations in fruits and in stem and leaves were calculated for each soil and dose.

It is important to point out that, in this study, the total Cu in soil that is bioavailable has been considered to be very similar to the total Cu concentration in soil. Although not realistic for aged contaminated soils, spiked soils realistically reflect the conditions in terms of contamination that can take place in agricultural soils as a result of different contamination processes. More specifically, they realistically reflect contamination processes and conditions associated with an excessive Cu-based pesticide and fungicide application, or due to spills [58] or intensive extractive activities nearby [59], where Cu is artificially added and is very bioavailable. In such cases, the values of total and bioavailable Cu content are very similar, so both concentrations can be used to analyse this type of contamination [39, 60].

3.5. Statistical analysis

After checking the distribution and homogeneity of variance, mean biomass produced for the different doses and soils was compared applying two-way ANOVAs and Turkey test, in order to elucidate differences amongst soils and doses. The influence of soil properties on biomass production and in the accumulation of Cu in the edible part of the plant was assessed by correlation analyses. Correlations were derived between each of the effective concentrations (EC_{50} and EC_{10}) calculated and the soil properties of the different soils sampled, and between the soil properties and the concentrations in plants at the different doses assayed. The correlation coefficients considered were Pearson's since the data had a normal distribution. All these statistical analyses were conducted using SPSS© version 19.3.

4. Results

Table 1 summarises the main properties of the seven soils assayed (Rojales, Sollana, Nules, Peníscola, Soil 1, Soil 2 and Soil 3). As it can be observed, a wide range of different soil properties was covered with the selected soils, enabling this way to analyse the influence of the different properties over the dynamics of Cu in soils and its transference to the plant.

Soil	pH	EC (dS/m)	SOM (%)	CCE (%)	CEC (cmol(+)/kg)	Sand (%)	Silt (%)	Clay (%)	Initial Cu (mg/kg)
Rojales	7.66	0.90	1.6	52	14.5	28	38	33	12.4
Sollana	7.48	2.38	3.8	53	27.6	12	41	47	30.9
Nules	7.72	3.26	8.7	39	37.1	19	34	48	58.5
Peníscola	7.72	1.86	2.7	45	16.8	49	25	25	17.4
Soil 1	5.36	1.10	3.7	0	4.2	10	10	80	7.6
Soil 2	5.67	1.34	4.6	0	13.1	26	36	38	17.6
Soil 3	7.41	2.05	3.5	0	36.5	42	43	15	15.5

EC, electrical conductivity; SOM, soil organic matter content; CCE, calcium carbonate equivalent content; CEC, cation exchange capacity

Table 1. Properties of the seven soils assayed [39–41].

	EC_{10} [a]		EC_{50} [b]		R^2 adj. (%) [c]	
soil	Lettuce	Tomato	Lettuce	Tomato	Lettuce	Tomato
Rojales	8.8 ± 0.9	32.9 ± 0.3	177 ± 2.1	500.7 ± 0.1	89	93
Sollana	46.2 ± 1.3	393.5 ± 0.2	680 ± 3.4	1223.8 ± 0.2	88	81
Nules	159 ± 3.4	491.4 ± 0.6	753 ± 2.9	1696.5 ± 0.4	97	50
Peníscola	–	358.4 ± 0.2	–	663.8 ± 0.2	–	98
Soil 1	49.0 ± 1.7	–	104.0 ± 2.0	–	90	–
Soil 2	106.9 ± 2.0	–	236.4 ± 2.4	–	94	–
Soil 3	443.1 ± 2.6	–	728.9 ± 2.9	–	93	–

– not assayed.
[a]Effective concentrations of added Cu that caused a 10% reduction in the biomass produced.
[b]Effective concentrations of added Cu that caused a 50% reduction in the biomass produced.
Percentage of variance accounted for by the log-logistic model.

Table 2. Toxicity threshold values (EC_{10} and EC_{50}, mg/kg) for Cu added to soil derived from the lettuce and tomato biomass tests in the seven soils assayed [39–41].

Table 2 shows and sums up toxicity threshold values (EC_{10} and EC_{50}) calculated for each soil and crop.

Dose Cu (mg/kg)	Rojales			Sollana			Nules		
	Total content of Cu (mg/kg) in soil	Total content of Cu in crop[a] (mg/kg)	CF[b]	Total content of Cu (mg/kg) in soil	Total content of Cu in crop[a] (mg/kg)	CF[b]	Total content of Cu (mg/kg) in soil	Total content of Cu in crop[a] (mg/kg)	CF[b]
0.01 (control)	8.9 ± 1.0	18 ± 3.0	2.03	22.1 ± 3.0	10.1 ± 2.0	0.46	44.4 ± 5.0	12.6 ± 2.0	0.28
65.9	77.0 ± 9.0	16.5 ± 3.0	0.21	70.0 ± 8.0	12.4 ± 2.0	0.18	83.4 ± 10.0	10.5 ± 2.0	0.13
659.0	365.1 ± 44.0	20.1 ± 3.0	0.06	403.2 ± 49.0	14.9 ± 3.0	0.04	359.4 ± 44.0	16.2 ± 3.0	0.05
1977.0	1549.7 ± 188.0	50.4 ± 9.0	0.03	1607.7 ± 195.0	24.0 ± 4.0	0.01	1622.9 ± 197.0	23.0 ± 4.0	0.01
3295.0	3271.2 ± 397.0	212.5 ± 36.0	0.06	3382.5 ± 410.0	74.6 ± 13.0	0.05	2738.7 ± 332.0	29.2 ± 5.0	0.01
6590.0	5831.0 ± 707.0	–	–	5853.0 ± 710.0	–	–	5850.0 ± 710.0	–	–

All the results are expressed in mg/kg in dry weight basis [39].
– no biomass produced.
[a]The conversion factors that have to be applied in order to calculate the content of metal in crop in fresh weight basis are the following: 11.2 for Rojales, 17.3 for Sollana and 17.6 for Nules.
[b]Concentration factor.

Table 3. Mean copper content in the edible parts of lettuces (mg/kg in dry weight basis), and mean total contents of copper in the European Mediterranean soils assayed.

Dose Cu (mg/kg)	Soil 1					Soil 2					Soil 3				
	Total content of Cu (mg/kg) in soil	Total content of Cu (mg/kg) in root	Total content of Cu (mg/kg) in leaves[a]	CF_{s-r}	CF_{r-l}	Total content of Cu (mg/kg) in soil	Total content of Cu (mg/kg) in root	Total content of Cu (mg/kg) in leaves[a]	CF_{s-r}	CF_{r-l}	Total content of Cu (mg/kg) in soil	Total content of Cu (mg/kg) in root	Total content of Cu (mg/kg) in leaves[a]	CF_{s-r}	CF_{r-l}
0.01 (control)	7.6 ± 4.5	10.6 ± 5.0	1.8 ± 1.0	1.3	0.2	17.6 ± 4.4	8.6 ± 2.4	1.4 ± 0.9	0.5	0.2	15.5 ± 1.0	7.9 ± 6.5	1.5 ± 0.2	0.5	0.2
65.9	53.7 ± 15.0	53.6 ± 17.9	3.0 ± 0.1	0.9	0.1	74.5 ± 20.7	96.3 ± 9.2	1.6 ± 0.2	1.2	0.1	39.6 ± 12.7	14.0 ± 6.9	3.5 ± 0.6	0.4	0.3
659.0	658.0 ± 72.0	–	–	–	–	670.6 ± 48.1	199.9 ± 21.3	10.1 ± 0.8	0.3	0.1	434.9 ± 57.5	60.5 ± 9.3	9.2 ± 0.1	0.1	0.2
1977.0	2281.7 ± 58.1	–	–	–	–	1808.5 ± 136.2	–	–	–	–	1985.0 ± 143.9	183.2 ± 15.9	35.5 ± 5.5	0.1	0.2
3295.0	3197.7 ± 498.2	–	–	–	–	3096.2 ± 540.0	–	–	–	–	3064.1 ± 146.5	–	–	–	–
6590.0	7227.8 ± 995.2	–	–	–	–	5338.0 ± 900.9	–	–	–	–	5604.5 ± 167.4	–	–	–	–

All the results are expressed in mg/kg in dry weight basis [41].
– no biomass produced.
CF_{s-r}: concentration factor, between soil and root; CF_{r-l}: concentration factor, between root and leaf.
[a]The conversion factors that have to be applied in order to calculate the content of metal in plant in fresh weight basis are the following: 8.2 for Soil 1, 8.8 for Soil 2, 9.9 for Soil 3.

Table 4. Mean copper content in the Australian Mediterranean soils assayed and mean copper content in roots and the edible part of lettuce.

Tables 3–5 show the results obtained in terms of Cu concentration in soils and in the different parts of the plants analysed, indicated previously.

	Total content of Cu (mg/kg)									
Dose Cu	Rojales					Sollana				
(mg/kg)	In soil	In plant[a]	In fruit[b]	$CF_{s\text{-}p}$	$CF_{p\text{-}f}$	In soil	In plant[a]	In fruit[b]	$CF_{s\text{-}p}$	$CF_{p\text{-}f}$
0.01 (control)	12.4 ± 1.7	23.8 ± 2.5	7.8 ± 1.9	1.92	0.33	30.9 ± 4.3	21.7 ± 2.2	8.2 ± 2.0	0.70	0.38
65.9	64.1 ± 8.9	28.8 ± 3.0	7.3 ± 1.7	0.45	0.25	79.1 ± 11.0	27.6 ± 2.9	8.1 ± 2.0	0.35	0.29
659.0	612.5 ± 84.9	31.9 ± 3.3	–	0.05	–	673.8 ± 93.4	26.3 ± 2.6	8.6 ± 2.0	0.04	0.33
1977.0	1879.9 ± 260.7	63.5 ± 6.6	–	0.03	–	2003.7 ± 277.8	27.8 ± 2.8	7.6 ± 1.8	0.01	0.27
3295.0	3670.0 ± 480.7	242.5 ± 25.1	–	0.07	–	2915.8 ± 404.3	28.5 ± 2.5	9.1±2.2	0.01	0.32
6590.0	6404.5 ± 888.1	641.3 ± 66.4	–	0.10	–	7080.0 ± 922.8	688.5 ± 71.2	–	0.10	–
Dose Cu	Nules					Peníscola				
(mg/kg)	In soil	In plant[a]	In fruit[b]	$CF_{s\text{-}p}$	$CF_{p\text{-}f}$	In soil	In plant[a]	In fruit[b]	$CF_{s\text{-}p}$	$CF_{p\text{-}f}$
0.01 (control)	58.1 ± 8.0	17.6 ± 1.8	6.6 ± 1.6	0.30	0.38	17.4 ± 2.4	20.4 ± 2.2	7.6 ± 1.8	1.17	0.37
65.9	108.5 ± 15.0	18.9 ± 2.0	8.8 ± 2.1	0.17	0.46	76.2 ± 10.6	23.7 ± 2.4	7.7 ± 1.9	0.31	0.32
659.0	683.2 ± 94.7	22.0 ± 2.3	6.8 ± 1.6	0.03	0.31	538.3 ± 74.6	31.4 ± 2.8	7.4 ± 1.8	0.06	0.24
1977.0	2023.1 ± 280.5	26.8 ± 2.8	7.8 ± 1.8	0.01	0.29	1658.2 ± 229.9	394.1 ± 40.8	8.3 ± 2.0	0.24	0.02
3295.0	2856.6 ± 396.1	44.4 ± 4.6	7.7±1.9	0.02	0.17	3185.6 ± 441.7	1187.5±122.9	9.9 ± 2.4	0.37	0.01
6590.0	6077.8 ± 842.7	1229.2 ± 127.2	8.7±2.1	0.20	0.01	6476.7 ± 897.9	–	–	–	–

All the results are expressed in mg/kg in dry weight basis [40].
– no biomass produced.
$CF_{s\text{-}p}$: concentration factor, between soil and plant; $CF_{p\text{-}f}$: concentration factor, between plant and fruit.
[a]The conversion factors that have to be applied in order to calculate the content of metal in plant in fresh weight basis are the following: 11.6 for Rojales, 10.2 for Sollana, 10.5 for Nules and 9.9 for Peníscola.
[b]The conversion factors that have to be applied in order to calculate the content of metal in fruit in fresh weight basis are the following: 16.7 for Rojales, 15.6 for Sollana, 14.8 for Nules and 18.5 for Peníscola.

Table 5. Mean copper content in the European Mediterranean soils assayed (mg/kg in dry weight basis), in plant (mg/kg in dry weight basis), and in the edible part of tomato (ripe fruit).

Regarding the definition of the critical limits, these could only be established for the European Mediterranean soils cropped with lettuce. For the Australian agricultural soils cropped with lettuce, the establishment of these limits was not possible due to the important toxic effect observed. On the other hand, for the European Mediterranean soils cropped with tomato, these limits could not be calculated due to the fact the Cu content in fruit kept constant, independently of the Cu dose assayed and type of soil. The results obtained are shown in **Table 6**.

Finally, regarding the statistical analysis, and as explained previously, different correlation analysis were carried out in order to determine which soil properties influence the dynamic of Cu in soil and were more significant in terms of biomass production and of Cu absorption. For further details regarding these analyses, please consult [39–41].

	Equation	**Critical limit**	**R^2 adj. (%)**
Rojales	$y = 0.0053x - 0.47$	1975	89
Sollana	$y = 0.0011x + 0.43$	8697	89
Nules	$y = 0.0003x + 0.75$	30817	88

Table 6. Critical limit for the soil studied.

4.1. European and Australian agricultural soils cropped with lettuce

As detailed previously, agricultural soils from two different Mediterranean areas of the world were considered. Different biomass assays having the same experimental design and crop were carried out in these areas, enabling to compare the results obtained and to draw different conclusions regarding the behaviour of Cu in soils and plants.

The analysis of the toxicity threshold values obtained for the Spanish and Australian agricultural soils and lettuce showed that biomass production is greatly influenced by Cu and that similar soil properties are relevant when analysing the effect of Cu and its mobility and bioavailabity. As it can be observed in **Table 2**, the range of toxicity thresholds established covered similar ranges in both Mediterranean areas, being of 8–753 mg_{Cu}/kg in the Spanish Region, and of 49–728 mg_{Cu}/kg in the Australian Region. In both cases, the maximum threshold value was obtained for the soil having the highest pH and clay content, independently of the soil type. Therefore, these two soil properties seem to be very relevant when analysing Cu mobility and availability in soils. The difference between the maximum thresholds obtained in each region can be linked to the fact that the soil of the Spanish region had a higher SOM content and a basic pH, which increases the retention capacity of soil.

The comparison of the results obtained in both areas also pointed out the relevance of pH when analysing the mobility and availability of Cu in agricultural soils, even in soils with medium clay contents. For the all soils assayed in the Spanish Mediterranean Region, whose pH values varied slightly and were all between 7 and 8, no biomass was produced after the fifth dose, while no biomass was produced after the second, third and fourth dose in the different soils of the Australian Region, increasing the toxic effect of Cu as pH decreased. In these latter soils, pH values varied amongst 5–7.5. The most important toxic effect was observed for one of the Australian soils assayed that had a low pH value (5.6) but a medium content of clay (38%).

Therefore, according to the results obtained, two different approaches have to be made when assessing Cu-contaminated agricultural soils, depending on the pH of these. In acidic soils (pH below 7), pH is the most relevant soil property and strongly influences the bioavailability of Cu, in spite of the contents and values obtained for other soils properties. Toxic effect of Cu increases as pH values decreased, and soil properties that we would expect to have some retention capacity are ineffective or have very little effect due to the influence of pH on their reactivity. In fact, at acid pH, the reactivity of SOM and clay is low or even null. Conversely, for basic soils (pH values exceeding 7), other properties have a more relevant effect, being clay/sand content, SOM and salinity the most relevant ones. Clay and SOM retain Cu by adsorption reactions, while salinity and sand content make Cu more bioavailable and increase the toxic effect.

Analysis of the transfer of Cu from soil to plant showed that it varied between these two areas. However, it is important to point out that comparison of results was difficult due to the important toxic effect observed in the Australian agricultural soils. No biomass was produced after earlier doses in the case of these soils, which made it complicated to compare absorption values and rates. In both areas, Cu content in the edible part of the plant increased as Cu concentration in soils also did, but no clear absorption pattern could be identified due to the limited data obtained in the Australian assays. However, the correlation analyses carried out between Cu contents and soils properties showed similarities between them and with the results obtained for biomass production. In this case, pH, salinity and sand content are the most determinant soil properties which enhance Cu transference from soil to lettuce, while SOM and clay content reduce this metals' transference to lettuce.

Concerning the critical limits, as commented previously, these could only be calculated for the European Mediterranean soils. When compared to with the Spanish soil quality standard, the results varied significantly. The critical value calculated for the non-saline soils (Sollana and Nules) was above 100 times the baseline value for Cu, being higher in the soil with the highest organic matter and clay content (Nules), whereas it was below in the soil with high salinity and low organic matter content (Rojales). It is important to point out that these values have to be interpreted carefully and considering they are only theoretical, especially the ones for Sollana and Nules. For these soils, no biomass would be produced if these concentrations were reached, as it has been proved in the assays carried out, where no biomass production was observed when the dose of Cu was 6590 mg/kg.

4.2. Lettuce and tomato cropped in different European Mediterranean

Within the same region, two different crops in different agricultural soils were assayed in order to analyse their different responses and behaviours to Cu in soil, in terms of biomass production and Cu absorption, and to evaluate the influence of soil properties on the mobility and availability of this metal to plants.

Toxicity threshold values obtained varied significantly between crops for the different soils assayed. For lettuce, as commented previously, effective concentration calculated varied between 8 and 753 mg_{Cu}/kg, while for tomato these concentrations varied between 33 and 1697 mg_{Cu}/kg. A more detailed analysis of these results indicate that, for EC_{10}, the values obtained for tomato are nearly twice the maximum value obtained for lettuce, except for one soil; and for EC_{50}, the lowest value obtained for tomato is very similar to the maximum concentration obtained for lettuce. This clearly indicates the different response of these two crops to the different Cu concentrations in soils, showing that tomato is more tolerant than lettuce to Cu-contaminated soils. In fact, according to [61] lettuce can be considered an accumulator crop, while tomato can be considered a non-accumulator crop.

The analysis of the influence of soil properties on the effect of Cu on plant biomass production led to similar results/conclusions in both assays. SOM, clay content and CEC are the most relevant properties affecting Cu soils dynamic [39, 40].

Regarding the metal accumulation in the plant, the concentrations determined both in tomato and lettuce shoots were also very similar, although this latter tends to accumulate slightly higher concentrations. The most important conclusion drawn is that in the case of tomato, low translocation rates to the edible part of the plant are observed, even in soils with high Cu concentrations, while Cu translocation and accumulation in the edible part of lettuce increase as soil Cu concentration increases. The results observed for tomato were particularly interesting, since Cu concentration in fruits kept low and constant, independently of the Cu concentration in soils and shoots. This indicates that these plants tend to accumulate Cu in shoots and roots, with very low translocation of it to fruit, pointing out its phytoremediation potential.

In both cases (lettuce and tomato), the increase in Cu concentration determined in plant was not proportional to the increase in Cu concentrations in soil, due to the fact that Cu accumulation in plant is limited. Since Cu concentration in tomato fruits kept constant, the critical limit of contaminant in soil for this crop could not be calculated and therefore cannot be compared with the critical limits calculated for lettuce.

The analysis of the influence of soil properties on the transfer and bioaccumulation of Cu in these crops also led to similar results/conclusions. Both salinity and sand content arised as soil characteristics that enhance the transfer of Cu from soil to plant; while SOM and clay content have the opposite effect.

Furthermore, it is important to point out that the maximum metal content in the edible part of the plant established by the identified legislations [19, 20] was not exceeded in any of the dose and soils assayed for tomato and by only one soil in the case of lettuce. This soil was the one having the highest salinity content, and therefore, it seems logical to observe this, due to the fact that, as explained previously, this soil property facilitates the transfer of Cu from soil to plant.

Finally, it is important to highlight that for both tomato and lettuce, and considering the results obtained for the effect of Cu and its interaction with soil properties on plant biomass production and metal bioaccumulation in plant, the soil quality standard established by the Spanish legislation is not valid from either approach. Toxicity threshold values calculated for both crops showed that this soils quality standard was too indulgent, and it indicated this approach as the most restrictive when establishing soil quality standards. Conversely, the critical limit calculated for lettuce (**Table 6**) and the results obtained for the accumulation of Cu in the edible part of the plant show that the soil quality standard established by the Spanish legislation was too restrictive, since this content would not be exceeded in any of the soils assayed. Only one critical limit established showed that this soil quality standard was too permissive and corresponded to the one calculated for the saline soil.

Therefore, the results obtained show that soil quality standards should be established considering the influence of the different soil properties and should be particular for each case and scenario.

Lastly, and since the baseline value considered and used in all the assays carried out is similar to those established in other Spanish Mediterranean regions [32, 33] and in other European Mediterranean regions [34, 35], it is important to highlight that the results obtained in this work could be used as guidance for all the European Mediterranean Region in order to propose adequate soil quality standards; and adequate and valuable phytoremediation strategies that could be applied to Cu-contaminated soils of this region.

5. Conclusions

Regarding the effect of Cu on biomass production, the toxicity values established for the different Mediterranean agricultural regions and soils considered cropped with lettuce covered similar ranges. In both cases, the maximum threshold value was obtained for the soil having the highest pH and clay content, independently of the soil type. This indicated that these two soil properties are relevant when analysing Cu mobility and availability in soils.

On the other hand, when analysing the toxicity values established for the Spanish Mediterranean soils but considering the two different crops assayed, significant differences were observed between crops, in terms of tolerance and response. These results indicated that tomato is more tolerant than lettuce to Cu-contaminated soils. However, the analysis of the influence of soil properties on the effect of Cu on plant biomass production led to similar results/conclusions in both assays. SOM, clay content and CEC are the most relevant properties affecting the dynamic of Cu in soil Cu.

Regarding the analysis of the Cu bioaccumulation results, assays carried out with lettuce showed significant differences between the Mediterranean regions considered. However, comparison of results was difficult due to the important toxic effect observed in the Australian agricultural soils.

Significant differences were also observed between crops when comparing the bioaccumulation rates and quantities established for each of them when cultivated in the Spanish Mediterranean Region. The most important result is related to the Cu accumulated in the edible part of the plant. While the concentration of Cu in this part of the plant increased as the concentration in soil also did for lettuce, it was not so for tomato, where the concentration kept constant for all doses and soils assayed.

However, in spite of the results obtained for the bioaccumulation of Cu in the edible part of the plant, the critical limit could only be calculated for lettuce grown in the Spanish Mediterranean agricultural soils. These critical limits showed that the soil quality standard established by the Spanish legislation was too indulgent for the non-saline soils, while it was too permissive for the saline ones.

Furthermore, and taking into account the maximum metal concentration established in the identified legislation [19, 20], this was only exceeded by lettuce grown in the saline soil of the Spanish Mediterranean Region, and only after the fourth dose. Therefore, special attention must be paid to soil with high salinity, since certain crops must not be cultivated in them due the potential accumulation of Cu in the edible parts of them.

Thus, and taking into account the influence of the soil properties on copper mobility and bioavailability in soil, it can be concluded that the influence of the different soil properties depends mainly on the pH of soils. In basic soils (pH > 7), soil organic matter content and clay content reduce the mobility and bioavailability of Cu through adsorption processes, while salinity and sand content enhance the absorption of this metal by plants. In acidic soils (pH < 7), the effect of low pH, increasing the mobility of Cu, is stronger and more significant than any other soil property.

So, considering the influence of soil properties on copper mobility and bioavailability in soil, soil quality standards for heavy metal contaminated soils should be defined/established considering the soil properties and the interaction of these with the heavy metal under analysis. In the case of Cu, the soil properties that should be considered when establishing these standards are as follows: pH, soil organic matter, clay content, sand content and salinity.

Moreover, for the type of crops considered, the effect of Cu on plant biomass production was the most relevant of those analysed, since it was the one that underwent a more severe impact. Therefore, this effect is one that should be considered when establishing adequate soil quality standards and proposing adequate soil management practices.

Finally, tomato showed an important phytoremediation potential, extracting Cu from not only low–medium but also from highly (>1700 mg/kg) Cu-contaminated basic agricultural soils, and having low translocation rates to fruits. However, soils with high Cu concentration underwent a noticeable reduction in terms of plant biomass production. Therefore, it is important to find an adequate balance between these two aspects, in order to propose this crop as a phytoremediation alternative in the appropriate soil conditions.

Acknowledgements

The first author would like to thank the Valencian Regional Government (Generalitat Valenciana) for the Doctoral grant (ACIF/2010/041) and the scholarship granted (BEFPI/2013/063), which allowed him to carry out his PhD and spend a period of six months in the Commonwealth Scientific and Industrial Research Organisation (CSIRO) Land and Water Division, respectively; and CSIRO Land and Water Division (Canberra, Australia) for enabling to conduct the experiments of this work. Both authors would like to strongly thank Dr. Luis

Recatalá for his direction and his unstoppable and uncountable help during his last year of life to achieve the results obtained, some of which have already been published and have been cited here.

Author details

Daniel Sacristán* and Ester Carbó

*Address all correspondence to: daniel.sacristan@uv.es

Land-Use Planning Department, Centro de Investigaciones Sobre Desertificación-CIDE (CSIC-University of Valencia-Generalitat Valenciana), Moncada, Valencia, Spain

References

[1] Holmgren GGS, Meyer MW, Chaney RL, Daniels RB. Cadmium, lead, zinc, copper and nickel in agricultural soils of the United States of America. Journal of Environmental Quality. 1993; 22: 335–348.

[2] Wong SC, LI XD, Zhang G, Qi SH, Min YS. Heavy metals in agricultural soils of the Pearl River Delta, South China. Environmental Pollution. 2002; 119: 33–44.

[3] Nicholson FA, Smith SR, Alloway BJ, Carlton-Smith C, Chambers BJ. An inventory of heavy metals inputs to agricultural soils in England and Wales. Science of the Total Environment. 2003; 31: 205-219.

[4] Romic M, Romic D. Heavy metals distribution in agricultural topsoils in urban area. Environmental Geology. 2003; 43: 795–805

[5] Mapanda F, Mangwayana EN, Nyamangara J, Giller KE. The effect of long-term irrigation using wastewater on heavy metal contents of soils under vegetables in Harare, Zimbabwe. Agriculture, Ecosystems and Environment. 2005; 107: 151–165.

[6] Luo L, Ma Y, Zhang S, Wei D, Zhu YG. An inventory of trace elements inputs to agricultural soils in China. Journal of Environmental Management. 2009; 90: 2524–2530.

[7] Wei B, Yang L. A review of heavy metal contaminations in urban soils, urban road dusts and agricultural soils from China. Microchemical Journal. 2010; 94: 99–107.

[8] Tóth G, Hermann T, Da Silva MR, Montanarella L. Heavy metals in agricultural soils of the European Union with implications for food safety. Environment International. 2016; 88: 299–309.

[9] CEC. Communication from the Commission to the council, the European parliament, the European Economic and Social Committee and the Committee of the regions:

Thematic Strategy for Soil Protection. Commission of the European Communities (CEC), Brussels, Belgium. 2006.

[10] Kong XB. China must protect high-quality arable land. Nature; 2014: 506 (7).

[11] Komárek M, Cadková E, Chrastny V, Bordas F, Bollinger JC. Contamination of vineyard soils with fungicides: a review of environmental and toxicological aspects. Environment International. 2010; 36: 138–151.

[12] Recatalá L, Micó C, Sánchez J, Boluda R. Approaches for characterising contaminated sites: an analysis considering the role of soil. In: Villacama Y, Brebbia CA and Usó JL, editors. Ecosystems and Sustainable Development III, Third International Conference on Ecosystems and Sustainable Development. Wessex Institute of Technology, Wessex, UK; 2001. p. 347–356.

[13] Visser WJF. Contaminated Land Policies in Europe, Chemistry and Industry Features. 1995.

[14] Page GW. Contaminated Sites and Environmental Clean-Up: International Approaches to Prevention, Remediation and Reuse. San Diego, CL, USA: Academic Press; 1997. 212 p.

[15] Rooney CP, Zhao FJ, McGrath SP (2006). Soil factors controlling the expression of copper toxicity to plants in a wide range of European soils. Environmental Toxicology and Chemistry. 2006; 25: 726–732.

[16] Sheldon AR, Menzies NW. The effect of copper toxicity on the growth and root morphology of Rhodes grass (*Chloris gayana* Knuth.) in resin buffered solution culture. Plant and Soil. 2005; 278: 341–349.

[17] Adriano DC. Trace Elements in Terrestrial Environments: Biogeochemistry, Bioavailability and Risks of Metals, 2nd edition. New York, USA: Springer; 2001. 796 p.

[18] Kachenko AG, Singh B. Heavy metals contamination in vegetables grown in urban and metal smelter contaminated sites in Australia. Water Air and Soil Pollution. 2006; 169: 101–123.

[19] Codex. Codex Alimentarius Commission, Joint FAO/WHO General Standard for Contaminants and Toxins in Foods. Rome, Italy. 1998.

[20] McLaughlin MJ, Hamon RE, McLaren RG, Speir TW, Rogers SL. Review: a bioavailability-based rationale for controlling metal and metalloid contamination of agricultural land in Australia and New Zealand. Australian Journal of Soil Research. 2000; 38(6): 1037–1086.

[21] Mediterranean Agriculture: Location and Characteristics [Internet]. 2016. Available from: http://www.yourarticlelibrary.com/agriculture/mediterranean-agriculture-location-and-characteristics-with-diagrams/25443/ (Accessed: 2016-05-28)

[22] International Cooperative for the Management of Mediterranean-Climate Ecosystems [Internet]. 2013. Available form: http://www.incomme.org/ (Accessed: 2016-05-28)

[23] Coccossis HN. Historical land use changes: Mediterranean regions in Europe. In: Brower FW, Thomas A, Chadwick MJ, editors. Land use Changes in Europe: Processes of Change, Environmental Transformations and Future Patterns. Dortrecht: Kluwer Academic Publishers; 1991. p. 441–461.

[24] Groom C, Parker J, Teller A. Agriculture. In: Stanners D, Bourdeau P, editors. Europe's Environment. The Dobris Assessment. Copenhagen, Denmark: European Environmental Agency (EEA); 1995. p. 447–463.

[25] Recatalá L, Segura M, Pastor A, Murillo R, Sánchez J (2009). Cambios de usos del suelo, procesos de desertificación, impactos ambientales e implicaciones en la calidad ambiental a nivel municipal. Capítulo 6. In: Recatalá L, editor and director. Indicadores e Índices Integrados en la Agenda 21 Local para la Evaluación de la Calidad Ambiental en Áreas Afectadas por Desertificación del Ámbito Mediterráneo. Valencia (Spain): CIDE-Centro de Investigaciones sobre Desertificación, Universitat de València, Fundación Biodiversidad; 2009. p. 160–196.

[26] FAO. FAO Soil map of Australia. Food and Agriculture Organisation of the United Nations, Rome, Italy. 1992.

[27] BOE. Royal Decree 9/2005, 14th January, where it is established the list of soil contaminant activities and criteria and standards to declare soils as contaminated. BOE nº 15 of the 18th of January 2005, Madrid, Spain. 2005.

[28] McLaughlin MJ, Parker DR, Clarke JM. Metals and micronutrients—food safety issues. Field Crops Research. 1999; 60: 143–163.

[29] ISO/DIS 19258. Soil quality. Guidance on the determination of background values. Geneve, Switzerland: International Standard Organisation (ISO). 2005.

[30] Micó C, Peris M, Recatalá L, Sánchez J. Baseline values for heavy metals in agricultural soils in an European Mediterranean region. Science of the Total Environment. 2007; 378: 13–17.

[31] Sánchez J, Recatalá L, Micó C, Peris M. Contenido de metales pesados en suelos agrícolas y su influencia en cultivos hortícolas. Valencia: Informe Anual de Proyecto. CIDE-Centro de Investigaciones sobre Desertificación (CSIC, Universitat de València, Generalitat Valenciana), Valencia, Spain; 2004.

[32] Pérez L, Moreno AM, González J (2000). Valoración de la calidad de un suelo en función del contenido y disponibilidad de metales pesados. Edafología 2000; 7(3): 113–120.

[33] Pérez C, Martínez MJ, Vidal J, Navarro C (2002). Proposed reference values for heavy metals in Calcaric Fluvisols of the Huera de Murcia (SE Spain). In: Proceedings of the International Symposium on Sustainable Use and Management of Soils in Arid and Semiarid Regions, 22–26 September 2002; Cartagena, Spain; 2002. p. 495–496.

[34] Facchinelli A, Sacchi E, Mallen L. Multivariate statistical and GIS-based approach to identify heavy metal sources in soils. Environmental Pollution. 2001; 114: 313–324.

[35] Mitsios IK, Golia EE (2005). Heavy metal concentrations in soils and irrigation waters in Thessaly region, Central Greece. Communications in Soil Science and Plant Analysis. 2005; 36: 487–501.

[36] SA EPA. South Australian biosolids guidelines for the safe handling, reuse or disposal of biosolids. Department of Environment and Natural Resources, South Australian Environment Protection Authority, Adelaide, SA, Australia; 1996.

[37] NSW EPA Environmental guidelines: Use and disposal of biosolids products. New South Wales Environmental Protection Authority, Sydney, NSW, Australia; 1997.

[38] Victorian EPA. Environmental guidelines for the management of biosolids (Draft). Environment Protection Authority, Melbourne, VIC, Australia; 1997.

[39] Recatalá L, Sacristán D, Arbelo C, Sánchez J. Can a single and unique Cu soil quality standard be valid for different Mediterranean agricultural soils under an accumulator crop? Water, Air and Soil Pollution. 2012; 223: 1503–1517.

[40] Sacristán D, Peñarroya B, Recatalá L. Increasing the knowledge on the management of Cu-contaminated agricultural soils by cropping tomato (*Solanum lycopersicum* L.). Land Degradation and Development. 2015a; 26: 587–595

[41] Sacristán D, Recatalá L, Viscarra-Rossel RA. Toxicity and bioaccumulation of Cu in an accumulator crop (*Lactucasativa* L.) in different Australian agricultural soils. Scientia Horticulturae. 2015b; 193:346-352.

[42] Recatalá L, Micó C, Peris M, Sáez E, Sánchez J (2002). Analysis of heavy metals in Mediterranean soils (Alicante, Spain) for soil quality policy making. In: Proceedings of the International Symposium on Sustainable Use and Management of Soils in Arid and Semiarid Regions, 22–26 September 2002; Cartagena, Murcia, Spain; 2002. p. 318–320.

[43] Recatalá L, Peris M, Micó C, Sánchez R, Sánchez J. Analysis of heavy metals in agricultural soils (Castellón, Spain). In: Proceedings of the Fourth International Conference on Land Degradation, 12–17 September 2004; Cartagena, Murcia, Spain; 2004. p. 325–326.

[44] WRB. World reference base for soil resources. A framework for international classification, correlation and communication. World soil resources report 103, 2nd edition. IUSS Working Group WRB. Food and Agriculture Organization of the United Nations (FAO), Rome. 2006.

[45] Tóth G, Montanarella L, Stolboboy V, Máté F, Bódis K, Jones A et al. Soils of the European Union. European Commission. Joint Research Centre. Institute for Environment and Sustainability. Office for Official Publications of the European Communities. Luxembourg; 2008.

[46] MAPA. Official methods for soil and water analysis. Volume III. Madrid, Spain: Ministry of Agriculture, Fisheries and Food. 1994.

[47] Rayment GE, Lyons DJ. Soil Chemical Methods—Australasia. Queensland, Australia: CSIRO Publishing; 2010. 520 p.

[48] Recatalá L, Micó C, Arbelo CD, Navarro J, Sánchez J (2008). Testing the usefulness of using a generic reference level for Cu for risk assessment in Mediterranean agricultural soils. In: Blum WEH, Gerzabek MH, Vodrazka M, editors. Book of Abstracts of the EUROSOIL 2008. Vienna, Austria; 2008. p. 318–320.

[49] Peñarroya B. Evaluation of a Cu soil quality standard for different representative Mediterranean agricultural soils. MSc Thesis. Valencia: Universitat de València; 2012.

[50] OECD. OECD Guidelines for the Testing of Chemicals. Section 2: Effects on Biotic Systems Test No. 208: Terrestrial Plant Test. OECD, Paris, France; 2006.

[51] Doelman P, Haanstra L. Short-term and long-term effects of heavy-metals on phosphatase-activity in soils—an ecological dose–response model approach. Biology and Fertility of Soils. 1989; 8: 235–41.

[52] Hinojosa MB, Carreira JA, Rodriguez-Maroto JM, Garcia-Ruiz R (2008). Effects of pyrite sludge pollution on soil enzyme activities: Ecological dose–response model. Science of the Total Environment. 2008; 396: 89–99.

[53] Micó C, Li HF, Zhao FJ, McGrath SP (2008). Use of Co speciation and soil properties to explain variation in Co toxicity to root growth of barley (*Hordeum vulgare L.*) in different soils. Environmental Pollution. 2008; 156: 883–890.

[54] Ruyters S, Salaets P, Oorts K, Smoldres E. Copper toxicity in soils under established vineyards in Europe: a survey. Science of the Total Environment. 2013; 443: 470–477.

[55] USEPA. Method 3052. Microwave Assisted Acid Digestion of Siliceous and Organically Based Matrices. Revision 0. United States Environmental Protection Agency (USEPA), Washington DC, USA; 1996

[56] Chen M, Ma LQ. Comparison of four USEPA digestion methods for trace metal analysis using certified and Florida soils. Journal of Environmental Quality. 1998; 27: 1294–1300.

[57] ANZFA. Food standards code, Amendment 35. Australia and New Zealand Food Authority, Canberra, ACT, Australia. 1997.

[58] Strawn DG, Baker LL. Speciation of Cu in a contaminated agricultural soil measured by XAFS, micro-XAFS, and micro-XRF. Environmental Science and Technology. 2008; 42: 37–42.

[59] Abreu MM, Matias MJ, Magalhães MCF, Basto MJ (2008). Impacts on water, soil and plants from the abandoned Miguel Vacas copper mine, Portugal. Journal of Geochemical Exploration. 2008; 96: 161–170.

[60] Micó C. Study of the contents of Heavy metals in soils with horticultural crops of the province of Alicante. PhD Thesis. Valencia (Spain): Universitat de València; 2005.

[61] Ross SM. Sources and forms of potentially toxic metals in soil-plant systems. In: Ross SM, editor. Toxic Metals in Soil-Plant Systems. Chichester: Wiley. 1994.

4

Treated Municipal Wastes: Are they Contaminating or Enriching the Soil?

Ahmed Al-Busaidi and Mushtaque Ahmed

Additional information is available at the end of the chapter

Abstract

Treated municipal wastes could be a mixture of treated sewage biosolids and green wastes (Kala compost) that can be applied for agricultural production. It can improve soil fertility and plant growth. However, long-term application of treated sewage biosolids could result in heavy metal accumulation and some health problems. The objective of this study is to evaluate the effect of different fertilizers, especially Kala compost, on the soil fertility and plant productivity. An open field was divided into nine plots and received either treated municipal wastes (Kala compost) or inorganic fertilizer, or a mixture of both fertilizers. The field was irrigated by drip system, and commercial cucumber, tomato, cabbage, lettuce, carrot, and potato were grown in each plot. Soil and plant were monitored continuously and samples were taken at different stages of the study. No symptoms of physical or chemical problems were observed in the open field and measured soil samples. Moreover, the soil had sufficient amount of different nutrients for plant growth and all measured micronutrients (heavy metals) were within the safe limit and below the allowable safe limit of the international standards. Good growth was observed in all grown crops and no symptoms of element toxicity were observed. Chemical analysis for fruit samples did not show any accumulation of heavy metals and all measured elements were within the safe limit for human consumption. It can be concluded that treated municipal wastes (Kala compost) were good media for plant growth that can enrich the soil with different elements needed for higher yield. However, more monitoring is needed with treated biosolid application and good management could be the key to avoid any adverse effect of any contaminant.

Keywords: treated wastewater, biosolids, Kala compost, heavy metals, plant growth

1. Introduction

Sewage sludge or biosolids, which are one of the final products from wastewater treatment plants, are considered the most promising waste that can be utilized in an effective and environmentally friendly manner. Sewage sludge creates very little to zero environmental impact if utilized properly. Throughout the world, the safe disposal of the sewage sludge is one of the major environmental concerns. However, opinions on the utilization of the sewage sludge vary due to the possible positive and negative points associated with the handling and treatment. In fact, sewage sludge is increasing annually as the population increases and it is a renewable product that will never stop forming. It can be treated differently with various methods depending on the purpose of the treatment [1, 2]. Sewage sludge is composed of organic compounds, macro- and micronutrients, trace elements including toxic metals, microorganisms, and micro-pollutants. Micro- and macronutrients serve as a source of plant nutrients, whereas organic constituents serve as soil conditioner. It contains high concentrations of N, P, Ca, and Mg. Potassium is, however, deficient in sewage sludge [3]. Sludge amendment improves soil properties such as porosity, bulk density, aggregate stability, and water-holding capacity. Sewage biosolids are often used as a fertilizer on farms to grow corn and cereal crops such as wheat. Using sewage biosolids as a nutrient source for field or forage crops or for improved pasture (1) improves soil fertility—offsetting the need for commercial fertilizers; (2) reduces production costs; (3) improves soil fertility; (4) enhances soil structure, moisture retention, and soil permeability; (5) adds organic matter—enhancing soil structure, moisture retention, and permeability, while reducing the potential for wind and water erosion [3].

Higher level of heavy metals in sewage sludge may be a cause for problems when applied in field used for agriculture. Whether any problem actually takes place will depend on soil pH, soil organic matter content, cation exchange capacity, movement of heavy metals in the soil profile, and changes that take place in the forms of heavy metals [4]. It is always advisable to use sewage sludge in low doses to reduce bioavailability of toxic heavy metals [5]. Sewage sludge amendment increases the production of a variety of plants including vegetables, cereals, grasses, and trees. The use of sewage sludge also results in more robust plants with faster development and greater biomass production [6]. It has been observed that crops contain heavy metals at concentrations harmful to human health when such crops were grown in soil amended with extremely high level of sewage sludge [7]. However, the metal concentrations in the sewage sludge depend on several factors such as (i) sewage origin, (ii) sewage treatment processes, and (iii) sludge treatment processes.

In Oman, "Haya Water" is a government company that is responsible for building, operating, and managing wastewater projects in Muscat Governorate. Haya Water has developed its pioneering Kala Composting Plant to enable the efficient reuse of sewage biosolids and green waste enabling their conversion to a compost product that can be used for agriculture, landscaping, and for individual gardens. The use of Kala compost (KALA) has various benefits such as farmers reusing a waste product, municipal authorities reducing their dependence on chemical fertilizers, as well as reducing greenhouse gas emission due to the use of environmental friendly waste management process [8]. However, high application of sewage biosolids could result in heavy metal accumulation and many health problems. Therefore, sewage biosolids applied to agricultural land must be well treated and continuously monitored to avoid any environmental risk problems. The objective of this study is to evaluate the effect of different fertilizers especially Kala compost on the quality of soil and crops. Specifically to (1) conduct research to assess the performance of the tested crops under different fertilizers, (2) to determine the changes in physicochemical properties of the soil treated by different fertilizers, (3) to determine the amount of water that can be saved using Kala compost, (4) to determine the effect of Kala compost on plant growth and find out any heavy metal accumulation in soil and plant, and (5) to monitor characteristics and yield components of crops grown and treated by different fertilizers.

2. Materials and methods

Research studies were carried out to achieve the set goals through detailed experimentation at Sultan Qaboos University (SQU), Agricultural Experiments Station (AES) open field.

New field at AES was prepared by removing rocks and big stones. The field was divided into nine plots and each plot (43.2 m^2) received either 216 kg of Kala compost or 4.5 kg of inorganic fertilizer (NPK) or a mixture of both fertilizers (MIX). Drip irrigation system was installed all over the field. Commercial cucumber, tomato, cabbage, lettuce, carrot, and potato were grown in each plot.

Soil salinity, moisture content, and temperature were monitored by using wet-sensor device. Moreover, direct soil samples were taken at depths of 0–15, 15–30, and 30–45 cm. Plant growth and yield of each crop treated by different fertilizers were observed. Fruits quality and quantity were assessed. Samples from soil and plants were taken for different physical, chemical, and biological analyses. All physicochemical analyses for soil and plants were done in soil and water labs (SQU) following standard methods and using inductively coupled plasma (ICP) machine for metal analysis, whereas biological analysis for plant samples was done in Muscat Municipality laboratories.

3. Result and discussion

3.1. Pure Kala compost

Pure Kala compost (saturated extract sample) was analyzed for physical, chemical, and biological properties. From **Table 1**, it was found that all measured parameters were within the acceptable level of the international standards and the compost can be applied to improve soil fertility. Actually, Kala compost is a mixture of different municipal wastes such as treated sewage sludge, plant materials, and cow manure. Therefore, it is expected to have low concentration of heavy metals and good values of different nutrients that can support plant growth. It was reported by the Ministry of Agriculture, Forestry, and Fisheries (MAFF) [9] that different compost will behave differently in the soil based on the processes used to generate waste materials.

Samples	pH	*EC (dS/m)	N (%)	OM (%)	TOC (%)	IC (%)	TC (%)	FC (FC media)
Kala 1	6.7	23.8	3.15	38.68	22.49	0.011	22.50	0
Kala 2	6.7	24.8	3.20	36.67	21.32	0.89	22.21	0
Kala 3	6.7	24.3	3.18	37.10	21.57	0.84	22.41	0
Samples	**Elements concentration (mg/l)**							
	Mn	**Cd**	**Cu**	**Fe**	**Zn**	**B**	**P**	**Al**
Kala 1	0.107	<0.001	0.160	0.590	0.068	0.451	2.506	0.127
Kala 2	0.082	<0.001	0.116	0.545	0.063	0.320	1.608	0.105
Kala 3	0.065	<0.001	<0.0004	0.394	0.001	<0.001	0.327	0.098
Samples	**Elements concentration (mg/l)**							
	Ba	**Ca**	**Cr**	**Co**	**Pb**	**Mg**	**Ni**	**Ti**
Kala 1	0.037	15.570	0.053	0.060	0.154	2.991	0.013	0.010
Kala 2	0.039	14.635	0.042	0.068	0.182	3.077	0.009	0.009
Kala 3	0.039	14.185	0.035	0.064	0.128	3.017	<0.001	0.008

*EC: Electrical conductivity; N: nitrogen; OM: organic matter; TOC: total organic carbon; IC: inorganic carbon; TC: total carbon; FC: fecal coliform bacteria.

Table 1. Chemical analysis for pure Kala compost.

3.2. Soil samples

Kala compost was a good source for organic matter and organic carbon that can support soil physical parameters. Organic matter could be the main reason in improving water-holding capacity of the soil amended by Kala compost (**Figure 1**). Moreover, it added more nitrogen to the soil and improved plant chlorophyll content (**Figure 2**). In addition, Kala compost reduced soil-compaction problem by improving soil bulk density where Kala compost gave 1.53 g/cm^3 and chemical fertilizer gave 1.72 g/cm^3. The good result for bulk density under Kala compost is supporting Kala application in which organic fertilizer can improve soil aggregate stability, soil structure, and support root growth.

Recent studies indicate that compost of biosolids in combination with woodchips or sawdust is used to grow horticulture crops under field or pots condition. It helps in improving soil

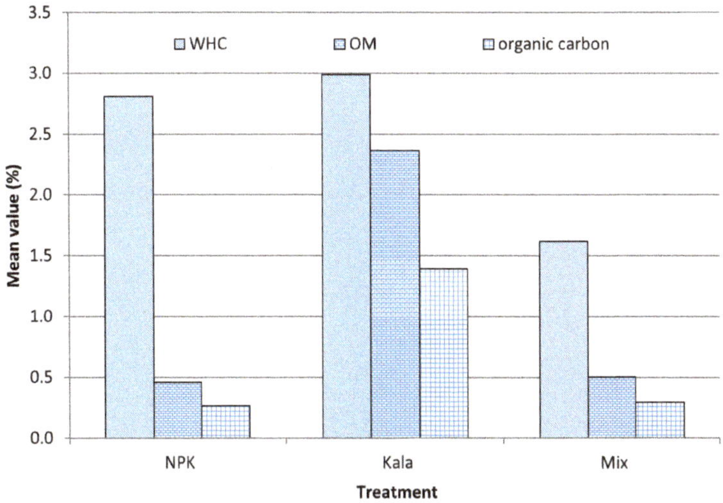

Figure 1. The effect of organic matter (OM) on soil organic carbon and water holding-capacity (WHC) at different treatments.

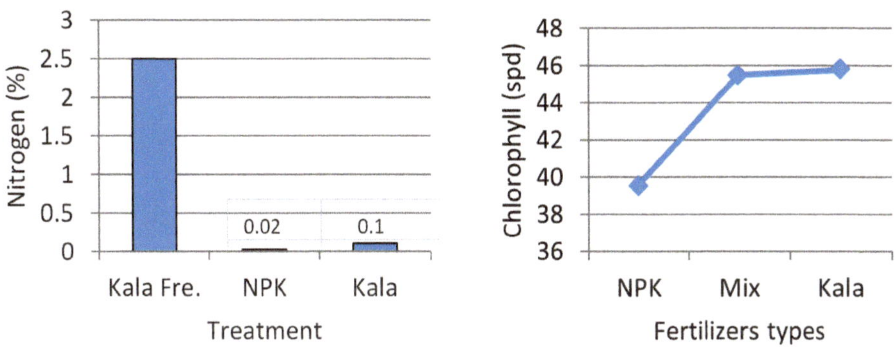

Figure 2. Soil nitrogen and plant chlorophyll values as affected by different treatments.

physical properties such as lowering bulk density, increasing water-holding capacity, increasing total soil porosity, and aggregate stability [10]. According to Wang et al. [11], sludge is shown to be efficient fertilizers as it improves soil physical properties such bulk density, porosity, aggregate stability and water retention and movement. Other properties also can be improved such as pH and contents of organic matter and nutrient contents as the raw sludge is rich in nutrients such as nitrogen, phosphorus, organic matter, and essential trace elements. A study showed clearly that water retention capacity was increased when 0.5% sewage sludge was added to soil. In fact, that increase was higher for raw sludge-amended soil than deposited sludge-amended soil [12].

3.3. Soil salinity

Soil salinity is a good indicator for soil fertility and salt toxicity. Using saturated paste extract method, it can be seen that chemical fertilizer (NPK) gave the highest value (3 dS/m) compared to other treatments (**Figure 3**). Whereas, Kala treatment gave reasonable value that was accepted by many crops. In all cases, salts could be added or diluted or leached down when the land is irrigated by good-quality water. Soil pH for all treatments was around 8. It was slightly affected by compost application.

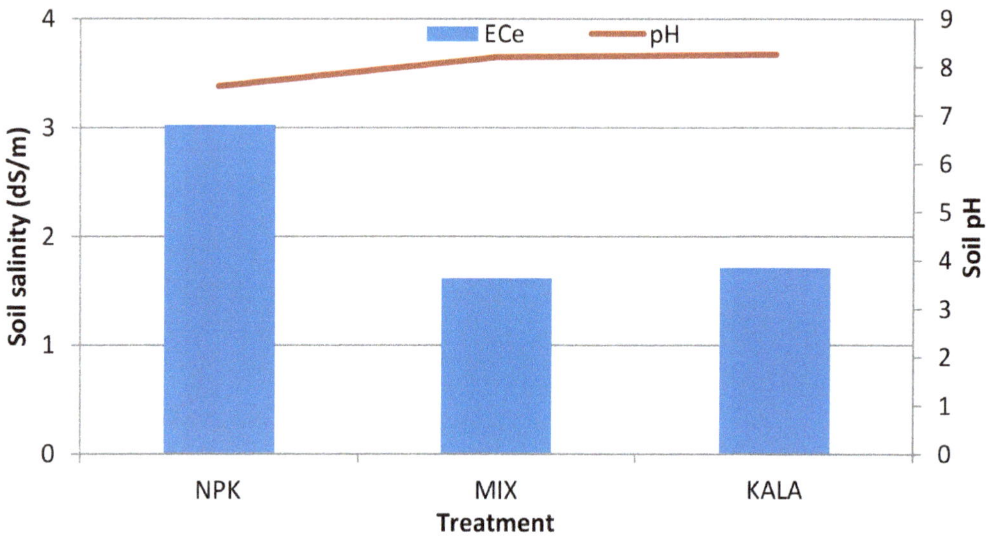

Figure 3. Soil electrical conductivity (ECe) and pH at the beginning of the study.

At the end of the study (**Figure 4**), the salinity value was almost similar to the result found in **Figure 3**. The main difference was that more salts or nutrients were released from Kala fertilizer. All salts found in each treatment were moved up or down the profile depending on air temperature for evaporation or amount of water added as irrigation or rainfall (**Figure 5**). Generally, Kala compost held less salts compared to NPK treatments, but at the same time all those salts were used to support plant growth and released slowly so they can be used as a source of nutrients without any problem of toxicity.

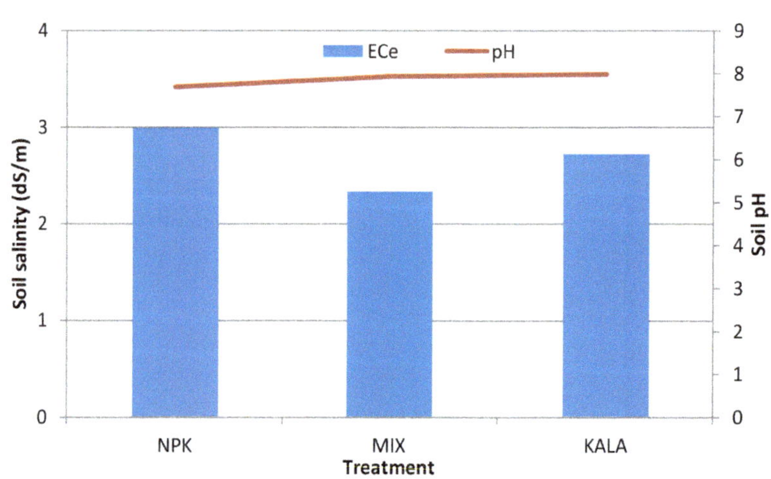

Figure 4. Soil electrical conductivity (EC_e) and pH at the end of the study.

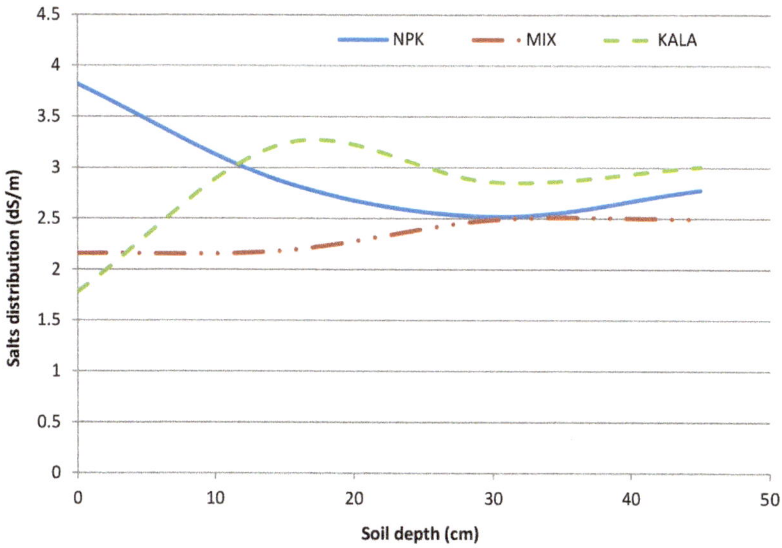

Figure 5. Salt distribution (EC_e) along the horizons.

Wet sensor is a good device for monitoring soil water content, temperature, and salinity. From **Table 2**, it can be seen that wet sensor confirms what was found in previous figures. Kala compost was maintaining much water that helped in reducing soil temperature with slow release of salts with time.

The application of organic amendment such as sewage sludge compost to agricultural field usually improves soil physiochemical properties through increasing the content of organic matter, the total nitrogen content, and the electrical conductivity, whereas it causes reduction in pH slightly [13]. Electrical conductivity could increase with sewage sludge compost application [14] as a result of acidification in combination with subsequent solubility of metallic elements.

Treatment	Time						
MC (% vol)	**05-Jan**	**12-Jan**	**26-Jan**	**02-Feb**	**09-Feb**	**16-Feb**	**23-Feb**
NPK	20.2	35.3	11.1	32.0	26.1	17.6	20.1
KALA	25.9	35.5	11.3	39.8	23.8	26.5	20.4
EC (dS/cm)							
NPK	176	170	211	178	135	97	122
KALA	133	137	149	211	130	133	115
Temp (°C)							
NPK	23.3	18.4	22.1	22.1	23.1	19.2	21.9
KALA	21.4	18.3	21.3	23.3	23.1	23.8	23.5

Table 2. Wet sensor readings for soil water salinity, moisture content (% vol), and temperature with time.

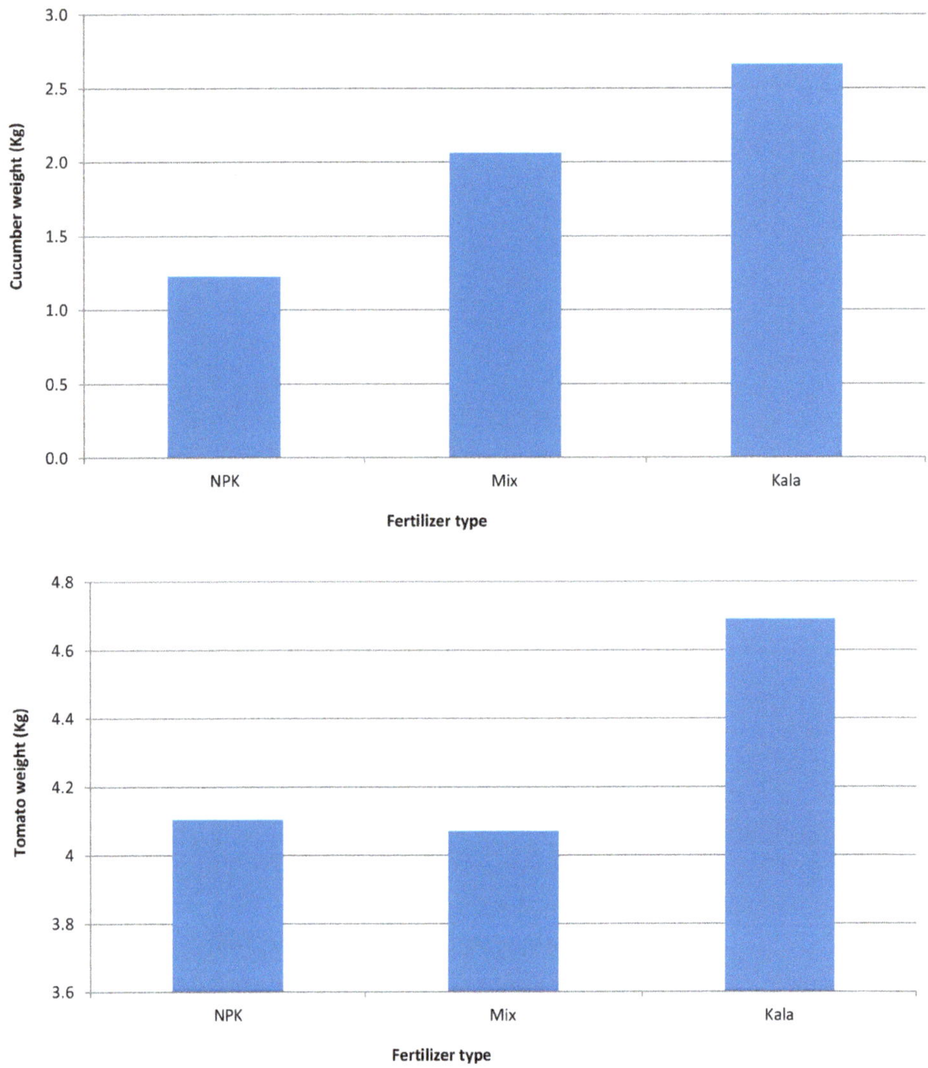

Figure 6. Average yield of cucumber and tomato.

3.4. Plant samples

From **Figures 6–8**, it can be seen that the best productivity of all tested crops was mostly with Kala compost followed by mix treatment and finally by NPK fertilizer. It does not mean that NPK treatment was bad but may be some plants did not get the right amount of fertilizer in the right time. The organic or mix of both organic and inorganic fertilizers usually is the best for consumer and surrounding environment. It seems that Kala compost was creating a good environment for plant by releasing multinutrients, reducing evaporation and keeping much water in the root zone compared to NPK treatment.

Good results were also found in Nielson et al. [15] study when the municipal biosolids were added to cultivate carrots and chard on irrigated soils. A significant increase in yield was found in plants growing biosolid-amended soil as compared to those grown in non-amended soil. In

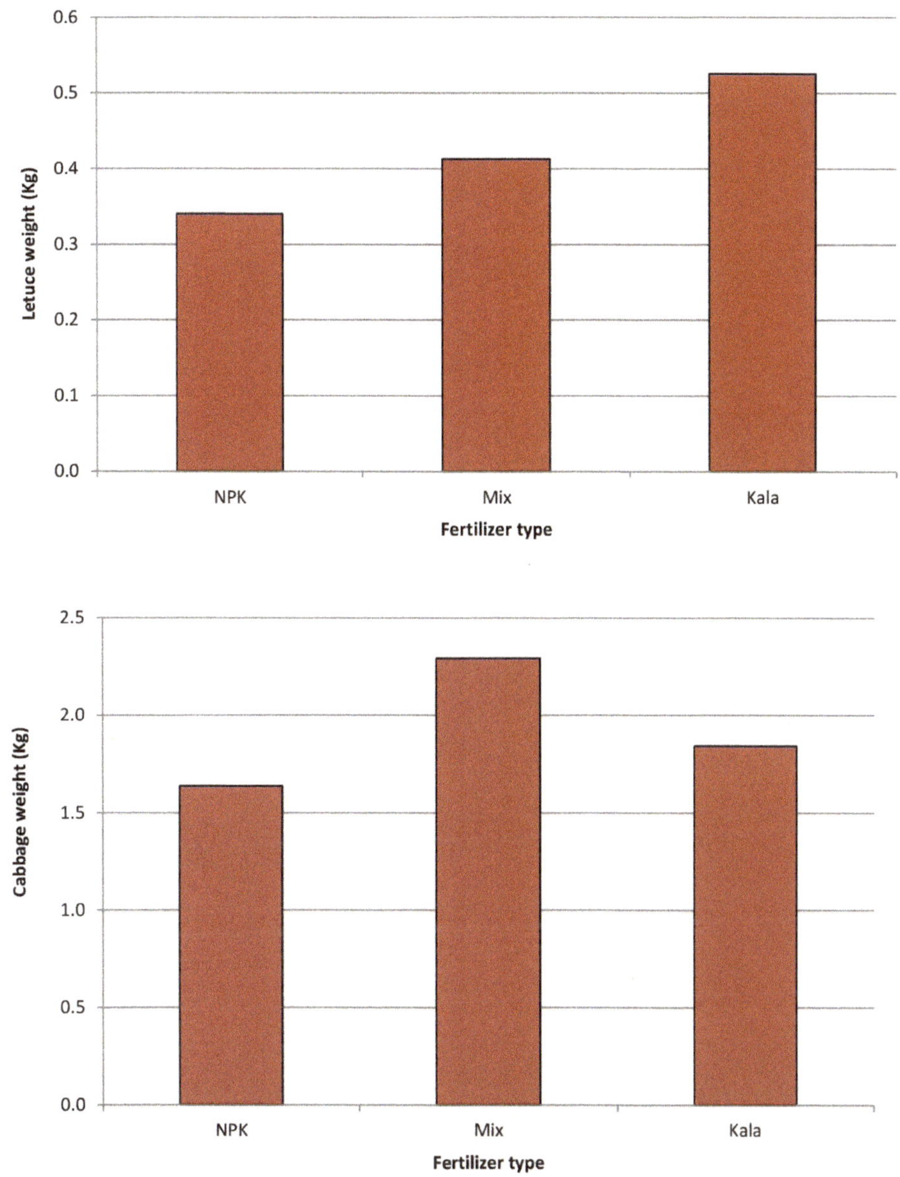

Figure 7. Average yield of lettuce and cabbage.

addition, a similar study with cotton (*Gossypium hirsutum*) also showed advancement of flowering and fruiting by 2–3 weeks under sludge-amended soil as compared to fertilizer-amended ones [16]. The grain yield of barley increased significantly under repeated sewage sludge application. The leaf protein concentration and dry matter accumulation in the plants grown in sludge-amended soil was higher from the beginning of development to ear emergence [17]. Moreover, it was found that the sludge amendment at the rate of 0.80, 160, and 320 t/ha dry wt. in soil increased the average dry weight of sunflower plants (*Helianthus annus* L.) [18]. Even in saline soil, Verlinden and McDonald [19] showed that compost amendment increased *Limonium sinuatum* and *Celosia argentea* yield. By supplying nutrients, particularly N and P, compost can improve the mineral-nutrient status and growth of plants in saline soils.

Finally, the faster development and greater biomass production in plants grown in sludge-amended soil may be responsible for an early reproductive cycle. Moreover, the complex organic and the inorganic compounds of sewage were broken down into simpler forms, and thus the final treated sludge became useful and beneficial to the seedling growth [20].

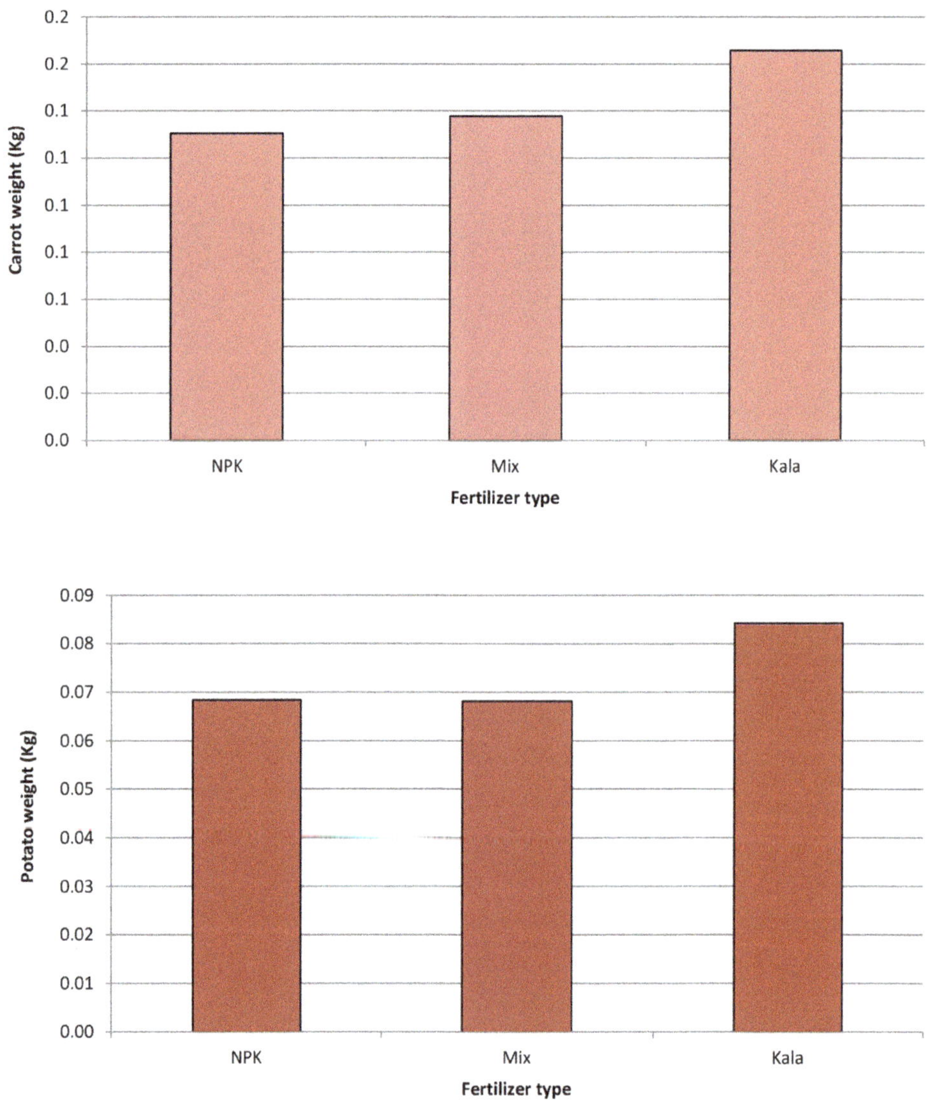

Figure 8. Average yield of carrot and potato.

3.5. Metal concentrations in soil samples

3.5.1. At the beginning of the study

From **Table 3**, it can be seen that all major cations were found in good amount for all treatments, whereas minor cations and heavy metals were detected in low concentrations. This means that all fertilizers had good concentrations of different nutrients in which they positively affected soil fertility. For heavy metals, all measured elements were within the acceptable level of international standards.

	Mg	K	P	Cd	Co	Cr	Cu
NPK	17.3957	92.9470	0.0300	<0.0010	0.0647	0.0613	<0.0004
KALA	9.0521	107.3446	3.4602	<0.0010	0.0696	0.0500	<0.0004
MIX	6.7443	91.3614	0.2732	<0.0010	0.0556	0.0456	<0.0004
	Fe	**Mn**	**Ni**	**Pb**	**Ti**	**Zn**	**B**
NPK	0.3477	0.0352	0.0100	0.2273	<0.0050	0.0211	0.0828
KALA	0.3602	0.0199	0.0410	0.2986	<0.0050	0.0302	0.0730
MIX	0.3397	0.0188	0.0050	0.2187	<0.0050	0.0010	0.0422

* Mg: magnesium; K: potassium; P: phosphorus; Cd: cadmium; Co: cobalt; Cr: chromium; Cu: copper; Fe: iron; Mn:4 manganese; Ni: nickel; Pb: lead; Ti: titanium; Zn: zinc; B: boron.

Table 3. Soil metal concentration (mg/l) in saturation extract at the beginning of the study.

3.5.2. *At the end of the study*

From **Table 4**, it can be seen that elements such as K, P, and Mg were found in good concentrations, which is good for plant growth, whereas microelements and heavy metals were in low concentrations and within the international standards for all treatments. As mentioned before, irrigation water was the main cause of releasing the nutrients to the root zones. However, the similarity in concentrations of most elements in NPK and Kala fertilizers means that original soil was a source for some elements (rock materials) and the added values came from each treatment.

	*Mg	K	P	Cd	Co	Cr	Cu
NPK	15.0237	85.3755	0.0300	<0.0010	0.0592	0.0654	<0.0004
KALA	38.5184	52.9741	0.4517	<0.0010	0.0566	0.0467	<0.0004
MIX	14.9962	60.2882	0.1451	<0.0010	0.0616	0.0484	<0.0004
	Fe	**Mn**	**Ni**	**Pb**	**Ti**	**Zn**	**B**
NPK	0.3356	0.0193	0.0050	0.2352	<0.0050	0.1488	0.1039
KALA	0.3376	0.0290	0.0114	0.2208	<0.0050	0.0232	0.2645
MIX	0.3340	0.0241	0.0050	0.2495	<0.0050	0.0656	0.0780

* Mg: magnesium; K: potassium; P: phosphorus; Cd: cadmium; Co: cobalt; Cr: chromium; Cu: copper; Fe: iron; Mn:4 manganese; Ni: nickel; Pb: lead; Ti: titanium; Zn: zinc; B: boron.

Table 4. Soil metal concentration (mg/l) in saturation extract at the end of the study.

Long-term fertilization of biosolids enhances soil condition and shows increase in land production and that increment confirms the potential of substantial revenue expansion [21]. A study was conducted in China by Wang et al. [11] to identify the effects of using sludge in agricultural lands. The study concluded that the biomasses of grass used in the experiment were increased as well as soil organic matter compared to control treatment where no sludge was added. Furthermore, the heavy metals Pb, Cu, and Zn were determined and found not exceeding the standards of acceptable levels of heavy metals. It is wise not to generalize how metals interact in soil and ultimately taken up by plants because many factors influence such interactions and uptake such as the type of metal, physical, and chemical properties of the soil and the type of crop. As it is difficult to take into consideration all such factors, the regulation of sewage sludge application is based on the total metal loading or concentration in soils. Kiekens et al. [22] observed much lower metal solubility in a calcareous clay soil than in sand (pH 6) regardless of whether the metals were added as salt or sludge form.

3.6. Metal concentrations in plant samples

To evaluate the nitrogen content for the tested crops (**Figure 9**), it can be seen that Kala treatment obtained the highest values which was expected due to the high content of nitrogen in Kala fertilizer compared to NPK. This value was clearly reflected in soil nitrogen and chlorophyll content shown in **Figure 2**. The high value of nitrogen could be one of the reasons for obtaining better productivity with Kala compost compared to NPK.

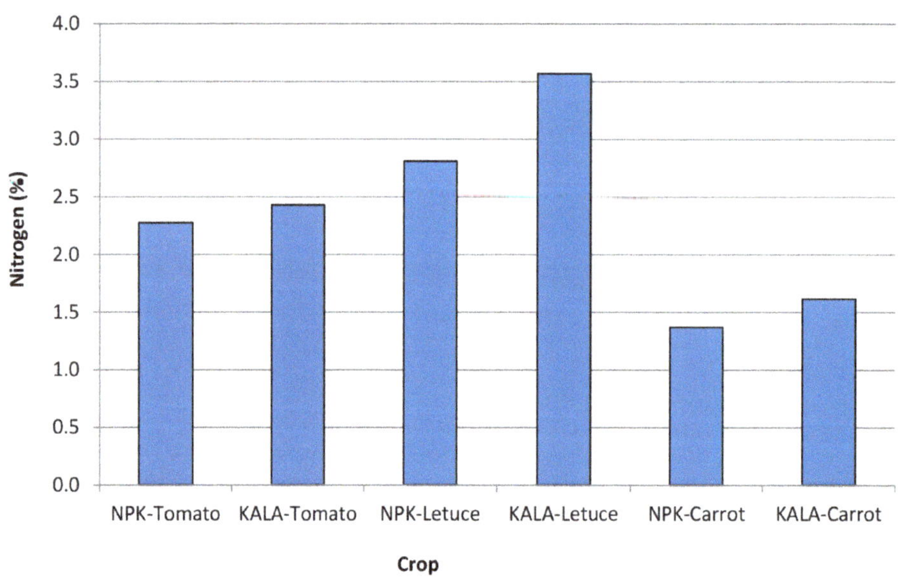

Figure 9. Nitrogen content in tested crops.

For microelement concentration in fruity plants, it can be seen from **Figures 10** and **11** that there were small changes between NPK and Kala treatments. For short-season plants such as cucumber (**Figure 10**), it can be seen that in some cases NPK gave higher values for some elements such as Mn, Pb, and Ni, whereas Kala gave higher values than NPK for others such as Fe, B, and Al.

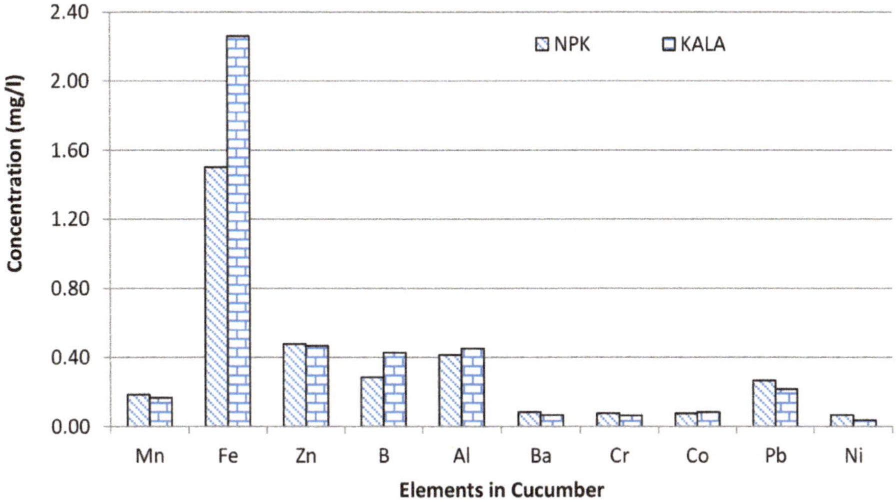

Figure 10. Heavy metal concentrations in cucumber.

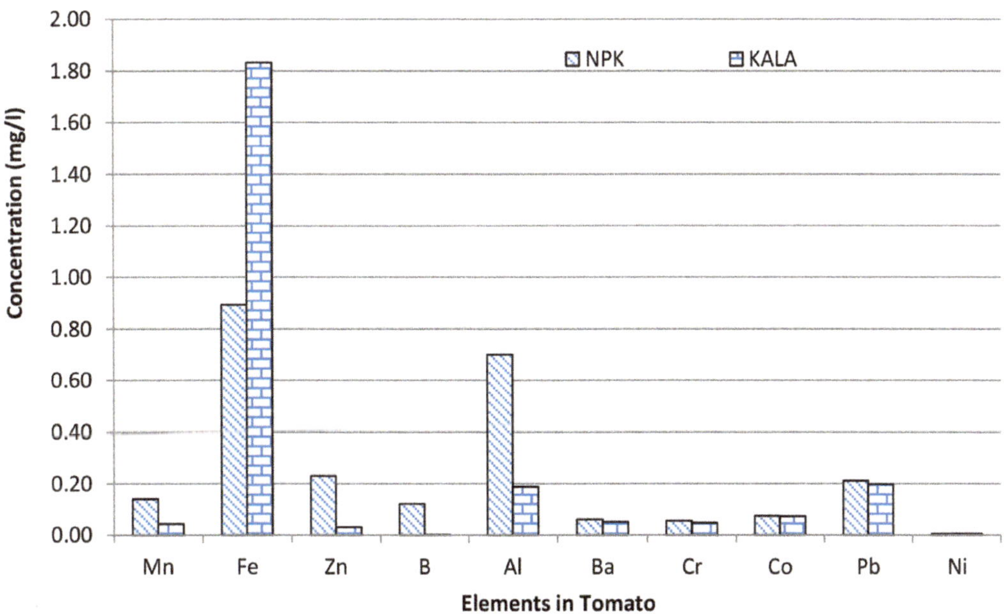

Figure 11. Heavy metal concentrations in tomato.

For long-season plants such as tomato (**Figure 11**), it can be seen that NPK was higher in all measured elements than Kala except for Fe.

For leafy plants, it can be seen from **Figures 12** and **13** that similar scenario was repeated and small variations were found between Kala and NPK fertilizers.

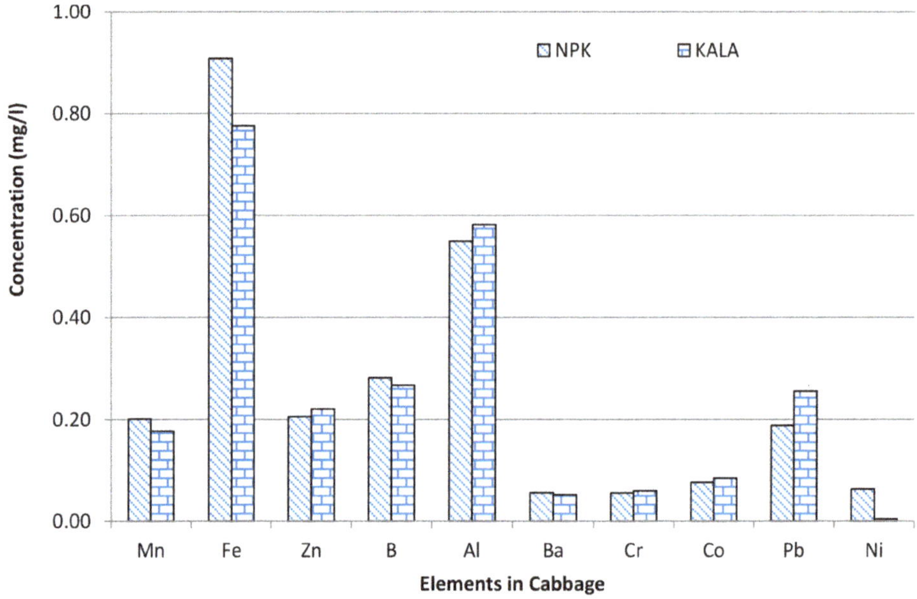

Figure 12. Heavy metal concentrations in cabbage.

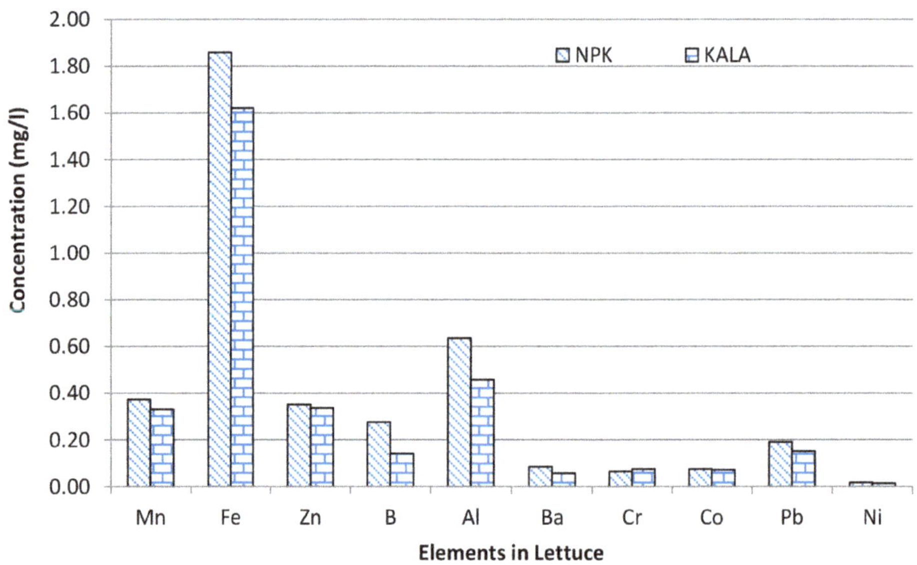

Figure 13. Heavy metal concentrations in lettuce.

For very short-season plant such as lettuce (**Figure 13**), it can be seen that all elements were in low concentrations with Kala compared to NPK. Iron (Fe) had the highest concentrations in all crops of both treatments.

For root crops such as carrot and potato (**Figures 14** and **15**), Iron (Fe) was high in both treatments of both crops. However, Kala compost was higher than NPK in some measured values.

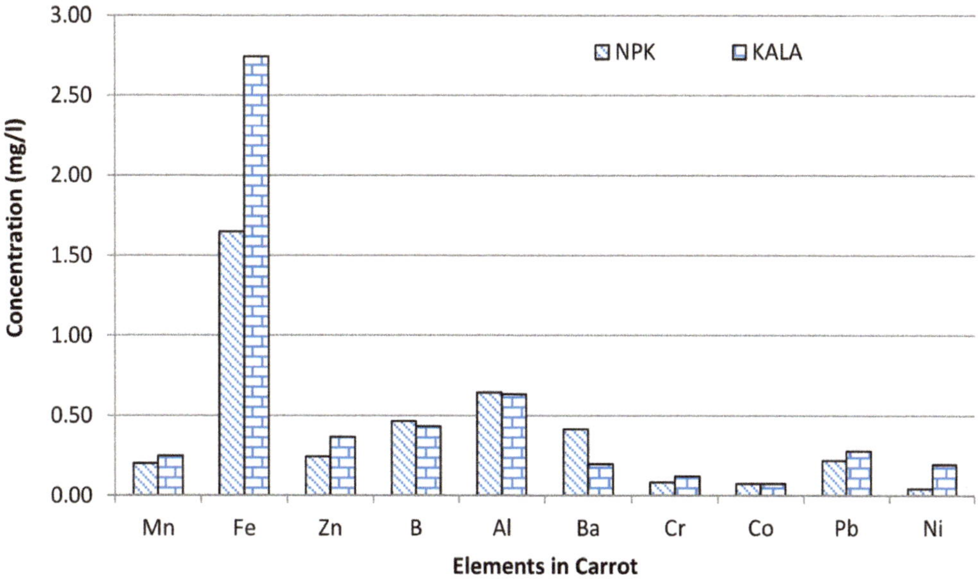

Figure 14. Heavy metal concentrations in carrot.

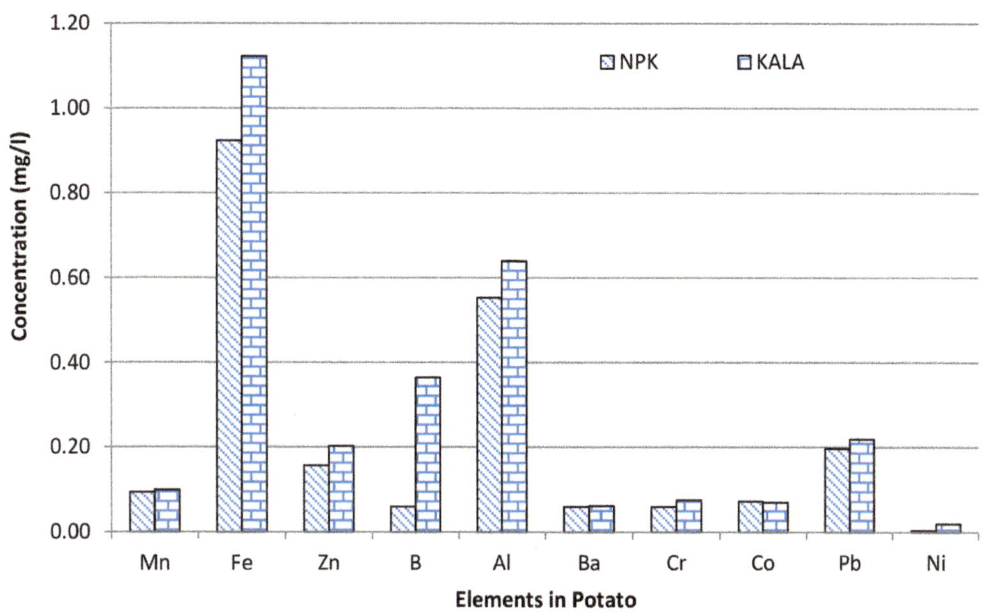

Figure 15. Heavy metal concentrations in potato.

For all treatments of all crops, cadmium (Cd) and copper (Cu) were found in very low concentration of <0.001 mg/l.

Several studies have evaluated the tissue concentrations of nutrients and heavy metals in plants when grown in the sewage sludge-amended soil. The accumulation pattern varied with soil type, plant species, phenology, and chelating effects of other metals [23]. Bonding of potentially toxic elements to sludge solids and soils can limit transfer to roots. Some metals, such as Cr and Pb, have very low solubility in soils and show a particularly strong barrier. Leafy crops tend to have less protection in the uptake of metals in comparison to root crops. Many experiments have shown the metals have lower concentrations in seeds and fruits compared to roots, stems, and leaves. For example, Mo is more concentrated in soybean seeds than in the leaves [24], and Tl concentrations in rapeseed are higher than in the leaves [25]. For slightly-moderately Cd-contaminated soils, the transfer of Cd to the seed of linseed (flax), sunflower, corn, and wheat can be sufficiently high to exceed health standards in some countries [26, 27], whereas Zn uptake by corn (maize) in a multiyear sewage sludge experiment on calcareous soils was within the safe limit [28].

For copper concentration in crops, results for Cu were observed in the long-term field sludge experiments of Hinesly and Hansen [29], Hinesly et al. [30], and Soon et al. [28]. It was observed that Cu concentration increased in maize stover when there was an increase in Cu loading in the soil through sludge application. But interestingly, the increase was not directly proportional to the amount of increased Cu application. Reasons for such behavior are Cu sorption by sludge and soil organic matter and plants' strong physiological barrier to Cu translocation [31].

Because of the complicated nature as to how metals behave in soils especially when they are added through sewage sludge, it is almost impossible to provide generalized guidelines. For any particular situation, various considerations should be given before setting metal application guidelines. Such concentration should include soil physical and chemical properties especially adsorption characteristics, crops to be grown, and usage of grown crops. Contamination of such land by metals should be regarded as irreversible and must be kept to the lowest practicable level [9].

3.7. Biological analysis

To evaluate microbial contamination, multisamples were sent to the Muscat Municipality laboratory from all crops. Different tests were done such as the total aerobic plate count, Coliform bacteria, *Escherichia coli, Staphylococcus aureus, Salmonella* spp., yeast, and mold. No harmful bacteria were found, and according to that all crops can be eaten safely.

Same finding was reported by Boswell [32], when he noticed that sewage sludge amendment increased the fruit yield significantly compared to the un-amended control and no toxic or detrimental effects on fescue were noted.

3.8. Water productivity

Water productivity factor can be calculated by comparing water used in this study with plant production (water productivity = total fruit weight, kg/water applied, m^3). The same amount of water was used to irrigate all crops, and as it was found in **Figures 6–8**, Kala compost gave better yield than NPK treatment, which means that water productivity of Kala compost was higher than NPK treatment.

Additions of organic fertilizers enhance soil fertility and improve soil structure. These improvements in soil physical properties increased water-holding capacity by promoting higher water retention in sludge-amended soils [33].

4. Conclusion

It can be concluded that treated municipal wastes (Kala compost) enriched the agricultural soil by improving soil physiochemical properties. Kala compost was a good conditioner for soil as it supported plants with many elements needed for high yield. Soil and plant chemical analysis did not show any problem of heavy metal accumulation. The application of Kala compost did not cause any environmental and human health problems. Therefore, it is safe to apply treated municipal wastes (Kala compost) in some agricultural crops if good management is practiced. Moreover, it is recommended that long-term records on application of treated municipal wastes (Kala compost) are reviewed, so clear findings can be generalized for future applications.

Acknowledgements

The authors would like to thank the staff from Sultan Qaboos University for their support in collecting and analyzing soil, water, and plant samples. Special thanks to Haya Water Company and USAID (FABRI) for their financial support.

Author details

Ahmed Al-Busaidi and Mushtaque Ahmed*

*Address all correspondence to: ahmed99@squ.edu.om

Department of Soils, Water and Agricultural Engineering, College of Agricultural and Marine Sciences, Sultan Qaboos University, Al-Khoud, Muscat, Oman

References

[1] Zabaniotou A, Theofilou C: Green energy at cement kiln in Cyprus—Use of sewage sludge as a conventional fuel substitute. Renew. Sustain. Energy Rev. 2008; 12 (2): 531–541.

[2] Singh RP, Agrawal M: Potential benefits and risks of land application of sewage sludge. Waste Manage. 2008; 28: 347–358.

[3] Logan TJ, Harrison BJ: Physical characteristics of alkaline stabilized sewage sludge (N-vitro soil) and their effects on soil properties. J. Environ. Qual. 1995; 24: 153–164.

[4] Korboulewsky N, Dupouyet S, Bonin G: Environmental risks of applying sewage sludge compost to vineyards: carbon, heavy metals, nitrogen, and phosphorous accumulation. J. Environ. Qual. 2002; 31: 1522–1527.

[5] Kulling D, Stadelmann F, Herter U: Sewage Sludge—Fertilizer or Waste? UKWIR Conference, Brussels. 2001.

[6] Ramulu US Sree: Reuse of Municipal Sewage and Sludge in Agriculture. Scientific Publishers, Jodhpur, India. 2002.

[7] Hue NV, Ranjith SA: Sewage sludges in Hawaii: chemical composition and reactions with soils and plants. Water Air Soil Pollut. 1994; 72: 265–283.

[8] Annual Haya Report for Kala, Haya Company, 2015 (Handout copy).

[9] MAFF: Review of the Rules for Sewage Sludge Application to Agricultural Land. Food Safety and Relevant Animal Health Aspects of Potentially Toxic Elements. MAFF Publications, London. 1993.

[10] Rosen CJ, Halbach TR, Swanson BT: Horticultural uses of municipal solid waste composts. Hort. Technol. 1993; 3 (2): 167–173.

[11] Wang QR, Cui YS, Liu XM, Dong YT, Christrie P: Soil contamination and plant uptake of heavy metals at polluted sites in China. J. Environ. Sci. Health. 2003; 38: 823–838.

[12] Epstein E: Effect of sewage sludge on some soil physical properties. J. Environ. Qual. 1975; 4: 139–142.

[13] Walid BA, Abdelbasset L, Noureddine G, Gijs DL, Marc V, Pascal B, Oswald VC,24 Naceur J, Tahar G: Accumulation and fractionation of trace metals in a Tunisian calcareous soil amended with farmyard manure and municipal solid waste compost. Journal of Hazardous Materials. 2010; 176: 99–108.

[14] Walker DJ, Bernal MP: Plant mineral nutrition and growth in a saline Mediterranean soil amended with organic wastes. Commun. Soil Sci. Plant Anal. 2004; 35: 2495–2514.

[15] Nielson GH, Hogue EJ, Nielson D, Zebarth BJ: Evaluation of organic wastes as soil amendments for cultivation of carrot and chard on irrigated sandy soils. Can. J. Soil Sci. 1998; 78: 217–225.

[16] Tsakou A, Roulia M, Christodoulakis NS: Growth of cotton plants (*Gossypium hirsutum*) as affected by water and sludge from a sewage treatment plant: I. Plant phenology and development. Bull. Environ. Contam. Toxicol. 2001; 66: 735–742.

[17] Bozkurt MA, Yarilga T: The effects of sewage sludge applications on the yield, growth, nutrition and heavy metal accumulation in apple trees growing in dry conditions. Turkish J. Agric. Forest. 2003; 27: 285–292.

[18] Morera MT, Echeverria J, Garrido J: Bioavailability of heavy metals in soils amended with sewage sludge. Can. J. Soil Sci. 2002; 82: 433–438.

[19] Verlinden S, McDonald L: Productivity and quality of statice (*Limonium sinuatum* cv. Soiree Mix) and cockscomb (*Celosia argentea* cv. Chief Mix) under organic and inorganic fertilization regiments. Sci Hortic-Amsterdam. 2007; 114: 199–206.

[20] Metcalf E: Wastewater Engineering: Treatment, Disposal, and Reuse, 4th ed. McGraw-Hill Publishing Company Ltd., New York, NY. 2003.

[21] Henry CL, Cole DW: Use of biosolids in the forest: technology, economics and regulations. Biomass Bioenerg. 1997; 13: 269–277.

[22] Kiekens L, Cottenie A, Van Landschoot G: Chemical activity and biological effect of sludge-borne heavy metals and inorganic metal salts added to soils. Plant Soil. 1984; 79: 89–99.

[23] Mahler RJ, Bingham FT, Sposito G, Page AL: Cadmium enriched sewage sludge application to acid and calcareous soils, relation between treatment, Cd in saturated extracts and Cd treatment. J. Environ. Qual. 1980; 9: 359–364.

[24] McBride MB, Richards BK, Steenhuis T, Spiers G: Molybdenum uptake by forage crops grown on sewage sludge-amended soils in the field and greenhouse. J. Environ. Qual. 2000; 29: 848–854.

[25] Tremel A, Masson P, Garraud H, Donard OFX, Baize D, Mench M: Thallium in French agrosystems. 2. Concentration of thallium in field-grown rape and some other plant species. Envir on. Pollut. 1997; 97: 161–168.

[26] Garrett RG, MacLaurin AI, Gawalko EJ, Tkachuk R, Hall GEM: A prediction model for estimating the cadmium content of durum wheat from soil chemistry. J. Geochem. Explor. 1998; 64: 101–110.

[27] Nan Z, Li J, Zhang J, Cheng G: Cadmium and zinc interactions and their transfer in soil–crop system under actual field conditions. Sci. Total Environ. 2002; 285: 187–195.

[28] Soon YK, Bates TE, Moyer JR: Land application of chemically treated sewage sludge: III. Effects on soil and plant heavy metal content. J. Environ. Qual. 1980; 9: 497–504.

[29] Hinesly TD, Hansen LG: Effects of using Sewage Sludge on Agricultural and Disturbed Land. Environmental Protection Agency, Washington, DC. 1983; EP A-600y2-83—113.

[30] Hinesly TD, Hansen LG, Bray DJ: Use of Sewage Sludge on Agricultural and Disturbed Lands. Environmental Protection Agency, Washington, DC. 1984; EP A-600y2-84-127.

[31] Poschenrieder C, Bech J, Llugany M, Pace A, Fenes E, Barcelo J: Copper in plant species in a copper gradient in Catalonia (North East Spain) and their potential for phytoremediation. Plant Soil. 2001; 230: 247–256.

[32] Boswell FC: Municipal sewage sludge and selected element application to soil: effect on soil and fescue. J. Environ. Qual. 1975; 4: 267–272.

[33] Ojeda G, Alcaniz JM, Ortiz O: Runoff and losses by erosion in soils amended with sewage sludge. Land Degrad. Dev. 2003; 14 (6): 563–573.

5

Edge of Field Technology to Eliminate Nutrient Transport from Croplands: Specific Focus on Denitrification Bioreactors

Michael Aide, Indi Braden and Sven Svenson

Additional information is available at the end of the chapter

Abstract

Tile drainage effluent from agriculture fields is beneficial to production agriculture; however, nitrate and phosphate transport from production fields to surface water resources is an environmental concern. The David M. Barton Agriculture Research Center (Cape Girardeau County, Missouri, USA) has a 40 ha controlled subsurface tile drainage/irrigation technology with associated denitrification bioreactor. Nitrate-bearing effluents from the controlled subsurface tile drainage/irrigation technology under a corn (*Zea mays* L.)-soybean (*Glycine max* L) rotation is sufficient to be an environmental concern. Nitrate-bearing effluent passage through the denitrification bioreactor typically promotes sufficient nitrate reduction (denitrification) that the bioreactor effluent water is less than 10 mg NO_3-N/L. Phosphorus, ammonium-N, and sulfate-S concentrations are not appreciably influenced by denitrification bioreactor passage.

Keywords: bioreactors, nitrate, controlled drainage, water quality, denitrification

1. Introduction

1.1. Impact of nutrient migration from cropland to fresh water

Hypoxia is considered as oxygen depletion in a water column to the point that living aquatic organisms may no long survive. Hypoxia in the northern Gulf of Mexico is defined as a dissolved oxygen concentration smaller than 2 mg/L. Hypoxia may be a naturally occurring phenomenon in selected marine environments (fjords, deep ocean basins, etc.); however, human activities are increasingly associated with the expanding of existing hypoxia zones [1]. One large hypoxia zone exists in the northern Gulf of Mexico, adjacent or superimposed on the Louisiana/Texas continental shelf.

Factors believed to be influencing the areal extent and the degree of oxygen depletion in the northern Gulf of Mexico include (i) nutrient concentrations flowing from the Mississippi River, (ii) eutrophication, (iii) microbial biomass respiration at the ocean floor, and (iv) water column stratification and attendant oxygen depletion. Mississippi River nutrient concentrations have increased in the twentieth century and continue to increase to the present time. The current concentrations of nitrogen and phosphorus in the Mississippi River and other rivers has been attributed to increased use of nitrogen and phosphorous fertilizers, the potential for nitrogen and phosphorus to become transported from crop fields to tributaries of the Mississippi River, and atmospheric deposition of oxidized nitrogen gases arising from the combustion of fossil fuels.

Eutrophication follows when aquatic systems receive these nutrients and increase primary production, including algae. The increased growth of phytoplankton exceeds the food web's capacity to consume the phytoplankton, permitting a portion of the phytoplankton to sink to the ocean bottom, supporting bacterial growth. Water column stratification isolates the reduced oxygen-bearing deep water layers. Organisms that are more predatory and higher in the food chain vacate the region, while other less mobile species perish. Disruption of commercial fishing is common. Hypoxia typically persists until weather patterns and storms remix the water column.

1.2. Review of controlled subsurface irrigation and drainage technologies

1.2.1. Nitrogen, phosphorus, and tile drainage

The United States Environmental Protection Agency (USEPA) maximum contamination level for nitrate-N is 10 mg NO_3-N/L and the scientific literature is replete with manuscripts addressing nitrate levels in groundwater and surface water exceeding this concentration [2–9]. Watersheds having N-fertilized row crop and metropolitan/suburban areas are known to contribute N runoff to tributaries, supporting hypoxia in the Gulf of Mexico [4].

Surface water runoff from intensively fertilized agricultural fields or urban landscapes, soil erosion, livestock and poultry operations, and effluent discharge from subsurface drainage technologies are important nutrient sources for freshwater contamination [10–15]. Nitrate concentrations emanating from subsurface drainage systems frequently exceed the USEPA maximum contamination levels [3, 5, 8, 16]. Phosphorus concentrations emanating from surface- or subsurface-drained landscapes are markedly most severe if the soils have a low P sorption capacity or have been heavily amended with phosphate manure/fertilizers [14, 17–23].

Dinnes et al. [4] reviewed the literature and noted that agricultural investigations aimed at reducing N losses from tile-drained soils include (1) properly adjust timing and rate of nitrogen fertilization, (2) quantify soil organic matter mineralization to reduce overapplication of nitrogen fertilizers, (3) using appropriate yield goals when making fertilizer recommendations, (4) encourage prescription fertilization practices, (5) employ nitrification and urease inhibitors, (6) employ remote sensing technologies to monitor crop nutrient status, (7) diversify crop rotations and cover crops, (8) manage plant residues, and (9) install riparian buffers and drainage control strategies. Drainage control strategies essentially manage soil water to promote anoxic soil conditions resulting in denitrification.

Kladivko et al. [7] effectively demonstrated that narrow-spaced lateral drainage lines have a greater capacity to promote nitrate removal. Fisher et al. [24] compared controlled subsurface drainage technologies with open drainage systems and documented that 30–75 cm water table depth maintenance reduced nitrate soil water concentrations and improved corn uptake of nitrogen. Randall et al. [8] investigated corn-soybean rotations in Minnesota and documented that nitrate leaching correlated with rainfall, that the soybean phase supported nitrate tile drain discharges because of residue mineralization and residual nitrate concentrations from the previous corn planting. Randall et al. [8] also observed that summer intervals exhibited the smallest nitrate leaching because the evapotranspiration rates exceeded the precipitation rates.

Phosphorus studies have centered on P runoff and P leaching [25–27]. Organic P and colloidal P may be mobile in controlled drainage systems [7, 17, 23]. Djodjic et al. [19] noted that dissolved reactive phosphorus was not effectively predicted by total P and that preferential water flow pathways did not allow for equilibrium assumptions. In a review, Hart et al. [28] noted that catchment studies typically show that 62–91% of surface runoff is associated with particulate P.

1.2.2. Denitrification bioreactors

Numerous ground and surface waters exceed the USEPA nitrate-N concentration of 10 mg NO_3-N/L drinking water limit [5, 6, 29, 30]. United States Environmental Protection Agency (USEPA) reports that a 45% reduction in nitrogen loads in the Mississippi River Basin is a goal to reduce water impact. In the Midwest, 15 million ha have artificial drainage capacities. The reported elevated nitrogen loads include 81 [31] and 88 kg N/ha [32], whereas more typical nitrogen loads are 25–35 kg N/ha, likely associated with nitrate-N effluent concentrations of 10–25 mg NO_3-N/L [33]. An emerging technology involves the design and construction of permeable reactive subsurface-packed beds having carbonaceous materials to support nitrate denitrification [34–36].

In a review by Christianson [37], denitrification bioreactors in the upper Midwest were effective in reducing nitrate-N effluent concentrations: 32.5 [38, 39], 40–65 [30], 50–60 [40], and 47% [41]. Denitrification bioreactors rely on microbial denitrification

$$5C + 4NO_{3^-} + 2H_2O = 2N_2 + 4HCO_{3^-} + CO_2.$$

The process requires a (i) carbon source (electron donor), (ii) low dissolved oxygen (DO) concentrations, (iii) denitrifying bacteria, and (iv) nitrate as an electron acceptor and results in either nitrogen gas (N_2) or nitrogen oxides (N_2O) production [40]. The microbial reaction pathway may be described as

$$NO_{3^-} \rightarrow NO_{2^-} \rightarrow NO \rightarrow N_2O.$$

Each step is catalyzed by nitrate reductase, nitrite reductase, nitric oxide reductase, and nitrous oxide reductase, respectively. The release of bicarbonate may modify the reactor pH. Low pH, low temperature, suboxic dissolved oxygen levels, and low C/N ratios act individually or collectively to support greater N_2O/N_2 ratios [37, 42, 43].

The reduction half reaction and associated log K_r and $E^o{}_H$ values for nitrate-dinitrogen gas couple may be written as [44]

$$1/5NO_{3^-} + 6/5H^+ e^- \rightarrow 1/10N_2 + ¾H_2OLogK_r = 21.1 \text{ and } E^o{}_H = 1.248 \text{ volts V.}$$

The IUPAC convention would list the reaction as

$$E_H(\text{voltsV}) = E^o + (RT/nF)\left\{\ln\left[NO_{3^-}\right]^{0.2}\left[H^+\right]^{1.2}/\left[N_2\right]^{0.1}\right\},$$

where $[H_2O]$ has unit activity Activity, R=001987 kcal/mole deg., T is temperature in Kelvin, and F is 23.061 kcal/volt g. eq. Given the partial pressure of nitrogen gas at 0.79 and a pH near neutrality, the E_H is a linear function of the nitrate concentration.

Denitrification reactor design is a complex function of reactor length and retention times suitable to reduce dissolved oxygen concentrations for the anaerobic process to facilitate nitrate reduction [42, 45]. Excessive retention times may promote sulfate-S reduction and mercury methanogenesis [37, 42]. Retention time is largely a function of reactor water flux, with greater water flow rates reducing the retention time. Chun et al. [41] observed that denitrification bioreactor nitrate reduction responded to first-order kinetics, whereas Schipper et al. [22] noted that field-scale bioreactors were better simulated using zero-order kinetics.

2. Materials and methods

2.1. Existing physical infrastructure

Located in Cape Girardeau County (Missouri, USA), the David M. Barton Agriculture Research Center hosts the Crop Science Unit. The Crop Science Unit has a controlled subsurface drainage and irrigation system. The controlled drainage system consists of a series of parallel 10 cm (4 in.) subsurface conduits having a parallel 10 m (30 ft) spacing collecting into 20 cm (8 in.) conducts for transport of surplus drainage water to field ditches. Irrigation and drainage are monitored by stop-log boxes fitted with adjustable baffles strategically arranged in the field to permit the restriction of water flow, allowing irrigation/drainage water to be added/removed throughout the system by gravity flow. The irrigation pumping system consists of five wells, each with capacity to pump 265 L/min (70 gal/min).

The denitrification bioreactor was constructed in June 2014. Sampling ports allow water sampling from the denitrification bioreactor at the influent and effluent tile lines. The denitrification bioreactor has dimensions of 10 m width, 20 m length, and 0.7 m thickness. The top of the denitrification bioreactor is approximately 0.6 m below the soil surface. Oak (*Quercus* sp.) wood chips having an approximately 5 cm (2 in.) equivalent circular diameter with 1 cm thickness constitute the denitrification bioreactor-packed bed fill **(Figure 1)**.

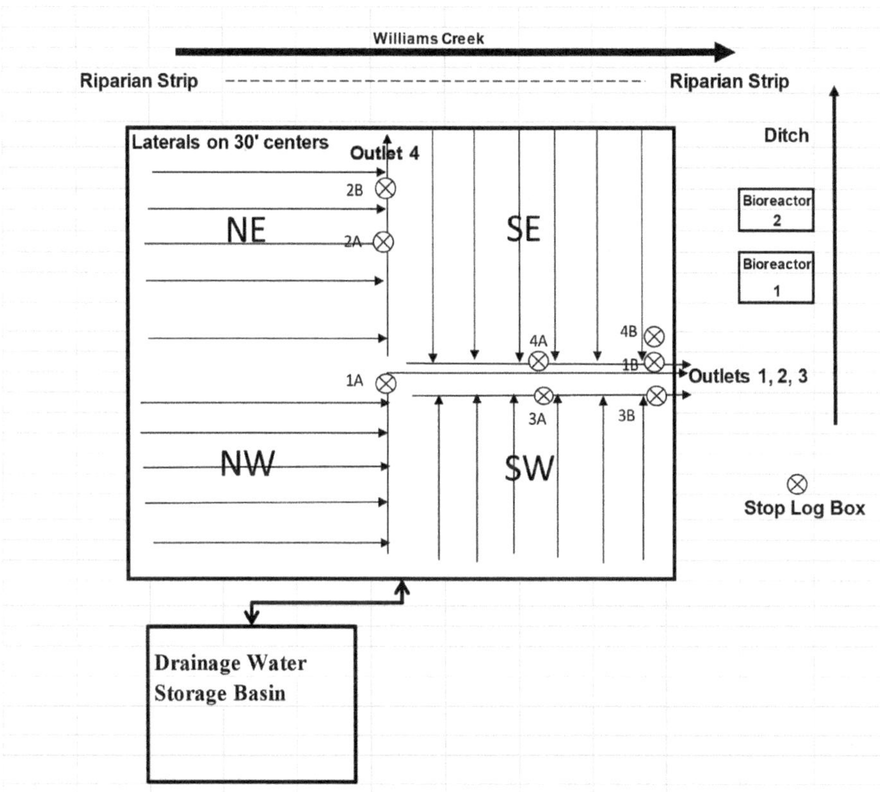

Figure 1. Technology development at the Crop Science Unit (40 ha or 100 acres).

2.2. Soil resources

The soils of the Wilbur series (USA Soil Taxonomy: coarse-silty, mixed, superactive, mesic Fluvaquentic Eutrudepts) consist of very deep, moderately well-drained soils that formed in alluvium. Six pedons show uniform silt loam textures throughout their soil profiles and display Ap-Bw-Cg horizon sequences [46]. Moderate medium platy structures in the near-surface horizons typically part to weak medium subangular blocky structures in the Bw horizons. The deeper Cg horizons generally show moderate coarse prismatic structures that part to weak medium subangular blocky structures. The dominant soil matrix colors are dark brown to dark yellowish brown in the Ap and Bw horizons, transitioning to light gray, gray, light brownish gray, and grayish brown in the Cg horizons. Iron-Mn accumulations and Fe depletions are evident throughout the soil profiles, especially in the Cambic and Cg horizons.

Soil pH generally ranges from slightly acid (pH 6.1–6.5) to neutral (pH 6.6–7.3) in the near-surface horizons to strongly acid (pH 5.1–5.5) and very strongly acid (pH 4.5–5.0) in the Bw and upper Cg horizons, whereas the deepest Cg horizons have moderate to slight acidity (pH 5.6–7.0). The soil organic matter contents are generally low (less than 2%) and decline with increasing soil depth. Soil phosphorus (extraction using Bray1-P) and sulfur (extraction using 2 M KCl) have their greatest concentrations in the near-surface horizons, showing a continuous P and S decline with increasing soil depth. The exchangeable cations are dominated by calcium (Ca), especially in the near-surface soil horizons. The total acidity is appreciable, particularly in the deeper soil horizons; however, some Wilbur pedons show reduced total acidity expressions in the deeper Cg horizons. The cation exchange capacity is low (<12 $cmol_{p(+)}$/kg) to medium (12–18 $cmol_{p(+)}$/kg) and roughly corresponds with the clay and soil organic matter contents.

Mechanical analysis indicates that silt is the dominant separate in all six pedons, with the sand separate being less than 10% and composed almost entirely of very fine sand. The clay mineralogy is mixed, with an abundance of hydroxyl Al-interlayered vermiculite, smectite, hydrous mica, and kaolinite. Smectite shows relatively greater abundances in the deeper soil horizons.

2.3. Soil water assessment

Field soil water measurements involve (i) water table height using piezometer tubes, (ii) irrigation water rates using flow meters, (iii) rainfall monitoring using a US Class A rain gauge, and (iv) volumetric soil moisture distribution using gravimetric samples and bulk density. Estimates of total tile drainage flow were obtained using electronic water elevation sensors in the stop-log boxes and box geometry to calculate water flow, where water was applied from Williams Creek with a centrifugal pump system. Levees were designed by field survey and established with a levee plow.

2.4. Crop production to assess nutrient uptake

Corn (*Zea mays* L.) was planted from 2008 to 2015 on 0.77 m (30 in.) row spacing. Phosphorus (P) and potassium (K) fertilization was applied using variable rate technology based on grid soil sampling. From 2012 to 2015, corn nitrogen fertilization rates were 378 kg N/ha (344 N lbs N/acre) as half of the urea was applied 1 week prior to planting and half applied 2 weeks after planting. Yield goals were 13,200 kg/ha at an established population of 85,000 plants/ha. Tissue testing (N, P, K, Ca, Mg, S, Na, Al, Fe, Mn, Zn, B, and Cu) and plant biomass accumulation were documented to assess nutrient uptake patterns at V7, R1, and R6 corn growth stages. Plant organ sampling includes biomass and nutrient accumulation in root, stem (culm), leaf, and seed, with total plant uptake and biomass accumulation based on the summation of the product of the plant organs biomass and concentration.

2.5. Field and laboratory protocols

Water sampling of tile drain and denitrification bioreactor influent and effluent was conducted weekly for the spring 2015 drainage season and daily for the denitrification bioreactor/ Williams Creek assessment. Water was collected in precleaned plastic collection bottles and stored in refrigeration cabinets until analyzed. Samples were analyzed for pH, NO_3-N, NH_4-N, H_2PO_4/HPO_4, SO_4-S, Ca, Mg, K, and Na at the University of Missouri's Fisher Delta Research Center using standard protocols. Nitrate concentrations were determined using an ion-specific electrode, ammonium concentrations were determined using colorimetric indophenol blue, phosphorus was determined using colorimetric ammonium molybdate, and sulfate-S was determined using the $BaCl_2$ turbidimetric method. Water pH was determined using a combination pH electrode. Exchangeable cations were extracted using 1 M ammonium acetate (pH 7) extraction. Water and soil calcium, magnesium, potassium, and sodium concentrations were determined using air-acetylene atomic absorption spectroscopy.

3. Research involving controlled subsurface irrigation and drainage at the David M. Barton Agriculture Research Center

This portion of the research project is a long-term assessment of controlled subsurface irrigation/drainage technologies with associated denitrification bioreactors. Tile drainage water chemistry and nitrate-ammonium concentrations available in soil from 2010 to 2013 are documented [47, 48]. These 4 years of investigation reveal soil nitrate concentrations generally showed an increase immediately after soil nitrogen fertilization practices and were sufficiently abundant to promote their transport from the soil resource to the tile drain effluent waters. The tile drainage chemistry data indicated (i) appreciable transport of nitrate-N in tile drain effluent waters (mean of 32 mg NO_3-N/L in 2008, mean of 80 mg NO_3-N/L in 2009, mean of 10 mg NO_3-N/L in 2010, and mean of 15 mg NO_3-N/L in 2012); (ii) denitrification soil pathways partially reduced a portion of the soil nitrate-N when the controlled drainage system establishes winter/early spring anoxic soil conditions, and (iii) the best strategy for reducing nitrate-N concentrations in tile drain effluent waters was adjusting: (i) N fertilization rates and (ii) the timing of their application.

Tile drainage from the 2014 soybean system illustrated pH levels near pH 6.5 ± 0.5 across all of the sampling sites for the duration of drainage. Greater nitrate sampling was performed in 2014 than 2013 because of the longer drainage interval; however, tile drainage effluent nitrate-N concentrations averaged from less than 10 mg NO_3-N/L for many of the sampling sites/times to more than 80 mg NO_3-N on at least four occasions (**Figure 2**). Ammonium concentrations in the tile drain effluents ranged from 0.25 mg NH_4-N/L to near 5 mg NH_4-N/L. The presence of appreciable nitrate and ammonium concentrations was reflective of a large nitrate pool remaining from the previous corn production and to a smaller extent soil organic matter mineralization. Phosphorus concentrations ranged from 0.1 to 3 mg P/L, thus phosphorus concentrations represent an environmental impact given they frequently exceed

0.2 mg PO_4-P/L. Sulfate-S concentrations ranged from 2.5 to 5.5 mg SO_4-S/L; however, these SO_4-S concentrations were not considered an environmental hazard.

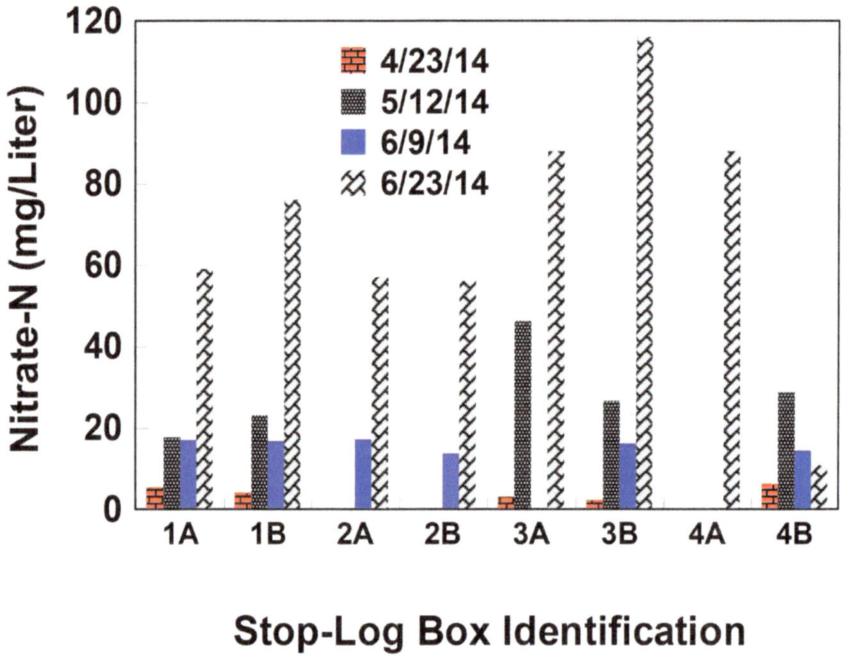

Figure 2. Drainage water nitrate-N concentrations from stop-log boxes in 2014.

Soil nitrate concentrations have been periodically monitored to estimate the soil nitrate pool for plant uptake and leaching potentials. The soil nitrate concentrations typically have fluctuated based on (i) the corn-soybean rotation stage, (ii) nitrogen (urea) fertilization rates and timing, (iii) soil denitrification (either intentionally establishing perched water tables by restricting drainage during the noncropping season and cropping season rainfall patterns), (iv) crop uptake (corn-soybean rotation and off-season cover crop establishment), and (v) soil mineralization and residue decomposition. Soil sampling established that nitrate-N concentrations were (i) greater after urea application for corn and (ii) dependent on rainfall patterns. Approximately 10–50% of the nitrate pool migrated from the upper 15 cm to the 15–30 cm layer within 1 month of application, with smaller portions of the nitrate pool ultimately percolating to deeper soil layers. As an example, 2013 (corn portion of the rotation) witnessed an April planting with urea (292 lbs/ac or 328 kg/ha) application just prior to 15 May 2013. On 15 May 2013, the majority of the urea was converted to ammonium with a portion of the ammonium converting to nitrate via nitrification reactions. On 7 June, the majority of the nitrogen application was nitrate, with a portion of the nitrate leaching into the 15–30, 30–45, and 45–60 cm deep soil layers (**Table 1**). Soil nitrate concentrations postcorn harvest (data not shown) and 24 March 2014 soil nitrate concentrations were comparatively smaller. A substantial portion of the field nitrogen pool was documented to be associated with grain and residue production (approximately 60%) and the remainder associated with the soil nitrate pool and lost from the soil system because of tile drainage effluents or soil denitrification reactions.

Depth (cm)	NO_3-N	NH_4-N
	mg N/kg	
15 May 2013 (corn)		
15	12.5–18.6	9.7–25.7
30	11.3–16.1	0.5–0.9
45	8.3–8.6	1.1–1.5
60	7.6–7.1	0.7–0.9
7 June 2013 (corn)		
15	14.4–21.4	3.2–5.8
30	15.0–16.7	3.4–7.9
45	14.5–15.6	2.2–3.4
60	14.4–14.6	0.7–3.4
24 March 2014 (soybean)		
15	6.9–7.9	0.3–1.4
30	7.0–7.7	0.2–0.6
45	6.9–7.9	0.2–0.7
60	7.0–7.7	0.2–1.4

Multiple replications.

Table 1. Soil nitrate and ammonium concentrations.

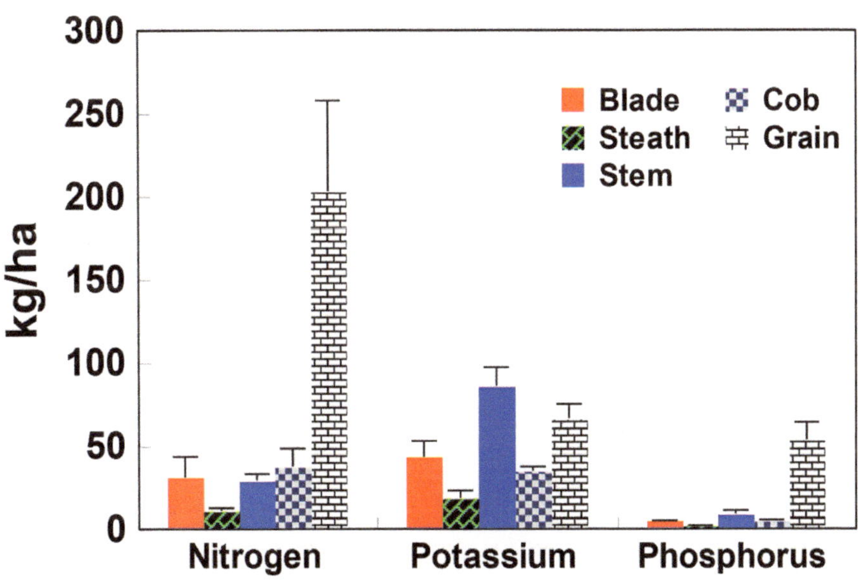

Figure 3. Concentrations of nitrogen, potassium, and phosphorus in 2014 corn by plant organs (Error bars are the standard deviations of three replicates.).

The associated corn biomass (**Figure 3**) demonstrates that nitrogen is primarily associated with grain (65%) and is thus removed from the soil landscape by harvest. Similarly, potassium (27% associated with grain) and phosphorus (74% associated with grain) demonstrate different harvest removals.

4. Research involving controlled subsurface irrigation and drainage with denitrification bioreactors at the David M. Barton Agriculture Research Center

4.1. Spring drainage water study: denitrification bioreactor inlet and outlet water chemistry for Spring 2015

The 2015 growing season was the first operational year for the denitrification bioreactor. Nitrate-bearing tile drainage water from land cultivated to corn (*Zea mays* L.) entered the denitrification bioreactor during the "drainage season." Mean phosphate, ammonium, nitrate, and sulfate concentrations and water pH are presented (**Table 2**) to illustrate the baseline chemistry and document that tile drainage effluent has sufficient nitrate-N to be considered as an environmental hazard.

Sampling sites	PO_4-P	NH_4-N	NO_3-N	SO_4-S	pH
			mg/L		
1A	0.3	0.9	21	2.7	6.8
1B	0.23	1.5	25.6	2.6	6.8
2A	0.19	1	16.4	2.7	6.6
2B	0.37	1	11.2	2	6.5
3B	0.2	0.7	15.8	1.6	6.9
4B	0.21	0.7	22	2.2	6.8
Bioreactor influx	0.23	0.8	59.1	4	6.7
Bioreactor effluent	0.25	0.9	38.6	2.1	6.6

Tile drainage sampling (1A, 1B, 2A, 2B, 3B, and 4B), mean of 12 sampling times from 20 March 2015 to cessation of drainage on 6 July 2015.

Table 2. Mean phosphorus, ammonium, nitrate, and sulfate concentrations and pH of tile drainage waters collected during the spring 2015 drainage season.

Nitrate-N concentrations were substantially reduced by passage through the denitrification bioreactor, except for 29 May 2015 that was postnitrogen fertilization and a heavy rain event with large water volumes migrating through the bioreactor (**Figure 4**). From March through early May, the influx of nitrate-N averaged 17 mg NO_3-N/L (standard deviation of 12 mg NO_3-N/L), whereas the effluent concentrations were 5 mg NO_3-N/L (standard deviation of 3 mg NO_3-

N/L). Nitrate concentrations from late May to mid-June and following nitrogen fertilization, the influx of nitrate-N averaged 69 mg NO_3-N/L (standard deviation of 31 mg NO_3-N/L), whereas the effluent concentrations were 21 mg NO_3-N/L (standard deviation of 40 mg NO_3-N/L).

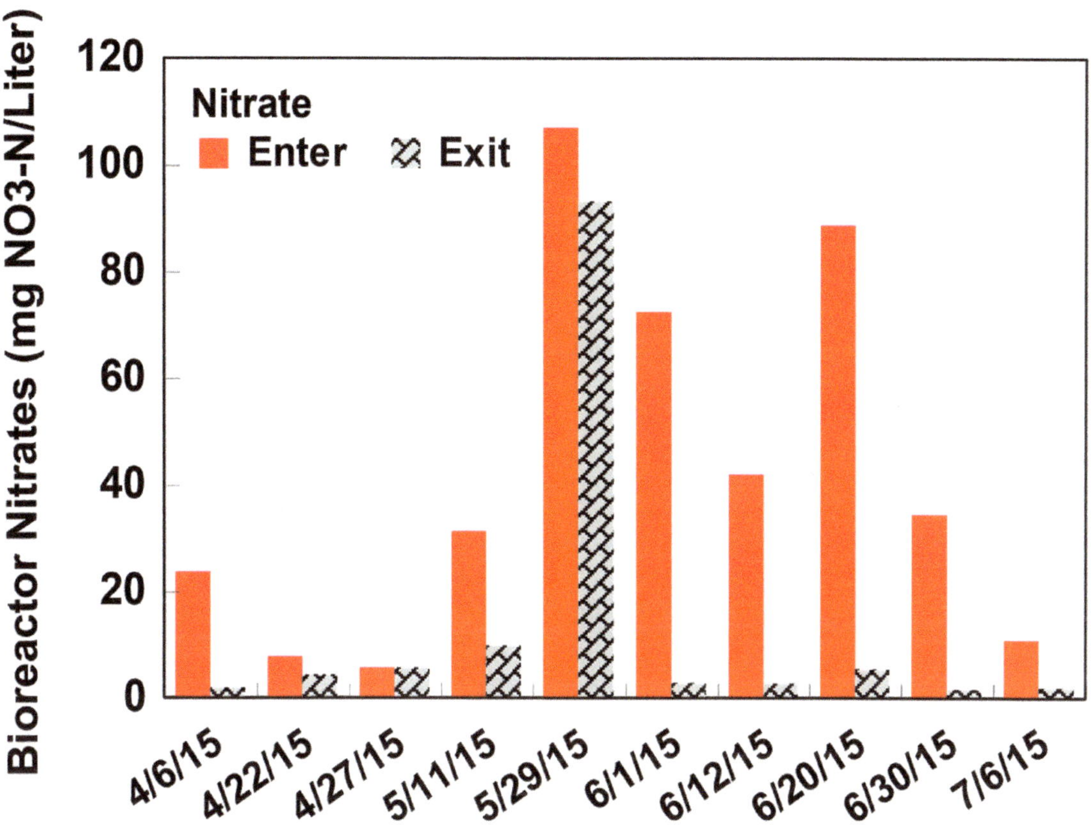

Figure 4. Denitrification bioreactor nitrate-N concentrations at the receiving and exiting terminals.

Ammonium concentrations were not appreciably influenced by bioreactor passage. Ammonium-N concentrations were generally less than 1 mg NH_4-N/L, except for 22 April 2015 (1.3 mg NH_4-N/L influx and 0.4 mg NH_4-N/L effluent) and 30 June 2015 (3.0 mg NH_4-N/L influx and 2.4 mg NH_4-N/L effluent). Phosphorus and sulfate concentrations and water pH were not appreciably influenced by fluctuations during the drainage season and were not significantly altered by denitrification bioreactor passage.

4.2. Williams Creek impoundment and denitrification bioreactor efficiency

In the winter of 2015, Williams Creek waters were pumped and impounded by a levee system and then allowed to infiltrate/percolate through the soil and entered the tile drainage system. Water captured by the controlled subsurface drainage technology was transported to the denitrification bioreactor.

4.2.1. *Williams Creek water and stop-log box 4B captured soil water*

Williams Creek water is classified as a calcium-carbonate type water with a pH range from 7.92 to 8.05, implying dissolved calcium carbonate was influencing pH. Soil water pH sampled from stop-log box 4B ranged from 6.36 to 7.15 with a mean near 6.75. Presumably, the soil's cation exchange complex buffered soil drainage water and reduced the pH of waters originating from Williams Creek.

The soil water comparisons for calcium, magnesium, potassium, and sodium (**Table 3**) reveal that calcium concentrations are greater in the Williams Creek impoundment trial than the spring 2015 drainage trial. The field was limed with calcite limestone in the winter of 2014–2015 and limestone requires a lengthy time interval to dissolve, perform cation exchange, and complete acid neutralization, thus increasing the calcium saturation of the cation exchange complex. Additionally, Williams Creek may be assumed to be a water solute calcium source.

ID	Ca (ppm)	Mg (ppm)	K (ppm)	Na (ppm)
6/12/2015				
4B	5.5	24	2.1	11.3
In	5.3	9.1	2.9	14.1
Out	8.9	11.6	2.8	13.8
12/13/2015				
4B	61	9.4	4.5	12.8
In	52	8.7	3.7	11.1
Out	54	8.6	3.5	11.2
12/14/2015				
4B	36	6	2.8	8.9
In	33.5	5.7	2.7	7.4
Out	35.5	5.8	2.6	8.3

Table 3. Soil water concentrations of calcium, magnesium, potassium, and sodium.

Williams Creek waters show elevated nitrate concentrations, ranging from 12.7 mg NO_3-N/L on 25 November 2015 to 672 mg NO_3-N/L on 4 December 2015 (**Figure 5**). Soil water shows a nitrate-N increase to 33.1 mg NO_3-N/L on 2 Dec 2015 and 44 mg NO_3-N/L on 3 December 2015, suggesting that the soil resource is influenced by nitrate-N originating from Williams Creek. Soil water nitrate-N concentrations are consistently smaller than the water from Williams Creek, implying that the soil resource is reducing nitrate-N concentrations by a combination of two processes: (i) dilution of Williams Creek nitrate-N concentrations with the preexisting soil water and (ii) denitrification soil processes.

Nitrate-N concentrations in soil water after 7 December 2015 show a gradual decline. Between 27 November and 29 November 2015, approximately 2.94 in. of rainfall occurred, inferring that rainfall acted to dilute the soil water nitrate-N concentrations. Williams Creek and soil water both demonstrated greater nitrate concentrations on 2 December 2015.

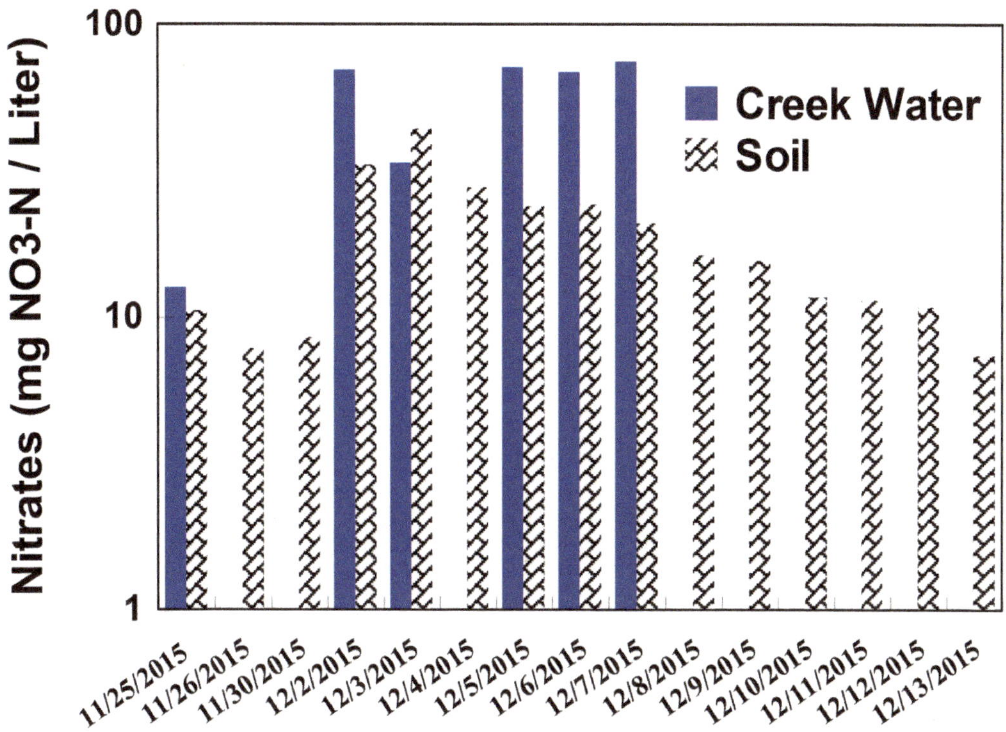

Figure 5. Nitrate concentrations from Williams Creek and stop-log box 4B. (Note: Log scale.) On 4 December 2015, Williams Creek showed 691 mg NO_3-N/L. (Data not shown on graph for graphics clarity.) Pumping from Williams Creek stopped on 8 December 2015.

Ammonium concentrations are generally small, less than 2 mg NH_4-N/L for Williams Creek and generally less than 1 mg NH_4-N/L for soil waters. Williams Creek water has the greatest ammonium concentration on 7 December 2015 (1.7 mg NH_4-N/L), approximately 3 days after the greatest nitrate-N concentrations, whereas soil water has the greatest ammonium concentration on 9 December 2015 (1.7 mg NH_4-N/L). Mean phosphorus concentrations are 0.36 mg PO_4-P/L for Williams Creek waters and 0.39 mg PO_4-P/L for the field sampling site waters, with the concentration differences being not significant. These phosphorus concentrations are considered sufficiently abundant to support water eutrophication. Sulfate concentrations were not significantly different between the Williams Creek waters (mean SO_4-S at 1.4 mg SO_4-S/L) and the field sampling site waters (mean SO_4-S at 1.2 mg SO_4-S/L).

4.3. Denitrification bioreactor nitrate reduction potential with Williams Creek source water

pH of the denitrification bioreactor inlet and effluent waters were not significantly different for each sampling date; however, the inlet water pH varied from a low pH of 6.33 (30 November 2015) to pH 7.07 (12 December 2015) and the effluent water pH varied from pH 6.31 (30 November 2015) to pH 7.18 (12 December 2015).

Denitrification bioreactor outlet nitrate-N concentrations were slightly too appreciably smaller than the corresponding inlet nitrate-N concentrations (**Figure 6**). The highest nitrate-N concentrations occurred on 2 December 2015, which corresponds with the nitrate-N concentration rise associated with stop-log box 4B. Nitrate-N concentrations from 2 December to 7 December 2015 ranged from 35.1 mg NO_3-N/L to 20.6 mg NO_3-N/L for the inlet concentrations and from 25.3 mg NO_3-N/L to 17.2 mg NO_3-N/L for the outlet concentrations. From 8 December to 13 December 2015, the inlet and outlet nitrate-N concentrations became increasingly smaller, and the outlet nitrate-N concentrations continued to be smaller than those of the corresponding inlet concentrations.

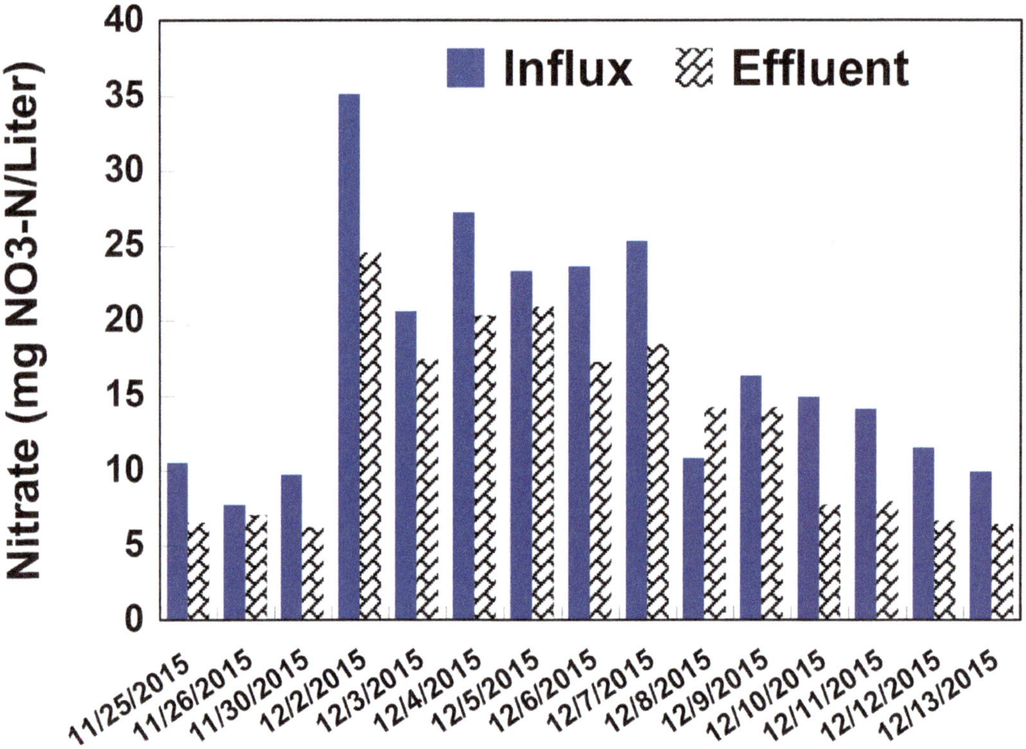

Figure 6. Water nitrate concentrations from the inlet (influx) and outlet (effluent) from the denitrification bioreactor.

Ammonium-N concentrations were substantially smaller than the corresponding nitrate-N concentrations. Ammonium-N concentration differences between the inlet and outlet waters suggest that the denitrification bioreactor sequestered ammonium-N or nitrification processes oxidized ammonium to nitrate (**Figure 7**). Denitrification bioreactor's mean phosphorus concentrations were smaller for the effluent (0.29 mg PO_4-P/L) than the inlet concentrations (0.38 mg PO_4-P/L); however, the concentration differences were not significant. Denitrification bioreactor's mean sulfate concentrations were greater for the effluent (1.1 mg SO_4-S/L) than the inlet concentrations (1.0 mg SO_4-S/L); however, the sulfate-S concentration differences were not significant.

Denitrification bioreactors in these field trials reduced effluent nitrate-N concentrations via denitrification pathways. Approximately 50% or greater nitrate-N reductions were observed when the flow volumes per unit time were sufficiently small for equilibrium attainment.

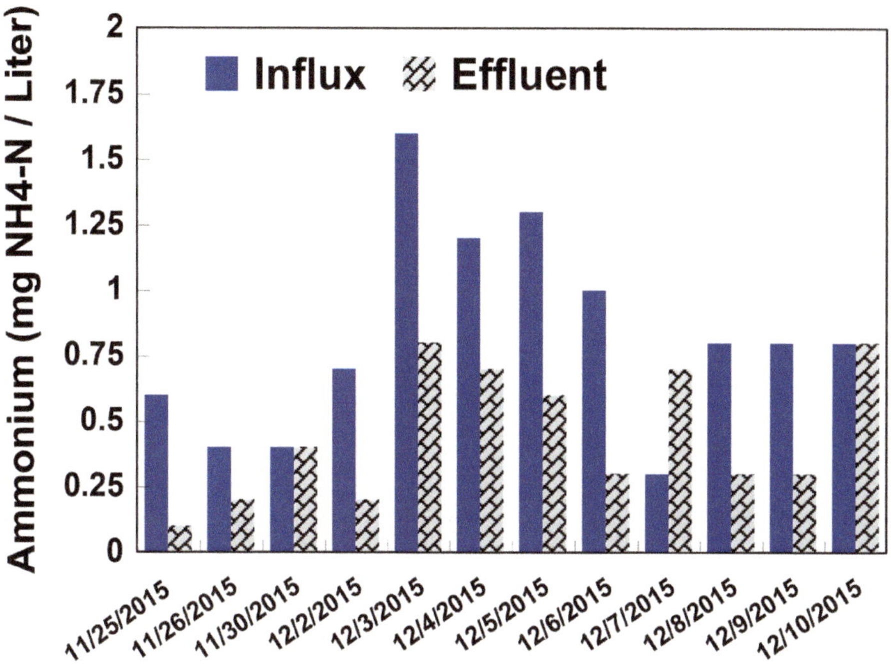

Figure 7. Water ammonium concentrations from the inlet (influx) and outlet (effluent) from the denitrification bioreactor.

5. Prospectus for future endeavors

(1) Development of effective crop nutrient management systems to improve crop uptake efficiency and reduce nitrate leaching.

(2) Development of "Soil Health" research initiatives to quantify soil structure attainment and carbon sequestration.

(3) Continue research on denitrification bioreactor design to reduce nitrate tile drainage. Engineering parameters based on reactor size, preferential bed packing materials, equilibrium thresholds, elimination of preferential flow path attainment, and pH maintenance require additional scrutiny.

Author details

Michael Aide*, Indi Braden and Sven Svenson

*Address all correspondence to: mtaide@semo.edu

Department of Agriculture, Southeast Missouri State University, Cape Girardeau, Missouri, USA

References

[1] Diaz, R.J. and Rosenberg, R. Marine benthic hypoxia: A review of its ecological effects and the behavioral responses of benthic macrofauna. Oceanogr. Mar. Biol. 1995; 33:245–303.

[2] Abrahamson, D., Radcliffe, D., Steiner, J., Cabrera, M., Hanson, J., Rojas, K., Schomberg, H., Fisher, D., Schwartz, L., and Hoogenboom, G. Calibration of the root zone water quality model for simulating tile drainage and leached nitrate in the Georgia Piedmont. Agron. J. 2005; 97:1584–1602.

[3] Abrahamson, D., Radcliffe, D., Steiner, J., Cabrera, M., Endale, D., and Hoogenboom, G. Evaluation of the RZWQM for simulating tile drainage and leached nitrate in the Georgia Piedmont. Agron. J. 2006; 98:644–654.

[4] Dinnes, D.L., Karlen, D.L., Jaynes, D.B., Kaspar, T.C., Hatfield, J.L., Colvin, T.S., and Cambardella C.A. Nitrogen management strategies to reduce nitrate leaching in tile-drained midwestern soils. Agron. J. 2002; 94:153–171.

[5] Jaynes, D., Dinnes, D., Meek, D., Karlen, D., Cambardella, C., and Colvin, T.S. Using the late spring nitrate test to reduce nitrate loss within a watershed. J. Environ. Qual. 2004; 33:669–677.

[6] Jaynes, D. and Colvin, T. Corn yield and nitrate loss in subsurface drainage from midseason nitrogen fertilizer application. Agron. J. 2006; 98:1479–1487.

[7] Kladivko, E., Frankenberger, J., Jaynes, D., Meek, D.W., Jenkinson, B.J., and Fausey, N.R. Nitrate leaching to subsurface drains as affected by drain spacing and changes in crop production system. J. Environ. Qual. 2004; 33:1803–1813.

[8] Randall, G., Vetsch, J., and Huffman, J. Nitrate losses in subsurface drainage from a corn-soybean rotation as affected by time of nitrogen application and use of nitrapyrin. J. Environ. Qual. 2003; 32:1764–1772.

[9] Selim, H.M., Davidson, J.M., and Rao, P.S.C. Transport of reactive solutes through multilayered soils. Soil Sci. Soc. Am. J. 1977; 41:3–10.

[10] Akhtar, M., McCallister, D., Francis, D., and Schepers, J. Manure source effects on soil phosphorus fractions and their distribution. Soil Sci. Soc. Am. J. 2005; 170:183–190.

[11] Davis, R., Zhang, H., Schroder, J., Wang, J., Payton, M., and Zazulak, A. Soil characteristics and phosphorus level effect on phosphorus loss in runoff. J. Environ. Qual. 2005; 34:1640–1650.

[12] Elliott, H., Brandt, R., Kleinman, P., Sharpley, A., and Beegle, D. Estimating source coefficients for phosphorus site indices. J. Environ. Qual. 2006; 35:2195–2201.

[13] Gburek, W., Sharpley, A., Heathwaite, L., and Folmar G. Phosphorus management at the watershed scale: A modification of the phosphorus index. J. Environ. Qual. 2000; 29:130–144.

[14] Langlois, J. and Mehuys, G. Intra-storm study of solute chemical composition of overland flow water in two agricultural fields. J. Environ. Qual. 2003; 32:2301–2310.

[15] Moore, P., Jr. and Edwards, D. Long-term effects of poultry litter, alum-treated litter, and ammonium nitrate on aluminum availability in soils. J. Environ. Qual. 2005; 34:2104–2111.

[16] Davis, D.M., Gowda, P.H., Mulla, D.J., and Randall, G.W. Modeling nitrate nitrogen leaching in response to nitrogen fertilizer rate and tile drain depth or spacing for southern Minnesota, USA. J. Environ. Qual. 2000; 29:1568–1581.

[17] Anderson, B.H. and Magdoff, F.R. Relative movement and soil fixation of soluble organic and inorganic phosphorus. J. Environ. Qual. 2005; 34:2228–2233.

[18] DeCamargo, O.A., Biggar, J.W., and Nielsen, D.R. Transport of inorganic phosphorus. Soil Sci. Soc. Am. J. 1979; 43:884–890.

[19] Djodjic, F., Borling, K., and Bergstrom, L. Phosphorus leaching in relation to soil type and soil phosphorus content. J. Environ. Qual. 2004; 33:678–684.

[20] Ilg, K., Siemens, J., and Kaupenjohann, M. Colloidal and dissolved phosphorus in sandy soils as affected by phosphorus saturation. J. Environ. Qual. 2005; 34:926–935.

[21] Page, T., Haygarth, P., Beven, K., Joynes, A., Butler, T., Keeler, C., Freer, J., Owens, P., and Wood, G. Spatial variability of soil phosphorus in relation to the topographic index and critical source areas: Sampling for assessing risk to water quality. J. Environ. Qual. 2005; 34:2263–2277.

[22] Schipper, L.A., Cameron, S.C., and Warnele, S. Nitrate removal from three different effluents using large-scale denitrification beds. Ecol. Eng. 2010; 36:1552–1557.

[23] Stampfli, N. and Madramootoo, C.A. Dissolved phosphorus losses in tile drainage under subirrigation. Water Qual. Res. J. Canada. 2006; 41:63–71.

[24] Fisher, M.J., Fausey, N.R., Subler, S.E., Brown, L.C., and Bierman, P.M. Water table management nitrogen dynamics, and yields of corn and soybeans. Soil Sci. Soc. Am. J. 1999; 63:1786–1795.

[25] Bruland, G. and Richardson, C. A spatially explicit investigation of phosphorus sorption and related soil properties in two riparian wetlands. J. Environ. Qual. 2004; 33:785–794.

[26] Cabrera, M.L. Modeling phosphorus in runoff: Basic approaches. *In* Radcliffe, D.E. and Cabrera, M.L. (eds), Modeling Phosphorus in the Environment. CRC Press, Boca Raton, FL. 2007; pp 65–81.

[27] Nelson, N.O. and Parsons, J.E. Basic approaches to modeling phosphorus leaching. *In* Radcliffe, D.E. and Cabrera, M.L. (eds), Modeling Phosphorus in the Environment. CRC Press, Boca Raton, FL. 2007; pp 81–104.

[28] Hart, M.R., Quin, B.F., and Nguyen, M.L. Phosphorus runoff from agricultural land and direct fertilizer effects: A review. J. Environ. Qual. 2004; 33:1954–1972.

[29] Jaynes, D.B., Colvin, T.S., Karlen, D.L., Cambardella, C.A., and Meek, D.W. Nitrate loss in substrate drainage as affected by nitrogen fertilizer rate. J. Environ. Qual. 2001; 30:1305–1314.

[30] Jaynes, D.B., Kasper, T.C., Moorman, T.B., and Parkin, T.B. In situ bioreactors and deep drainpipe installation to reduce nitrate losses in artificially drained fields. J. Environ. Qual. 2008; 37:429–436.

[31] Kaspar, T.C., Jaynes, D.B., Parkin, T.B., and Moorman, T.B. Rye cover crop and gamagrass strip effects on NO_3 concentration and load in tile drainage. J. Environ. Qual. 2007; 36:1503–1511.

[32] Lawlor, P.A., Helmers, M.J., Baker, J.L., Melvin, S.W., and Lemke, D.W. Nitrogen application rate effect on nitrate-nitrogen concentration and loss in subsurface drainage for the corn-soybean rotation. Trans. ASABE. 2008; 51:83–94.

[33] Jaynes, D.B., Hatfield, J.L., and Meek, D.W. Water quality in Walnut Creek watershed: Herbicides and nitrate in surface waters. J. Environ. Qual. 1999; 28:45–59.

[34] Greenan, C.M., Moorman, T.B., Kaspar, T.C., Parkin, T.B., and Jaynes, D.B. Comparing carbon substrates for denitrification of subsurface drainage water. J. Environ. Qual. 2006; 35:824–829.

[35] Greenan, C.M., Moorman, T.B., Kaspar, T.C., Parkin, T.B., and Jaynes, D.B. Denitrification in wood chip bioreactors at different water flows. J. Environ. Qual. 2009; 38:1664–1671.

[36] Kovacic, D.A., David, M.B., Gentry, L.E., Starks, K.M., and Cooke, R.A. Effectiveness of constructed wetlands in reducing nitrogen and phosphorus export from agricultural tile drainage. J. Environ. Qual. 2000; 29:1262–1274.

[37] Christianson, L.E.,"Design and performance of denitrification bioreactors for agricultural drainage"(2011). Graduate Theses and Dissertations. Paper 10326. http://lib.dr.iastate.edu/etd/10326

[38] Van Driel, P.W., Robertson, W.D., and Merkley, L.C. Upflow reactors for riparian zone denitrification. J. Environ. Qual. 2006; 35:412–420.

[39] Van Driel, P.W., Robertson, W.D., and Merkley, L.C. Denitrification of agricultural drainage using wood-based reactors. Trans. ASABE. 2006; 49:565–573.

[40] Moorman, T.B., Parkin, T.B., Kaspar, T.C., and Jaynes, D.B. Denitrification activity, wood loss, and N_2O emissions over 9 years from a wood chip bioreactor. Ecol. Eng. 2010; 36:1567–1574.

[41] Chun, J.A., Cooke, R.A., Eheart, J.W., and Cho, J. Estimation of flow and transport parameters for woodchip-based bioreactors: II. Field-Scale Bioreactor. Biosyst. Eng. 2010; 105:95–102.

[42] Christianson, L., Bhandari, A., and Helmers, M. Emerging technology: Denitrification bioreactors for nitrate reduction in agricultural waters. J. Soil Water Conserv. 2009; 64:139A–141A.

[43] Elgood, Z., Robertson, W.D., Schiff, S.L., and Elgood, R. Nitrate removal and greenhouse gas production in a steam-bed denitrifying bioreactor. Ecol. Eng. 2010; 36:1575–1580.

[44] Langmuir, D. 1997. Aqueous Environmental Chemistry. Prentice-Hall, Upper Saddle River, NJ.

[45] Robertson, W.D., Ptacek, C.J., and Brown, S.J. Rates of nitrate and perchlorate removal in a 5-year-old wood particle reactor treating agricultural drainage. Ground Water Monitor. Remediat. 2009; 29:87–94.

[46] Aide, M., Braden, I., Hermann, N., Mauk, D., Mueller, W., Svenson, S., and Weathers, J. Assessment of a large subsurface controlled drainage and irrigation system: I. Design, Soil Properties, and Attendant Crop Performance. Trans. Missouri Acad. Sci. 2013; 44:1–7.

[47] Aide, M., Braden, I., Hermann, N., Mauk, D., Mueller, W., Svenson, S., and Weathers, J. Assessment of a large subsurface controlled drainage and irrigation system: II. Corn performance. Trans. Missouri Acad. Sci. 2013; 44:8–11.

[48] Aide, M., Braden, I., Hermann, N., Mauk, D., Mueller, W., Svenson, S., and Weathers, J. Assessment of a large subsurface controlled drainage and irrigation system: III. Water chemistry of the tile effluent and its potential impact on surface water resources. Trans. Missouri Acad. Sci. 2013; 44:11–17.

6

The Molecular-Based Methods Used for Studying Bacterial Diversity in Soils Contaminated with PAHs

Anna Gałązka and Jarosław Grządziel

Additional information is available at the end of the chapter

Abstract

Soil contamination could adversely affect microbial diversity, and perhaps also above- and below-ground ecosystem functioning. It is important to study microbial diversity not only for basic scientific research, but also to understand the link between diversity and community structure and function in the pollution site. The study of microbial diversity and their function in contaminated soil creates a serious problem because they observed significant limitations in methodology and taxonomy of this group. Methodology for the determination of bacterial diversity does not include their function in the soil and other environment areas. Microbes are known for their catabolic activity in bioremediation, but changes in microbial communities are still unpredictable. The bioremediation of a pollutant and its rate depend on the environmental conditions, number and type of the microorganisms, nature and chemical structure of the chemical compound being degraded. However, molecular methods have been used to study soil bacterial communities. While many anthropogenic activities, such as city development, agriculture, and use of pollution, can potentially affect soil microbial diversity, it is unknown how changes in microbial diversity can influence below-ground and above-ground ecosystems. There are problems associated with studying bacterial diversity in soil. These arise not only from methodological limitations, but also from a lack of taxonomic knowledge. Methods to measure microbial diversity in soil can be categorized into two groups: biochemical-based techniques and molecular-based techniques. But more common for studying microbial diversity in soil contaminated with polycyclic aromatic hydrocarbons are the molecular methods.

Keywords: bacterial diversity, soil contamination, PAHs, trace elements, molecular methods, DGGE, NGS

1. Introduction

The oil refinery industry is involved in the global processes of exploration, extraction, transporting (often with oil tankers and pipelines), and marketing petroleum products. The products of largest volume of the industry are oil and gasoline [1, 2]. Crude oil and petroleum are also the raw materials for many chemical products, including pharmaceuticals, solvents, fertilizers, pesticides, and plastics. Oil and its derivatives (such as polycyclic aromatic hydrocarbons, PAHs) are among very significant and dangerous sources of ecosystem contaminants. Oil derivatives that contaminate soil are a threat to human health as well as a hazard to all living beings [2].

Polycyclic aromatic hydrocarbons (PAHs) are a large group of carcinogenic compounds emitted into the atmosphere by incomplete combustion of fossil fuel or biomass. As semi-volatile chemicals, PAHs can be transported over long distances in the atmosphere. In general, 3-4-5 ringed PAHs are largely predominant in air wherever the sampling was established, whether in rural, suburban, or urban areas. PAHs can pass from air to water, soil, and vegetation, through dry gaseous, dry particle-bound, and wet depositions [3]. They are persistent in various environmental media and can subsequently enter the food chains. Nowadays, it is well known that human exposure mainly occurs by ingestion of contaminated agricultural and natural food [4, 5]. Using plant in bioremediations is more popular and common. Plants are capable of accumulating PAHs from the soil, water, and air. In the ryzosphere of plants, we have a very higher activity of microorganism capable of using PAHs as the only source of carbon and energy [2, 5, 6].

The main source of hydrocarbons (PAHs) is incomplete combustion of organic different material. Polycyclic aromatic hydrocarbons are colorless, white, or yellow solids. They present low solubility in water and also low vapor pressure [7]. They arise mainly from anthropogenic sources (forest fires, oil seeps, and volcanic eruptions). Other sources of PAHs are burning of fossil fuel, coal tar, wood, oil derivatives, petroleum spills, and discharge. These substances are very toxic, mutagenic, and carcinogenic [8]. The remediation and bioremediation of PAHs are very longer and technically hard. Their persistence in soil increases with increase in molecular weight of PAHs. It is estimated that more than 90% of the total burden of oil derivatives such as PAHs reside in the surface layer of soils where they accumulate the most. Recent determinations of PAHs in agricultural soils in Poland indicate that the content of these contaminants in the majority of the soils is low but in long-term contaminated soils, this content is very higher [9, 10].

Several techniques of remediation of PAHs are known: volatilization, photooxidation, chemical oxidation, adsorption on soil particles, and microbial biodegradation. The main popular techniques are expensive and very time-consuming. Otherwise, the effect of that remediation in many cases transfers the pollutant from one phase (soil, water, or air) to another [2, 4].

Bioremediation process is much less dangerous, and the results (products) of this process are safe for the environment such as inorganic minerals, H_2O, CO_2 (aerobic), or CH_4 (anaerobic) [1, 11].

Microbes are known for their catabolic activity in bioremediation, but changes in microbial communities are still unpredictable [1, 11]. The most popular PAH-degrading microorganisms are bacteria and fungi. The bioremediation of PAHs very often depends on the environmental conditions (climates, number and type of the microorganisms, soil structure, plants). The extent of biodegradation process depends on many biotic and abiotic factors, including pH, temperature, oxygen, microbial population, degree of acclimation, accessibility of nutrients, chemical structure of the compound, cellular transport properties, and chemical partitioning in growth medium [2, 4, 12].

Overall, PAHs are immobile and persistent in soil and also more difficult to extract. PAHs are less accessible to living organisms (microorganism) when they come in contact with the aggregate soil structure [2]. There are many methods used to clean up PAH and oil derivatives in contaminated soils, but bioremediation using bacteria and fungi consortium is most popular [1, 2, 8].

However, molecular methods have been used to study soil bacterial communities in contaminated soil with PAHs and oil derivatives. While many anthropogenic activities, such as city development, agriculture, and use of pollution, can potentially affect soil microbial diversity, it is unknown how changes in microbial diversity can influence below-ground and above-ground ecosystems. The study of bacterial diversity in soil contaminated with PAHs has some problems. The problems arise not only from the methodological limitations, but also from a lack of taxonomic knowledge. This studies focuses on whole groups of microorganism (bacteria and fungi) and its function in in the contaminated sites.

2. Bacteria and nitrogen fixation microorganisms in bioremediation of contaminated soil

Microorganisms have some potential as an effective and inexpensive mean to remediation of contaminated soils [13]. The successful application of bioremediation techniques (bioaugmentation, phytoremediation) is largely dependent on the some capacity of plant growth-promoting microorganisms to efficiently colonize growing plants roots [14].

Bacteria are the class of microorganisms actively involved in the degradation of organic pollutants from contaminated sites especially from soils rhizosphere [13, 14]. A number of bacterial species are known to degrade PAHs (shown in **Table 1**). These bacteria very often are isolated from contaminated soil and have special potential to degradation of oil derivatives. The most carcinogenic and toxic from PAHs is benzo(a)pyrene. This hydrocarbon is a model contaminate in bioremediation study. Bacteria which can degrade benzo(a)pyrene grow well on alternative carbon source in liquid culture experiments [19–21].

Other authors [22] observed a 5 % decrease in benzo(a)pyrene concentration after 168 h during incubations with *Sphingomonas paucimobilis* strain of bacteria. They also noticed that resting cells of *S. paucimobilis* grown on nutrient agar supplemented with glucose resulted in significant evolution of 14 CO_2 (28%), indicating higher hydroxylation and ring cleavage. Some

Bacteria	Plant	Contaminant	Role of bacteria	Ref.
Azospirillum lipoferum	Wheat	Crude oil	Promoted development of wheat root system enhanced level of oil degradation	[14]
Azospirillum brasilense	Tall fescue	Polycyclic aromatic hydrocarbons (PAHs)	Increased plant tolerance to PAHs Promoted plant growth under stress	[15]
Azospirillum spp. *Pseudomonas stutzeri*	Meadow fescue Maize Winter rye	Polycyclic aromatic hydrocarbons (PAHs) Crude oil	Promoted development of plant root system enhanced level of oil and PAH degradation Increased plant tolerance to PAHs	[9, 10]
Enterobactor cloacae	Tall fescue	Total petroleum Hydrocarbons (TPHs)	Promoted plant growth in the presence of environmental contaminants such as TPHs	[14]
Pseudomonas fluorescens	Wheat	Trichloroethylene (TCE)	Degraded TCE with toluene o-monooxygenase	[16]
Pseudomona fluorescens	Alfalfa	Polychlorinated biphenyls(PCBs)	More effectively metabolized PCBs with bph gene cloned	[17]
Pseudomonas putida	Arabidopsis	PCBs	Utilized plant secondary metabolites	[18]

Table 1. Examples of bioremediation of organic contaminants in soil with bacteria species.

authors [14, 19] isolated 11 strains from a variety of contaminated sites (oil, motor oil, refinery derivatives) with the ability to degrade benzo(a)pyrene. Bacteria capable to PAHs degradation and using as the only source of carbon and energy belong to the main species *Pseudomonas, Agrobacterium, Bacillus, Burkholderia, Sphingomonas, and Phanerochaete chrysosporium* [23]. Other authors reported PAH degradation using other bacteria including *Rhodococcus* sp., *Mycobacterium,* and mixed culture of *Pseudomonas* and *Flavobacterium* species [20]. In study of Heitkamp et al. [24], the authors described about bacterial isolated from oil-contaminated soil which was capable of mineralizing the pyrene. *Pseudomonas aeruginosa* isolated from a stream heavily polluted by a petroleum refinery was very effective in degradation of phenanthrene [25]. *Pseudomonas aeruginosa* actively grow over high doses of phenanthrene with complete removal of the pollutant in a period of 30 days of the experiment. Other authors report that *Mycobacterium* species isolated from a PAH-contaminated soil were able to utilize pyrene as the only sole source of carbon and energy (up to 60% of the pyrene added (0.5 mg ml^{-1}) within 8 days at 20°C of temperature) [26]. Some products of this degradation pathway were analyzed (Cis-4,5-pyrene dihydrodiol, 4-5-phenanthrene dicarboxylic acid, 1- hydroxy-2-naphthoic acid, 2-carboxybenzaldehyde, phthalic acid, and protocatechuic acid). In the study of Yuan et al. [11], the authors isolated strains of bacteria from a petrochemical waste which having the capacity of degrading acenaphthene, fluorene, phenanthrene, anthracene, and pyrene by 70–100% in a period of 40 days of the experiment. This bacteria belong to the *Pseudomons fluoresens* and *Haemophilus* species. Dean-Ross et al. [15] isolated two bacterial strains (*Mycobacterium flavescens* and *Rhodococcus* spp.) from some sediments. This bacteria were found to be capable of PAH degradation (pyrene mineralization by *M. flavescens* and anthracene mineralization by *Rhodococcus species*) [27]. The study also proposed the degradation pathway of fluoranthene. In both strains, metabolism of fluoranthene occurred on the fused ring of fluoranthene molecule, producing 9-fluorenone-1-carboxylic acid.

Microbial degradation is the mean to remove PAHs from contaminated soils, especially using strains of bacteria which are able to degrade PAHs and using them as a source of carbon and energy and fix free nitrogen such as the strains of *Azospirillum* spp. and *Pseudomonas stutzeri*. These strains are the diazotrophic bacteria capable of free nitrogen fixing, hydrocarbon degradation as an only source carbon, and energy and biosurfactant production. Bacteria of the genus *Pseudomonas* are known in the literature as the most active degraders of hydrocarbons in natural biotopes of polluted sites and within biotechnological preparations [9, 10, 69].

Diazotrophic bacteria such as *Azospirillum* spp. and *Pseudomonas stutzeri* are also using in bioremediation of crude oil derivatives in soils naturally and artificially polluted [9, 10]. Gałązka et. al. reported the study with three soils artificially polluted with PAHs (anthracene, phenanthrene, and pyrene) at the doses of 100, 500, and 1000 mg kg^{-1} d.m. of soil and diesel fuel at the doses of 0.1%, 0.5%, and 1% (v/v). In study was also used soil naturally contaminated with crude oil (brown soil). Grasses were inoculated with the mixture of bacteria strains *Azospirillum* and *Pseudomonas stutzeri* and applied in the bioremediation process in the amount of 1 ml per 500 g of soil.. The amounts of anthracene, phenanthrene, and pyrene were determined in soils artificially polluted and Σ15 PAHs in soils artificially polluted with diesel fuel, as well as in brown soil aged polluted with crude oil. It was found that the inoculation of plants with *Azospirillum* spp. and *Pseudomonas stutzeri* had a positive effect on bioremediation process either in soils artificially polluted with PAHs (decrease from 25–60% of the primary concentration comparing to the control) or in soils polluted with diesel fuel (decrease from 2–25%) [9, 10]. The slime of *Azospirillum* spp. and *Pseudomonas stutzeri* introduced to soil did not limit the development of indigenous bacteria consortia in the polluted soil; instead, progressive biodegradation of PAHs enabled major growth of total number of bacteria, *Actinomycetes* and their biological groups. The ability of *Azospirillum* spp. and *Pseudomonas stutzeri*, populating rhizosphere and the inside of grass roots, to free nitrogen fixing and the use of PAHs (phenanthrene, anthracene, and phyrene) as the only source of carbon and energy suggests that in the future, after the series of detailed analysis, it will be possible to invent preparation based on these species, suitable for bioremediation of soils polluted with PAHs, with very limited supplementation of environment with nitrogen fertilizers. The successful results were observed (an important decrease in the content of PAHs in soils) in soil inoculated with *Azospirillum* and *Pseudomonas stutzeri* after grass growth (maize, meadow fescue). This processes were especially effective in calcareous rendzina artificially polluted with PAHs and in soil long-term contaminated with crude oil [28, 29].

2.1. Bacterial diversity in soil contaminated with PAHs

Soil microorganisms play a big roles in various biogeochemical cycles and are responsible for the cycling of organic compounds especially oil derivatives and polycyclic aromatic hydrocarbons. Also they influence above-ground ecosystems by contributing to plant nutrition, plant health, soil structure, and soil fertility. Our knowledge on soil microbial diversity is limited in part by our inability to study soil microorganisms. It is known that in 1 g of soil there are 10^{30} different soil microorganisms [30]. Only 1% of this soil bacterial population can be cultured by classical methods. About 99% is unknown, and this group of microorganism is possible to measure only in using molecular methods [31, 32].

Various molecular methods have been used to study soil bacterial communities. Many biotic and abiotic factors play a big role to changes in microbial diversity (contamination, anthropogenic activities, plant growth). It is not known how changes in microbial community structure influence ecosystem functions. Study of microorganisms function is the need for reliable and accurate mechanisms of understanding their diversity and taxonomic [33–35].

Typically, diversity studies include the relative diversities of communities across a gradient of stress, disturbance, or other biotic or abiotic difference [35]. It is difficult with current techniques to study true diversity since we do not know what is present and we have no way of determining the accuracy of our extraction or detection methods. Species diversity consists of species richness, the total number of species present, species evenness, and the distribution of species [32].

Methods to measure microbial diversity in soil can be categorized into two groups: biochemical-based techniques and molecular-based techniques. But more common for studying microbial diversity in soil contaminated with polycyclic aromatic hydrocarbons are the molecular methods.

2.2. Limitations of molecular methods to study bacterial diversity in contaminated soils

Molecular techniques based on polymerase chain reaction (PCR) have been used to overcome the limitations of culture-based methods; however, they are not without their own limitations [32, 34].

Soil microorganisms (especially bacteria) are located between soil aggregates. There is a very big problem with separating these from micro- and macro-components of soil structure. The study bacterial biodiversity requires isolated genomic DNA from bacterial cells [35]. This process is dependent on bacterial cells (gram-negative or gram-positive bacterial cells). Gram-negative cells would be lysed when the cell extraction is sensitive, but the gram-positive cells may be lysed in stronger conduction, but in this case DNA may be disintegrated [32]. The special method of DNA or RNA extraction from bacterial cells used can also bias biodiversity studies. The harsh and drastic DNA extraction methods (bead beating) can shear the nucleic acids, leading to some problems in subsequent PCR detection products [36]. With soil samples, it is necessary to remove some inhibitory substances (fulvic acids, humic acids). These substances can be coextracted and can strongly interfere with subsequent PCR and analysis. Second step of analysis can lead to loss of DNA or RNA inhibitory of PCR. The most popular in bacterial biodiversity studies are primers which targeted typical regions coding genes present in all organisms such as 16S rRNA or ITS (internal transcribed spacer). This genes have well-defined regions for taxonomic classification of bacteria and are not subject to horizontal transfer and have sequence databases available to researchers.

Many authors [32, 34, 36, 37] discussed some issues surrounding differential PCR amplification including different affinities of primers to templates, different copy numbers of target genes, hybridization efficiency, and primer specificity. In addition, some sequences with lower G+C content are thought to separate more efficiently in the denaturing step of polymerase chain reaction and therefore could be preferentially amplified [32, 34]. There are known a few important points in optimalization of PCR such as amplification including different affinities of primers to templates, different copy numbers of target genes, hybridization efficiency, and primer specificity. The above discusses a few limitations of molecular-based methods, which can influence the analysis and interpretation of their community analysis. Molecular-based methods provide valuable information about the microbial community as opposed to only culture-based techniques.

3. Molecular techniques based on PCR methods to study bacterial diversity

The molecular methods of study bacterial diversity include some methods profiling of soil microbial communities, based upon culture-independent techniques (cloning, fingerprinting techniques, automated ribosomal intergenic spacer analysis (ARISA), or terminal/restriction fragment length polymorphism (TRFLP, RFLP) (**Table 2**) [32, 34, 35, 74, 73].

Application of these techniques yields information that can be used to assess how environmental factors contribute to changes in microbial community structure. Although a considerable amount is known about how culturable bacteria respond to anthropogenic agents, little is known about how organic compounds influence the structure of soil microbial communities in situ. It has been suggested that microbial community structure in polluted environments is influenced by the complexity of chemical mixtures present and time of exposure and is thought generally to lead to a reduction in microbial diversity. We do not know why the amount of PAH contamination together with the PAH compound present significantly affected microbial community structure in PAH-contaminated soils [35, 37].

DNA hybridization is a measure of genetic complexity of the microbial/bacterial community and has been used to estimate diversity in soil contaminated. The similarity between communities of two different samples can be studied by measuring the degree of similarity of DNA through hybridization kinetics [39]. Nucleic acid hybridization using specific probes is an important qualitative and quantitative tool in molecular bacterial ecology. These hybridization techniques can be done on extracted DNA or RNA, or *in situ*.

Method	Advantages	Ref.
G+C content	Not influenced by PCR biases	[38]
	Includes all DNA extracted	
	Quantitative	
	Includes rare members of community	
DNA hybridization	Same as nucleic acid hybridization Thousands of genes can be analyzed	[39]
	If using genes or DNA fragments, increased specificity	
Denaturing and temperature gradient gel electrophoresis (DGGE/TGGE)	Large number of samples can be analyzed simultaneously	[40, 41]
	Reliable, reproducible, and rapid	
Restriction fragment length polymorphism (RFLP)	Detect structural changes in microbial community	[42]
Terminal restriction fragment length polymorphism (T-RFLP)	Simpler banding patterns than RFLP	[42]
	Can be automated; large number of samples	
	Highly reproducible	
	Compare differences in microbial communities	
Ribosomal intergenic spacer analysis (RISA) and automated ribosomal intergenic spacer analysis (ARISA)	Highly reproducible community profiles	[43]

Table 2. Advantages of some molecular-based methods to study soil microbial diversity.

The known sequences of some oligonucleotide/polynucleotide probes ranging in specificity from domain to species can be tagged with markers at the 5'-end of DNA. The most popular markers are fluorescent markers that include derivatives of fluorescein or rhodamine. Quantitative dot-blot hybridization methods are used to measure the relative abundance of the special group of microorganisms (bacteria). In these methods, samples (bacterial culture) are lysed to release all nucleic acids. In dot-blot hybridization with specific and universal oligonucleotide primers, the rRNA sequences are quantified relative to total rRNA [32, 34, 35]. The changes in the activity and hence the amount of rRNA content or changes in the abundance in the population may represent the relative abundance is samples. Hybridization methods of studying bacterial biodiversity can also be conducted at the cellular level and can be done in situ (valuable spatial distribution information on microorganisms in environmental sample) [34]. The method, known as fluorescent in situ hybridization or FISH (fluorescence in situ hybridization), has been used successfully to study the spatial distribution of bacteria in biofilms [39]. The lack of sensitivity is the most limited point in the methods such as in situ hybridization or hybridization of nucleic acids extracted directly from soil samples. The some unless sequenced are present in very high copy and there are not detected in this methods. Polymerase chain reactions the methods which there is no this problem. DNA extracted directly from soil samples can act as a template for PCR or mRNA and can be reverse-transcribed into cDNA and then amplified using standard PCR methods [31, 32]. The use of mRNA in biodiversity studies will allow a snapshot of the active bacterial population in contaminated soil, whereas DNA extracted directly from this samples can represent active as well as dormant bacteria. The amplified PCR product can be hybridized with either oligonucleotide probes to provide specific information on the bacterial community in contaminated soil or with other samples to which bacterial community similarity is compared [35]. The PCR targeting the 16S rDNA has been used extensively to study prokaryote (bacteria) diversity and allows identification of prokaryotes as well as the prediction of phylogenetic relationships [26]. Initially, molecular-based methods for ecological studies relied on cloning of target genes isolated from environmental samples [44]. Although sequencing has become routine, sequencing thousands of clones is cumbersome [45].

3.1. The denaturing gradient gel electrophoresis methods to study bacterial diversity

The property of double-stranded DNA molecules allowing their separation in an electric field is used in many electrophoretic techniques. A standard electrophoresis consists in separating the DNA molecules by size. For this purpose, the agarose gel is prepared with the appropriate concentration, typically from 0.5 to 2%, and is connected to constant electric field. The DNA molecules pass through the small spaces within the gel and migrate at different rates depending on their size [46]. As a result, towards the end of the gel we observe DNA fragments of smaller sizes (less base pairs), and the large fragment will move slower, remaining closer to the top. In this way, it is possible to know the approximate size of the analyzed fragments [See **Figure 1**, gel on the left]. However, this method cannot be used to distinguish between each of the DNA molecules of the same size, differing only in the nucleotide sequence. The solution was developed in 1987 (See [47]). Method called denaturing gradient gel electrophoresis(DGGE) is based on the fact that only double-stranded DNA fragments move in the electric field, whereas single-stranded not have such ability, or at least their mobility is strongly reduced. Denaturation of the double-stranded structure of DNA into single strands is accomplished by treatment DNA using high temperature and denaturing agents, usually a mixture of formamide and urea [48]. The specific temperature and concentration of denaturant in which the DNA is denatured, also known as the melting point of DNA, are dependent on nucleotide sequence. This correlation means that even a single base mutation can change the melting point of DNA. What is important in understanding the phenomenon, it is not only the influence of bonds between paired bases, but also the interaction between neighboring pairs [49, 70]. This makes it possible to distinguish DNA fragments of the same size but with different nucleotide sequence [See **Figure 1**, gel on the right].

DGGE electrophoresis is usually performed at a constant temperature (usually 60°C) in the presence of two denaturing agents: formamide and urea, the concentration of which depends on the experiment and analyzed fragments. The analysis is carried out in polyacrylamide gel (6–12%), which consists of a mixture of acrylamide and bis-acrylamide, usually in a 37.5:1 ratio [50]. This polymer is resistant to high temperatures and denaturing agents, and also creates the appropriate pores through which DNA can easily migrate. It is also characterized by a much higher resolving power with respect to agarose [51].

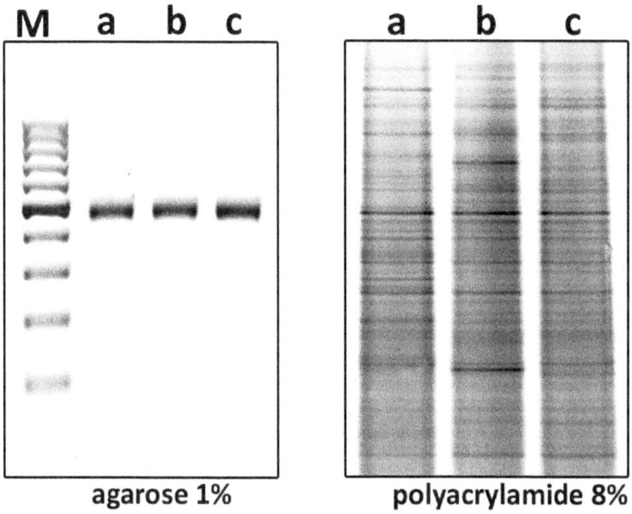

Figure 1. Comparison of agarose electrophoresis and DGGE. The letter M represents size marker of the DNA; the letters a–c are designations of samples. The same PCR products were placed on both gels for comparison.

Gel preparation and electrophoresis are in a vertical orientation, where the top of the gel is the lowest concentration of denaturing agents (usually from 0 to 30%) and the bottom of the gel fills the highest concentration (usually 50–80%). Between the extreme values, the concentration of denaturing agents creates an increasing gradient. Throughout the run electrophoresis is supplied a constant voltage, typically about 60V for 16 h [52]. In some cases, it can be applied a higher voltage of 130–150 V for 3–6 h, while the bands are then more blurred [53, 54]. This affects the image of electrophoresis. Electrophoresis in the gradient of denaturant allows the rapid identification of the different variants of genes (alleles), detection of mutations in medicine, and an overview of genetic diversity in any environment. Many studies using DGGE method is used for rapid diagnosis of disorders of human microbiota [55, 56] or to analyze the change in the composition of the bacteria in the fermenters or other dynamic biological systems [57]. DGGE limitation is the selection of appropriate fragments of DNA for analysis. This method keeps its resolving power in fragments size between 100 and 500 bp. The analyzed DNA fragments are always PCR products–amplicons, typically including the hypervariable regions of the 16S rDNA gene (in the case of bacteria) or ITS (internal transcribed spacer) in the case fungi. The ITS regions are situated between the small and large subunits of the ribosomal rDNA. The advantage of choosing these regions is the presence of both conservative and those highly variable sequences [58, 71, 72].

DGGE method has been known for more than 30 years but is continually improving. The first enhancement was the introduction of the GC-clamp. This is 20- to 60-nt-long DNA fragment that is added to one of the primers for PCR and contains only the G and C bases. It has been found to increase resolving power of the method by maintaining a small fragment of double-stranded structure, even at high temperatures (almost 100°C) and in high concentrations of denaturant [59].

Another improvement of the method is the use of specific markers (as a references). This involves selecting the reference strains of known origin and certified taxonomy, and then isolating the DNA. The next step is to prepare DGGE-PCR amplicons. Appropriately prepared amplicons are placed in an empty well of the polyacrylamide gel as a reference. Taking advantage of markers, it is possible to normalize gels and then compare different experiments with each other. The second application is to compare the quality and the quantity of bands in the analyzed wells, with those in the well marked as a reference in order to classify and the species composition in the sample, as well as their abundance [60].

It should be noted that this method has a broad spectrum of applications, from medicine to the currently developing metagenomics, and provides a complementary tool to traditional classical methods of exploring the composition of microorganisms. Although it does not provide as comprehensive and complete results as sequencing, the costs of its implementation and the time in which you can get to know the preliminary results are much smaller. This is a very good method for the presumptive identification of microorganisms as well as continuous monitoring of changes in the composition of microbial communities such as contaminated soil, water, bioreactors, or the composition of the human microflora.

It is worth mentioning also the limitations of DGGE. First of all, this method is based on PCR; therefore, the selection of appropriate conditions but also suitable polymerase is a key issue. Most of the problems with this method stems from mistakes at this stage. Polymerase chain reactions is always associated with the possibility of introducing errors by altering the genetic profile in the investigated samples. Occasionally, PCR products from different organisms, despite differing nucleotide sequences, may also have the same melting point. This causes the risk of missing some of the bands on the gel. On the other hand, there is a risk of nonspecific products in PCR (e.g., as a result of amplification of the chloroplast or mitochondrial DNA) to give false results. Often, in order to avoid such a situation there can be applied several-step PCRs (e.g., nested PCR), as well as touchdown PCR which is known to increase the specificity of the reaction [61].

3.2. Next-generation sequencing

Next-generation sequencing (NGS), otherwise high-throughput sequencing, resulted in a breakthrough in the automation and commercialization of the sequencing process.. In 2000, the company Lynx Therapeutics launched the first fully automated sequencing apparatus, the principle of which was still based on the Sanger method. In 2004, the company 454 Life Sciences has developed and successfully launched the sale of second-generation sequencer, which used discovered in 1996 pyrosequencing method. In addition to the huge success in the prevalence of the device, the cost of sequencing decreased sixfold in comparison with the device from 2000 [62, 63].

High-throughput sequencing is probably the fastest growing method used in the biology and biotechnology. To date emerged a series of modifications which resulted in the development of equipment relatively cheap and efficient.

On the market, there is a large selection of sequencing systems introduced by many other companies, but this chapter focuses on Illumina sequencing system. It is the most common method in the study of metagenomes different environments. Due to the a very dynamic development of the technology described herein, performance data and bandwidth become outdated several times a year.

DNA prepared for sequencing must meet several requirements. First of all, it must be free from contamination and PCR inhibitors such as humic acids, ethanol, and phenol compounds. A very important and crucial step in the preparation of biological samples is appropriate for DNA extraction and its purification. Commercially available kits provide high-performance elution of DNA, contain enzyme (such as DNase) inhibitors, and allow getting rid of impurities.

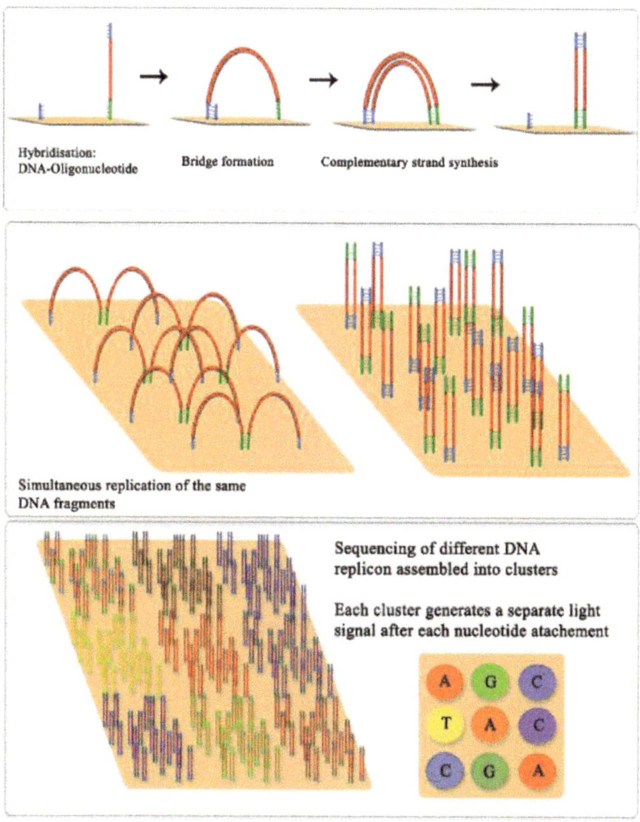

Figure 2. Cluster formation in Illumina NGS sequencing.

An important advantage is the ability to simultaneously sequencing of many samples at the same time. This is done by marking samples by attaching specific, short DNA fragments of known sequence treated as barcodes. The principle of the sequencing uses fluorescently labeled nucleotides. During the attachment of one nucleotide, generation of a light signal occurs and the reaction is temporarily blocked. After registration signal, a fluorescent label is cleaved enzymatically allowing the connection of the next nucleotide. Each of the nucleotides (A, T, C, G) has a different type of fluorescent label recognized as a different wavelength. DNA is immobilized on the surface of the flow cell, which allows direct and equal access of polymerase to each of the each DNA molecule [64]. At a distance of less than one micron, there are more than a thousand copies of the same DNA fragments to form one cluster. Different DNA fragments form separate clusters, allowing for simultaneous sequencing of millions of DNA fragments [**Figure 2**].

The parameters of current devices are extremely high. Within 24 h, around 5 Gb (giga bases) of reads can be obtained, when reading 200–300 bp fragments (V3-V4 hypervariable regions for example). With exceptionally large genomic projects, there can be used the device with the highest performance (HiSeq series) allowing to generate up to 1 Tb of data within a few days [65].

Next-generation sequencing in combination with other molecular methods (including DGGE) is a very complex and indispensable method of testing microbiomes and the ecological. Metagenomic approach to the knowledge of the biodiversity present in difficult conditions, such as contaminated soil or sewage, sells out all other known methods, allowing the examination of not only a fraction of microorganisms, but also discovering new, previously unknown species [66–68].

4. Summary

The better understanding of the link between bacterial diversity and their community structure and function is very important to study microbial diversity in contaminated soil. This is not only important for basic scientific research but also to study biodiversity in soil contaminated with PAHs. Significantly higher amounts of 16S rRNA have been found in all microbial groups analyzed in fields that have never been cultivated than agricultural fields and also in soil contaminated with PAHs. This suggests a decrease in bacterial biomass or activity in cultivated fields. However, it is unknown what these reductions in diversity mean to ecosystem functioning, and it is important for the sustainability of ecosystems to examine and better understand the link between diversity and function. There are some limitations associated with studying organisms in contaminated soil. There are some taxonomic and methodological limitations. The methods to study bacterial diversity (numerical, taxonomic, structural) are improving for some group of bacteria and fungi. It is generally thought that a diverse population of microorganisms will be more resilient to biotic and abiotic stress and more capable of adapting with environmental changes (contamination). The knowledge of plant–microbe–soil-contaminant interactions is increasing, but the complexity of interacting biological, chemical, and physical factors means that much remains to be understood.

As new techniques are developed, our level of understanding increases and our knowledge expands.

Acknowledgements

The research was conducted within the frames of Task 1.4. Evaluation and formation of biodiversity of soil and microbial activity of soil with regard to habitat conditions and management system. Multi—Annual Programme IUNG—PIB 2016—2020.

Author details

Anna Gałązka* and Jarosław Grządziel

*Address all correspondence to: agalazka@iung.pulawy.pl

Department of Agricultural Microbiology, Institute of Soil Science and Plant Cultivation—State Research Institute, Pulawy, Poland

References

[1] Andreoni V., Gianfreda L. Bioremediation and monitoring of aromatic-polluted habitats. Appl. Microbiol. Biotechnol. 76 (2007) 287–308.

[2] Gan S., Lau E.V., Ng H.K. Remediation of soils contaminated with polycyclic aromatic hydrocarbons (PAHs). J. Hazard. Mat. 172 (2009) 532–549.

[3] Leglize P., Alain S., Jacques B., Corinne L. Adsorption of phenanthrene on activated carbon increases mineralization rate by specific bacteria. J. Hazard. Mat. 151 (2008) 339–347.

[4] Ma B., He Y., Chen H.-H. , Xu J.-M., Rengel Z. Dissipation of polycyclic aromatic hydrocarbons (PAHs) in the rhizosphere: Synthesis through meta-analysis. Environ. Pollut. 158 (2010) 855–861.

[5] Ma B., Chen H., He Y., Wang H., Xu J. Evaluation of toxicity risk of polycyclic aromatic hydrocarbons (PAHs) in crops rhizosphere of contaminated field with sequential extraction. J. Soils Sediments 10 (2010) 955–963.

[6] Cheem S.A., Khan M.I., Shen C., Tang X., Farooq M., Chen L., Zhang L., Chen Y. Degradation of phenanthrene and pyrene in spiked soils by single and combined plants cultivation. J. Hazard. Mat. 177 (2010) 384–389.

[7] Cofield N., Banks M.K., Schwab A.P. Liability of polycyclic aromatic hydrocarbons in the rhizosphere. Chemosphere 70 (2008) 1644–1652.

[8] Samanta S.K., Singh O.V., Jain R.K. Polycyclic aromatic hydrocarbons: environmental pollution and bioremediation. Trends Biotechnol. 20 (2002) 243–248.

[9] Gałązka A., Król M., Perzyński A. The efficiency of rhizosphere bioremediation with *Azospirillum* sp. and *Pseudomonas stutzeri* in soils freshly contaminated with PAHs and diesel fuel. Pol. J. Environ. Stud. 21 (2012) 345–353.

[10] Gałązka A., Gałązka R. Phytoremediation of polycyclic aromatic hydrocarbons in soils artificially polluted using plant-associated-endophytic bacteria and Dactylis glomerata as the bioremediation plant. Pol. J. Microbiol. 64 (2015) 239–250.

[11] Yuan S.Y., Shiung L.C., Chang B.V. Biodegradation of polycyclic aromatic hydrocarbons by inoculated microorganisms in soil. Bull. Environ. Conatm. Toxicol. 69 (2002) 66–73.

[12] Barac T., Taghavi S., Borremans B., Provoost A., Oeyen L., Colpaert J.V., Vangronsveld J., van der Lelie D. Engineered endophytic bacteria improve phytoremediation of water-soluble, volatile, organic pollutants. Nat. Biotechnol. 22 (2004) 583–588.

[13] Greene E.A., Kay J.G., Jaber K., Stehmeier L.G., Voordouw G. Composition of soil microbial communities enriched on a mixture of aromatic hydrocarbons. Appl. Environ. Microbiol. 66 (2000) 5282–5289.

[14] Huang X.D., El-Alawi Y., Gurska J., Glick B.R., Greenberg B.M. A multi-process phytoremediation system for decontamination of persistent total petroleum hydrocarbons (TPHs) from soils. Microchem J. 81: (2005) 139–47.

[15] Dean-Ross D., Moody J., Cerniglia C.E. Utilisation of mixtures of Polycyclic aromatic hydrocarbons by bacteria isolated from contaminated sediment. FEMS Microbiol. Ecol. 41 (2002) 1–7.

[16] Yee D.C., Maynard J.A., Wood T.K. Rhizoremediation of trichloroethylene by a recombinant, root-colonizing Pseudomonas fluorescens strain expressing toluene ortho-monooxygenase constitutively. Appl. Environ. Microbiol. 64 (1998) 112–118.

[17] Villacieros M., Whelan C., Mackova M., Molgaard J., Sanchez-Contreras M., Lloret J. Polychlorinated biphenyl rhizoremediation by Pseudomonas fluorescens F113 derivatives, using a Sinorhizobium meliloti nod system to drive bph gene expression. Appl. Environ. Microbiol. 72 (2005) 2687–2694.

[18] Narasimhan K., Basheer C., Bajic V.B., Swarup S. Enhancement of plant-microbe interactions using a rhizosphere metabolomics-driven approach and its application in the removal of polychlorinated biphenyls. Plant. Physiol. 132 (2003) 146–153.

[19] Aitken M.D., Stringfellow W.T., Nagel R.D., Kazunga C., Chen S.H. Characteristics of phenanthrene-degrading bacteria isolated from soils contaminated with polycyclic aromatic hydrocarbons. Can. J. Microbiol. 44 (1998) 743–752.

[20] Janbandhu A., Fulekar M.H. Biodegradation of phenanthrene using adapted microbial consortium isolated from petrochemical contaminated environment. J. Hazard. Mat. 187 (2011) 333–340.

[21] MacNaughton S.J., Stephen J.R., Venosa A.D., Davis G.A., Chang Y.-J., White D.C. Microbial population changes during bioremediation of an experimental oil spill. Appl. Environ. Microbiol. 65 (2001) 3566–3574.

[22] Ye B., Siddiqi M.A., Maccubbin A.E., Kumar S., Sikka H.C. Degradation of polynuclear aromatic hydrocarbons by Sphingomonas paucimobilis. Environ. Sci. Technol. 30 (1996) 136–142.

[23] Bumpus J.A. Biodegradation of polycyclic aromatic hydrocarbons by Phanerochaete chrysosporium. Appl. Environ. Microbiol. 61 (1989) 2631–2635.

[24] Heitkamp M.A., Franklin W., Cerniglia C.E. Microbial metabolism of polycyclic aromatic hydrocarbons: isolation and characterization of a Pyrene-degrading bacterium. Appl. Environ. Microbiol. 54 (1998) 2549–2555.

[25] Romero M.C., Cazau M.C., Giorgieri S., Arambarri A.M. Phenanthrene degradation by microorganisms isolated from a contaminated stream. Environ. Pollut. 101 (1998) 355–359.

[26] Kim S.J., Kweon O., Jones R.C., Ricky D.E., Cerniglia C.E. Genomic analysis of polycyclic aromatic hydrocarbon degradation in Mycobacterium vanbaalenii PYR-1. Biodegradation 19 (2008) 859–881.

[27] Walter U., Beyer M., Klein J., Rehm H.J. Degradation of pyrene by Rhodococcus sp. UW1. Appl. Microbiol. Biotechnol. 34 (1991) 671–676.

[28] Gałązka A., Król M., Perzyński A. Bioremediation of crude oil derivatives in soils naturally and artificially polluted with the use of maize as the test plant. Part I. PAHs degradation. Acta Sci. Pol. Agricultura. 9 (2010) 13–24.

[29] Gałązka A., Król M., Perzyński A. Bioremediation of crude oil derivatives in soils naturally and artificially polluted with the use of maize as the test plant. Part II. Crop yeld. Acta Sci. Pol. Agricultura. 9 (2010) 25–36.

[30] Desai C., Pathak H., Madamwar D. Advances in molecular and "-omics" technologies to gauge microbial communities and bioremediation at xenobiotic/anthropogen contaminated sites. Bioresour. Technol. 101 (2010) 1558–1569.

[31] Guo C., Sun W., Harsh J.B., Ogram A. Hybridization analysis of microbial DNA from fuel oil-contaminated and noncontaminated soil. Microb. Ecol. 34 (1997) 178–187.

[32] Łyszcz M., Gałązka A. Selected molecular methods used in assessing the biodiversity of soil organisms. Postępy Mikrobiologii 55 (3) (2016) 309–319.

[33] Siciliano S.D., Germida J.J., Banks K., Greer C.W. Changes in microbial community composition and function during a polyaromatic hydrocarbon phytoremediation field trial. Appl. Environ. Microbiol. 69 (2003) 483–489.

[34] Theron J., Cloete T.E. Molecular techniques for determining microbial diversity and community structure in natural environments. Crit. Rev. Microbiol. 26 (2000) 37–57.

[35] Yrjälä K., Keskinen A.-K., Lkerman M.-L., Fortelius C., Sipila T.P. The rhizosphere and PAH amendment mediate impacts on functional and structural bacterial diversity in sandy peat soil. Environ. Pollut. 158 (2010) 1680–1688.

[36] Nocker A., Burr M., Camper A. Genotypic microbial community profiling: a critical technical review. Microb. Ecol. 54 (2007) 276–289.

[37] Roling W.F.M., van-Breukelen B.M., Braster M., Goeltom M.T., Groen J. Analysis of microbial communities in a landfill leachate polluted aquifer using a new method for anaerobic physiological profiling and 16S rDNA based fingerprinting. Microb. Ecol. 40 (2000) 177–188.

[38] Clegg C.D., Ritz K., Griffiths B.S. %G+C profiling and cross hybridisation of microbial DNA reveals great variation in below-ground community structure in UK upland grasslands. Appl. Soil Ecol. 14 (2000) 125–134.

[39] Cho J.-C., Tiedje J.M. Bacterial species determination from DNA–DNA hybridization by using genome fragments and DNA microarrays. Appl. Environ. Microbiol. 67 (2001) 3677–3682.

[40] Muyzer G. DGGE/TGGE a method for identifying genes from natural ecosystems. Curr. Opin. Microbiol. 2 (1999) 317–322.

[41] Muyzer G., Smalla K. Application of denaturing gradient gel electrophoresis (DGGE) and temperature gradient gel electrophoresis in microbial ecology. Antonie van Leeuwenhock. 73 (1999) 127–141.

[42] Raikos V., Vamvakas S.S., Kapolos J., Koliadima A., Karaiskakis G. Identification and characterization of microbial contaminants isolated from stored aviation fuels by DNA sequencing and restriction fragment length analysis of a PCR-amplified region of the 16S rRNA gene. Fuel 90 (2011) 695–700.

[43] Ranjard L., Brothier E., Nazaret S. Sequencing bands of ribosomal intergenic spacer analysis fingerprints for characterization and microscale distribution of soil bacterium populations responding to mercury spiking. Appl. Environ. Microbiol. 66 (2000) 5334–5339.

[44] Nyyssonen M., Piskonen R., Itavaara M. A targeted real-time PCR assay for studying naphthalene degradation in the environment. Microb. Ecol. 52 (2006) 533–543.

[45] Schmalenberger A., Schwieger F., Tebbe C.C. Effect of primers hybridizing to different evolutionarily conserved regions of the small-subunit rRNA gene in PCR-based microbial community analyses and genetic profiling. Appl. Environ. Microbiol. 6 (7) (2001) 3557–3563.

[46] Sambrook J., Russel D.W. Molecular Cloning: A Laboratory Manual. 3rd Ed., Cold Spring Harbor, NY: Cold Spring Harbor Laboratory Press; 2001.

[47] Myers R.M., Maniatis T., Lerman L.S. Detection and localization of single base changes by denaturing gradient gel electrophoresis. Methods Enzymol. 155 (1987) 501–527.

[48] Muyzer G., de Waal E.C., Uitterlinden A.G. Profiling of complex microbial populations by denaturing gradient gel electrophoresis analysis of polymerase chain reaction-amplified genes coding for 16S rRNA. Appl. Environ. Microbiol. 59 (1993) 695–700.

[49] SantaLucia J. A unified view of polymer, dumbbell, and oligonucleotide DNA nearest-neighbor thermodynamics. Proc. Natl. Acad. Sci. U.S.A. 95 (1998) 1460–1465.

[50] Green S.J., Freeman S., Hadar Y., Minz D. Molecular tools for isolate and community studies of Pyrenomycete fungi. Mycologia. 96 (2004) 439–451.

[51] Stellwagen NC. Electrophoresis of DNA in agarose gels, polyacrylamide gels and in free solution. Electrophoresis. 30 (Suppl. 1) (2009) 188–195.

[52] Yergeau E., Filion M., Vujanovic V., St-Arnaud M. A PCR-denaturing gradient gel electrophoresis approach to assess Fusarium diversity in asparagus. J. Microbiol. Methods 60 (2005) 143–154.

[53] Mohammadi P., Hamidkhani A., Asgarani E. Comparative analysis of denaturing gradient gel electrophoresis and temporal temperature gradient gel electrophoresis profiles as a tool for the differentiation of Candida species. Jundishapur J. Microbiol. 8 (2015) e22249.

[54] Schabereiter-Gurtner C., Piñar G., Lubitz W., Rölleke S. An advanced molecular strategy to identify bacterial communities on art objects. J. Microbiol. Methods 45 (2001) 77–87.

[55] Neves A.B., Lobo L.A., Pinto K.C., Pires E.S., Requejo M., Maia L.C., Antonio A.G. Comparison between clinical aspects and salivary microbial profile of children with and without early childhood caries: a preliminary study. J. Pediatr. Dent. 39 (2015) 209–214.

[56] Sagheddu V., Patrone V., Miragoli F., Puglisi E., Morelli L. Infant early gut colonization by Lachnospiraceae: high frequency of Ruminococcus gnavus. Front. Pediatr. 4(57) (2016) 1–6.

[57] Wei Q., Kawagoshi Y., Huang X., Hong N., Van Duc L., Yamashita Y., Hama T. Nitrogen removal properties in a continuous marine anammox bacteria reactor under rapid and extensive salinity changes. Chemosphere. 148 (2016) 444–451.

[58] Liu J., Yu Y., Cai Z., Bartlam M., Wang Y. Comparison of ITS and 18S rDNA for estimating fungal diversity using PCR-DGGE. World J. Microbiol. Biotechnol. 31 (2015) 1387–1395.

[59] Sheffield V.C., Cox D.R., Lerman L.S., Myers R.M. Attachment of a 40-base-pair G + C-rich sequence (GC-clamp) to genomic DNA fragments by the polymerase chain reaction results in improved detection of single-base changes. Proc. Natl. Acad. Sci. 86 (1989) 232–236.

[60] Liang Z., Drijber R.A., Lee D.J., Dwiekat I.M., Harris S.D., Wedin D.A. A DGGE-cloning method to characterize arbuscular mycorrhizal community structure in soil. Soil Biol. Biochem. 40 (2008) 956–966.

[61] Li W., Han L., Yu P., Ma C., Wu X., Xu J. Nested PCR-denaturing gradient gel electrophoresis analysis of human skin microbial diversity with age. Microbiol. Res. 169 (2014) 686–692.

[62] Schuster S.C. Next-generation sequencing transforms today's biology. Nat. Met. 5 (2008) 16–18.

[63] Brenner S., Johnson M., Bridgham J., et al. Gene expression analysis by massively parallel signature sequencing (MPSS) on microbead arrays. Nat. Biotechnol. 18 (2000) 630–634.

[64] Hodkinson B.P., Grice E.A. Next-generation sequencing: a review of technologies and tools for wound microbiome research. Adv. Wound Care 4 (2015) 50–58.

[65] Illumina. An Introduction to Next-Generation Sequencing Technology. 2016. Available from: http://www.illumina.com/content/dam/illumina-marketing/documents/products/illumina_sequencing_introduction.pdf

[66] Samarajeewa A.D., Hammad A., Masson L., Khan I.U., Scroggins R., Beaudette L.A. Comparative assessment of next-generation sequencing, denaturing gradient gel electrophoresis, clonal restriction fragment length polymorphism and cloning-sequencing as methods for characterizing commercial microbial consortia. J. Microbiol. Methods 108 (2014) 103–111.

[67] Gorgé O., Bennett E.A., Massilani D., Daligault J., Pruvost M., Geigl E.M., Grange T. Analysis of ancient DNA in microbial ecology. Methods Mol. Biol. 1399 (2016) 289–315.

[68] Frisli T., Haverkamp T.H., Jakobsen K.S., Stenseth N.C., Rudi K. Estimation of metagenome size and structure in an experimental soil microbiota from low coverage next-generation sequence data. J. Appl. Microbiol. 114 (2013) 141–151.

[69] Muratova A.Y., Turkovskaya O.V., Antonyuk L.P., Makarov O.E., Pozdnyakova L.I., Ignatov V.V. Oil-oxidizing potential of associative rhizobacteria of the genus *Azospirillum*. Microbiology 74 (2005) 210–215.

[70] Smalla, K., Wieland, G., Buchner, A., Zock, A., Parzy, J., Kaiser, S., Roskot, R., Heuer, H., Berg, G. Bulk and rhizosphere soil bacterial communities studied by denaturing gradient gel electrophoresis: plant-dependent enrichment and seasonal shifts revealed. Appl. Environ. Microbiol. 67 : (2001) 4742–4751.

[71] Liu H., Yang C.-Y., Tian Y., Lin G.-H. Using population dynamics analysis by DGGE to design the bacterial consortium isolated from mangrove sediments for biodegradation of PAHs. Int. Biodeterior. Biodegrad. 65 (2011) 269–275.

[72] Maarit-Niemi R., Heiskanen I., Wallenius K., Lindstrom K. Extraction and purification of DNA in rhizosphere soil samples for PCR-DGGE analysis of bacterial consortia. J. Microbiol. Methods 45 (2001) 155–165.

[73] Liu W.-T., Marsh T.L., Cheng H., Forney L.J. Characterization of microbial diversity by determining terminal restriction fragment length polymorphisms of genes encoding 16S rRNA. Appl. Environ. Microbiol. 63 (1997) 4516–4522.

[74] Osborn A.M., Moore E.R.B., Timmis K.N. An evaluation of terminal-restriction fragment length polymorphisms (TRFLP) analysis for the study of microbial community structure and dynamics. Environ. Microbiol. 2 (2000) 39–50.

7

Simulation of Phosphorus Transport in Soil Under Municipal Wastewater Application Using Hydrus-1D

Ali Erfani Agah, Patrick Meire and Eric de Deckere

Additional information is available at the end of the chapter

Abstract

Today, wastewater irrigation is one of the best options to reduce the stress on limited availability of fresh water and to meet the nutrient requirements of crops. In the present study, the simulation accuracy and performance of the HYDRUS-1D model to predict phosphorus leaching have been evaluated and compared to lysimeter data. More specifically, the effects of irrigation using four types of water (wastewater, effluent, mixture of freshwater and effluent, and freshwater) on three types of soil (sandy loam, loam, and clay loam) have been investigated both experimentally and numerically. Barley was planted as a common agricultural crop. The leachates from lysimeters have been collected and sampled at the beginning, middle, and end of the growing season. These samples have then been analyzed for phosphorous. The results show that the trend of change in nutrient concentration (P) was a function of plant requirement. Maximum process of leaching occurred concurrent with minimum plant requirement. The average phosphorus leaching into the root depths turns out to be insignificant, as it amounts to only 0.65–1.65%. This reassuring result means that wastewater with high concentrations of phosphorus compounds (up to 5–10.3 PO_4-P mgl^{-1}) can just be treated through an intermittent application to the land surface. Overall, a good agreement between experimental- and numerical-model results is obtained, wherefore the model overestimates the mean phosphate leaching during the growing season of the crop slightly. On the basis of these results, soil with loamy texture was considered to be the most suitable type for irrigation with wastewater and effluent. The results of this research indicate that with a proper management program in regard to the types of soil to be used, crops to be cultivated, water quality, and timing maneuver, the negative impacts of low quality water on soil/plant/groundwater systems can be minimized.

Keywords: irrigation, wastewater, phosphorous, barley, HYDRUS-1D

1. Introduction

Besides wastewater usage and their environmental impact, water shortages are a severe problem in several parts of the world. Many parts of the world are threatened by water scarcity. In the Middle East, the threat of water scarcity is particularly important as it is an arid region with limited fresh water sources. Therefore, seeking for unconventional sources of water is inevitable in this area. The use of treated sewage water for irrigation ensures the reuse of water resources. Municipal wastewater not only offers an alternative water irrigation source, but also the opportunity to consider as low price fertilizer because of its high nitrogen (N), phosphorus (P), and potassium (K) content [1].

Phosphorus is a valuable nutrient contained in wastewater [2]. There is potential for these nutrients present in recycled water to be used as a fertilizer source when the water is recycled as an irrigation source for agriculture [3]. Phosphorus (P) is commonly found in municipal and agricultural waste and wastewater, originating from the digestion of phosphorus-containing food sources. Municipal wastewaters may contain 5–20 mgl^{-1} of total phosphorus, of which 1–5 mgl^{-1} is organic and the rest is inorganic. Phosphorus in natural waters is usually found in the form of phosphates (PO_4^{3-}). During irrigation with wastewater, phosphorus may be leached from or retained in the soil or taken up by plants. Too much phosphorus in the water causes algae to grow faster than the ecosystems can handle.

Phosphorus can move into surface water bodies by runoff or erosion and cause water quality problems such as eutrophication. Phosphates are not toxic to people or animals unless they are present in very high levels. The phosphate in wastewater is initially quite soluble and available [4]. Movement of phosphate is slow but may be increased by rainfall or irrigation water flowing through the soil. Due to erosion of soil and when the sediment reaches a body of water it may act as a sink or a source of P in solution. Therefore, to develop effective management practices, there is a need to improve the understanding of P transport in the soil profile through percolation or matrix flow. In the case of blue-green algae, toxic by-products can be produced, which create health issues if a lake or reservoir would be used as a source of drinking water. For this reason, phosphorus removal is an essential role of wastewater treatment plants and testing for phosphorus in the plant effluent is critical. Controlling phosphorus discharged from municipal and industrial wastewater treatment plants is a key factor in preventing eutrophication of surface water bodies. The objectives of this study were, using HYDRUS-1D model [5], as a tool, to develop an understanding of vertical distribution and transport processes PO_4 leaching in soil lysimeter condition. Calibration and validation of HYDRUS-1D model was based on the experimental results.

2. Material and methods

2.1. Experimental site

The experiment was carried out in the field of lysimeters at the Mashhad research station site, (36°13′ latitude, 59°38′ longitude) in northern east Iran during growing season (2004–2005).

This research was done to investigate the soil capacity to remove impurities when it is irrigated with wastewater and effluent and to study the potential impacts on groundwater quality. For this purpose, the effects of irrigation with four types of water (wastewater, effluent, mixture of freshwater and effluent, and freshwater) on three types of soil (sandy loam, loam, and clay loam) were investigated. A randomized completely blocked design was performed with three replications. The experiment was carried out, using 36 lysimeter (2 × 1.5 m) as experimental units. The number of lysimeters was equal to the number of experimental treatments × replicates (i.e., 4 × 3 × 3 = 36). Barley was planted as a common agricultural crop. A layer of gravel was placed at the bottom of each lysimeter to facilitate drainage. The leachates from lysimeters were collected and sampled at the beginning, middle, and end of the growing season. The samples were analyzed for chemical oxygen demand (COD) [6], phosphate, and nitrate [7]. Physicochemical characteristics of irrigation water, wastewater, and soil used in this study are summarized in **Tables 1** and **2**, respectively.

Parameter	Unit	Irrigation water			Standard value	
		Wastewater	Effluent	Well water	FAO[a]	IDE[b]
PH	–	8.3	7.9	8.2	6.5–8.4	6–8.5
EC	dSm^{-1}	1.7	1.4	0.6	<3	–
SAR	$(meql^{-1})^{1/2}$	3.8	4.7	0.24	<3	–
TSS	mgl^{-1}	254*	101*	3	–	100
Na^+	$meql^{-1}$	8.07	8.35	0.4	–	–
K^+	$meql^{-1}$	0.1	–	–	–	–
Ca^{2+}	$meql^{-1}$	3.7	2.6	1.8	–	–
Mg^{2+}	$meql^{-1}$	5.3	3.7	3.8	–	8.2
Cl^-	$meql^{-1}$	6.6*	5.3	1.5	<4	6
Hco_3^-	$meql^{-1}$	6.7	5.6	3.9	<8.5	–
So_4^-	$meql^{-1}$	2.9	3.5	0.5	–	5.2
NO_3-N	mgl^{-1}	3.1	23.4	108	5–30	10
NH_4-N	mgl^{-1}	29	3.4	0.2	–	–
Total-N	$meql^{-1}$	53.6	29	3.4	2.5–43	–
PO_4-P	mgl^{-1}	5.9**	3.4	0.13	4.1	–
COD	mgl^{-1}	384.6	27	20	–	100
BOD	mgl^{-1}	252	13.3	0	–	200

[a]Food and Agriculture Organization of the United Nations.
[b]Iranian Department of Environment.
*The standard is higher than the range of Iranian Department of Environment.
**The standard is higher than the range of FAO.

Table 1. Physicochemical characteristics of water and treated wastewater.

Soil sample*	Parameters												
	Anions solution saturation extract (meql^{-1})				Total anions	Cations solution saturation extract (meql^{-1})				Total cations	EC (dsm^{-1})	pH (–)	SAR (meql^{-1})$^{1/2}$
	CO_3^{2-}	HCO_3^-	SO_4^{2-}	Cl^-		Ca^{2+}	Mg^{2+}	Na^+	K^+				
C	–	1.8	35	9	45.7	14	20	12	0.1	45.9	4.2	7.4	0.87
L	–	2.5	4.5	6.3	13.2	4.1	7.2	2.1	–	13.1	1.2	7.7	0.9
S	–	2.2	5.3	6.1	13.6	4.2	7.3	2.1	–	13.5	1.4	7.8	2.9

*C: clay loam, L: loam, and S: sandy loam.

Table 2. Some chemical properties of soil layers at the experimental field site at initial condition.

2.2. Data collection

In this model, some physical and soil hydraulic properties, concerning soil moisture retention characteristics, θ (h), and saturated hydraulic conductivity, K_{sat}, were measured in the field. The parameters of van Genuchten's [8] model were evaluated by fitting on θ (h) data using the curve RETC code. The average values of van Genuchten parameters for lysimeter study at different soil types are given in **Table 3**.

Soil sample*	Particle fraction (%)			Texture (–)	Bulk density (kgcm^{-3})	θ_r (cm^3 cm^{-3})	θ_s (cm^3 cm^{-3})	a (cm^{-1})	n (–)	l (–)	K_{sat} (cm day^{-1})
	Clay	Silt	Sand								
S	22.09	19.19	58.72	Sandy loam	1.51	0.065	0.41	0.075	1.89	0.5	106.1
L	20.30	39.68	40.02	Loam	1.43	0.078	0.43	0.036	1.56	0.5	24.96
C	48.65	28.75	22.6	Clay loam	1.3	0.095	0.41	0.019	1.31	0.5	6.24

*C: clay loam, L: loam, and S: sandy loam.

Table 3. Physical properties and van Genuchten parameters for soil sample with θ_r, residual water content (cm^{3}cm^{-3}); θ_s, saturated water content (cm^{3}cm^{-3}); a (cm^{-1}) and n(–), empirical parameters; l(–), pore-connectivity and tortuosity factor and K_{sat}, saturated hydraulic conductivity (cm h^{-1}).

2.3. The HYDRUS-1D-flow and transport model

In this study, HYDRUS-1D software, version 4.14, was used to conduct numerical simulations of one-dimensional water flow and phosphorous transport in vertical profiles of unsaturated soil to simulate the phosphorous transport in the different soil types under municipal wastewater application. The total depth of each soil profile was 200 cm with one soil type in each profile. Raw sewage then passes through the filter mesh, effluents-treated municipal wastewater, obtained daily from the Parkanabad wastewater treatment plants, mixture of 50% effluents and 50% well water, and well water was used as the influent. Irrigation water was applied to the lysimeters at a flow of 0.78–0.21 m^3 m^{-2} day^{-1} in 2004 and 2005, respectively. Each soil profile was oriented vertically, so that the irrigation water flowed in a vertical direction.

Plant	h_o	h_{opt}	h_{2H}	h_{2L}	h_3
			cm		
Barley	-15	-30	-325	-600	-8000

*Water uptake is assumed to be zero close to saturation (i.e., wetter than some arbitrary "anaerobiosis point" h_o). Root water uptake is also zero for pressure heads less than the wilting point (h_3). Water uptake is considered optimal between pressure heads h_{opt} and h_2, whereas for pressure heads between h_2 and h_3 (or h_o and h_{opt}), water uptake decreases (or increases) linearly with pressure head.

Table 4. Effective root depth, root water uptake parameters, and root distribution*.

The initial condition for volumetric soil water content was between 0.1 and 0.2 for different soil types in all simulations. In case of water flow, the upper water flow boundary condition was atmospheric boundary condition with surface layer, given by the following equation:

$$-K\left(\frac{\partial h}{\partial x}+\cos(\alpha)\right)=q_0(t)-\frac{dh}{dt} \qquad at\ x=L(Soil\ surface) \tag{1}$$

where q_0 is the net infiltration rate (precipitation minus evaporation).

Irrigation treatments	Data of sampling						
	22.6.2004	29.6.2004	8.7.2004	18.7.2004	29.7.2004	6.8.2004	Mean
Total nitrogen (mg l^{-1})							
Wastewater	43	46.5	45.9	47.6	60.8	77.6	53.6
Effluent	36.8	22.6	20	30	29.1	35.2	29
Well water	3.41	3.41	3.41	3.41	3.41	3.41	3.41
Ammonia (mg l^{-1})							
Wastewater	24.5	27.6	24.3	27.3	33.3	36.9	29
Effluent	1.37	2.2	2.4	5.5	3.1	5.9	3.41
Well water	0.21	0.21	0.21	0.21	0.21	0.21	0.21
Nitrate (mg l^{-1})							
Wastewater	11.5	8.3	11.9	10	9.4	13.4	10.8
Effluent	34.1	19.3	16.2	22	23	26	23.4
Well water	3.1	3.1	3.1	3.1	3.1	3.1	3.1
Phosphate (mg l^{-1})							
Wastewater	3.6	5.4	3.2	5.8	7.5	10.3	6
Effluent	2.7	5	2.6	2.4	1.9	6	3.4
Well water	0.13	0.13	0.13	0.13	0.13	0.13	0.13

Table 5. The amount of nitrogen and phosphate in different irrigation water (mg l^{-1}).

In this study, the lower water flow boundary condition was free drainage. The minimum allowed pressure head at soil surface is the wilting value and was set at the value of 100,000 cm provided by HYDRUS-1D. The root water uptake by plants is described by the macroscopic approach of Feddes et al.'s [9] model. Information on root water uptake with compensation is available in Ref. [5]. The coefficients of Feddes et al.'s [9] model are presented in **Table 4** [5]. The maximum root depth, seeding depths, and the root growth ratio of barley were 100, 5, and 5 cm, respectively.

To investigate the concentration of nitrogen and phosphate in wastewater, effluent, and well water, at any time of sampling from the Parkanabad wastewater treatment plants, quality of the water/wastewater in terms of total nitrogen, ammonia, nitrate, total phosphate, and chemical oxygen demand (COD) were tested based on standard methods [6]. Mean concentration of nitrogen and phosphate in different irrigation water are presented in **Table 5**.

Irrigation water**	Soil sample***	Data of sampling											
		22.6.2004		29.6.2004		8.7.2004		18.7.2004		29.7.2004		6.8.2004	
		Chemical oxygen demand (*COD*) (mgl^{-1})											
		1*	2*	1	2	1	2	1	2	1	2	1	2
W_1	S	20	13	20	30	20	30	20	20	20	27	20	24
W_1	L	20	12	20	20	20	28	20	18	20	21	20	19
W_1	C	20	10	20	28	20	30	20	20	20	28	20	21
W_2	S	26	18	27	24	29	33	25	29	27	37	24	34
W_2	L	26	17	27	26	29	35	25	28	27	31	24	39
W_2	C	26	13	27	27	29	37	25	27	27	38	24	31
W_3	S	35	17	45	35	25	42	30	28	29	35	27	40
W_3	L	35	20	45	38	25	35	30	25	29	28	27	33
W_3	C	35	35	45	37	25	40	30	25	29	27	27	34
W_4	S	400	25	430	47	380	53	385	38	392	51	381	45
W_4	L	400	27	430	48	380	57	385	40	392	50	381	41
W_4	C	400	25	430	50	380	52	385	37	392	48	381	42

1*Input *COD* in terms of milligrams per liter, the pollution load of wastewater, and water used in irrigation.
2*Drainage *COD* in terms of milligrams per liter, contamination of water is drained from the lysimeters.
**W_1: freshwater, W_2: mixture of and effluent, W_3: effluent, W_4: wastewater.
***S: sandy loam, L: loam, C: clay loam.

Table 6. The amount of chemical oxygen demand (COD) in different irrigation water (mgl^{-1}).

As shown in **Table 5**, about 42% of phosphate in raw wastewater is removed during the treatment process. According to Mojid et al. [10], the maximum permissible level of phosphate in wastewater for irrigation should not be more than 4.1 mg l^{-1}. In our study, the amount of phosphate in raw wastewater was more than FAO's standard. About effluent, however, the average of phosphate was less than 4.1 mg l^{-1} [11], but in some samples, its concentration was higher than the standard amount. Results of the analysis of chemical oxygen demand (COD) and irrigation water are presented in **Table 6**. This table includes the average results from three similar lysimeters in each irrigation (irrigation water and the type of soil) and through this we can observe the relative change transfer of contamination by COD index into the deep soil during the irrigation season.

The HYDRUS-1D model was also used to simulate PO_4 transport under different irrigation treatments and soil types in one-dimensional vertical lysimetrs. The HYDRUS-1D was run for the main processes of water flow and general solute transport. No hysteresis was considered in the simulations. A total of three simulations (one for each soil types) were performed. Each simulation modeled one-dimensional unsaturated water flow, root water uptake, and phosphate transport. In each simulation, the precipitation and irrigation water were applied to the soil surface of lysimeter. The soil surface in each simulation was covered with barley crop. The initial values for the longitudinal dispersivity (λ) were derived from HYDRUS-1D dataset and from a study done by [12, 13]. HYDRUS-1D model was then calibrated manually by using these initial values for the λ parameter. The λ parameter was calibrated against the concentration of PO_4-P in drainage water from lysimeters throughout the experiment. The final value of λ was determined by using several iterations when the mass balance errors were minimized to <1%. We assumed the molecular diffusion coefficient in free water (DW) was set to zero, therefore the transport of solute through diffusion was considered negligible. The initial water conditions were specified in terms of water content between 0.1 and 0.2 for different soil types in all simulations. The upper water flow boundary condition at the surface ($x = L$) was specified as the atmospheric boundary condition with a surface layer. This boundary condition imposed time-dependent conditions to specify the atmospheric conditions at the top of the lysimeter. Initial concentration of PO_4 on the top node of the lysimeter was specified equivalent to the amount of PO_4 wastewater added on top of the lysimeter before running the experiment. The lower water flow boundary conditions were prescribed using gravitational free draining. As for solute (PO_4) transport, concentration flux boundary conditions were implemented at the upper boundary, and a zero gradient boundary condition was set at the lower solute boundary condition. The reaction parameters required by the HYDRUS-1D model were derived from the adsorption experiment reported by Abou Nohra et al. [14]. The reaction parameters ($\boldsymbol{k_d}$ and $\boldsymbol{\beta}$) required by the HYDRUS-1D model were derived based on Eq. (2):

$$s = K_d \log c^{\beta} \tag{2}$$

where s is the concentration of PO_4 adsorbed to the soil ($M\ M^{-1}$), c is the concentration of PO_4 in solution ($M\ L^{-3}$), k_d is the equilibrium constant ($L^3\ M^{-1}$), and β is a shape-fitting parameter [15]. The solute transport and reaction parameters considered in the simulations for different soil samples are listed in **Table 7**. The HYDRUS-1D models were run for phosphorous transfer into two stages: calibration and validation. Results obtained from 2004 were used to calibrate the parameters to improve the fit between the simulated and measured data. Similarly, the results obtained from 2005 were used to validate the output from the model.

Model parameter	Soil sample*		
	S	L	C
Soil bulk density, g cm^{-3}	1.51	1.43	1.35
Longitudinal dispersivity, cm	1	1.15	1.23
Equilibrium constant-adsorption isotherm coefficient, $cm^3 mg^{-1}$	1	1.25	1.35
Shape fitting parameter-adsorption isotherm coefficient, –	1.35	1.45	1.6

*S: sandy loam, L: loam, C: clay loam.

Table 7. Transport and reaction parameters for different soil samples.

3. Results and discussion

3.1. Model calibration and validation

Predicted and measured values of cumulative deep percolation (DP) for different soil types are presented in **Figure 1**. Comparing linear relationship between the predicted and measured values of DP with the 1:1 line, the measured values of DP matched well with the predicted values. This indicated that the HYDRUS-1D model is capable to predict DP at different irrigation treatments. The slopes of the linear relationship are statistically equal to 1.0 and the values of NRMSE and "d" are 0.12–0.15, 0.21–0.991, and 0.987–0.976 for sandy loam, loam, and clay loam, respectively. These indicated a high accuracy of the prediction of DP by HYDRUS-1D model for barley crop.

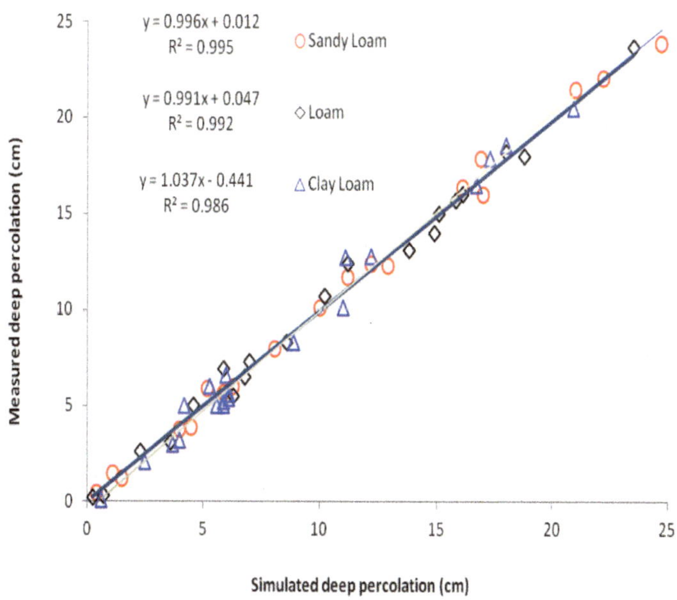

Figure 1. Relationship between predicted and measured values of deep percolation for barley.

Values of measured and predicted leached PO_4 for barley crop during the growing season at different soil lysimeters and for different irrigation water are shown in **Figure 2**. The linear

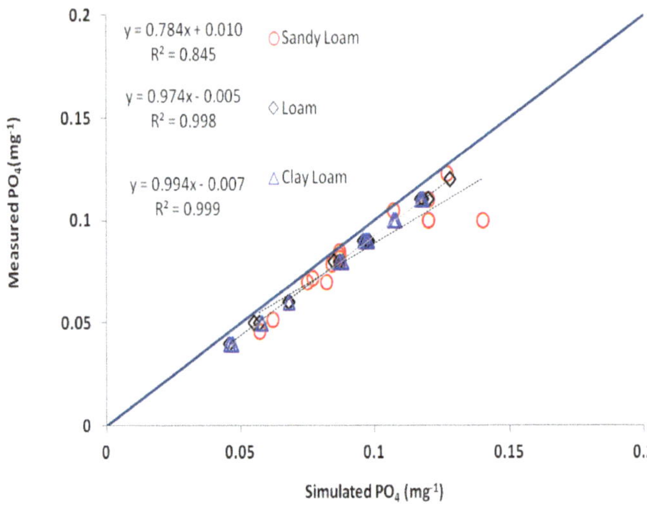

Figure 2. Relationship between predicted and measured phosphate leaching for barley.

relationship between the measured and predicted values of leached PO_4 were compared with the 1:1 line and the slope and intercept values were calculated. Ideally, the slope and intercept should be one and zero, respectively, indicating a perfect match between predicted and measured values. However, this is a very strict requirement and rarely met in practice. In this study, the slopes of the linear relationship for PO_4 is statistically equal to 1.0 and intercept values were 0. 216, 0.870, and 0.036 for sandy loam, loam, and clay loam, respectively. The close similarity between the measured and predicted PO_4 content at different soil profile depths over

Irrigation water*	Soil sample**	NO_3	PO_4	NO_3	PO_4	NO_3	PO_4	NO_3	PO_4
		AE (–)		RMSE (mg l^{-1})		NRMSE (–)		d (–)	
W_1	S	0.017	0.052	0.029	0.129	0.015	0.214	0.991	0.870
W_1	L	0.077	0.081	0.133	0.173	0.073	0.223	0.990	0.881
W_1	C	0.043	0.081	0.075	0.171	0.048	0.271	0.987	0.872
W_2	S	-0.040	0.052	0.069	0.129	0.030	0.237	0.993	0.778
W_2	L	0.003	0.038	0.006	0.091	0.003	0.271	0.994	0.891
W_2	C	0.007	0.017	0.012	0.047	0.007	0.271	0.991	0.887
W_3	S	0.070	0.087	0.121	0.107	0.016	0.211	0.989	0.859
W_3	L	0.127	0.210	0.219	0.189	0.040	0. 247	0.982	0.792
W_3	C	0.180	0.290	0.312	0.202	0.053	0.258	0.987	0.897
W_4	S	0.073	0.013	0.127	0.142	0.015	0.219	0.990	0.919
W_4	L	0.003	0.019	0.006	0.021	0.001	0.284	0.985	0.903
W_4	C	0.030	0.020	0.052	0.087	0.012	0.253	0.984	0.898

[1]AE, the average error; RMSE, the root mean square error; NRMSE, normalized root mean square error; and d, the index of agreement.

*W_1: freshwater, W_2: mixture of and effluent, W_3: effluent, W_4: wastewater.

**S: sandy loam, L: loam, C: clay loam.

Table 8. Statistical indexes for calibration and validation of HYDRUS-1D[1].

time resulted in a high correlation coefficient (0.991), high index of agreement (0.984), low average error (0.077), low root mean square error (0.312 mg l^{-1}), and low normalized root mean square error (9%), demonstrating a very good calibration of the model (**Table 8**). These indicated a high accuracy of the prediction of leached PO_4 by HYDRUS-1D model for barley crop in different soil types. The model overestimated the measured phosphate leaching in all soil types used in the model simulation. Correlation coefficient values were at around 0.914, index of agreement at around 0.907, average error at around 0.305, root mean square error values at around (0.0298 mg^{-1}), and normalized root mean square error at around 11% for all lysimeter soil. Overall, the values calculated for phosphate leaching demonstrate a good correlation of the model to field data.

3.2. PO4 1 leaching to depth

The findings of phosphor concentration in different kinds of irrigation and drainage water are displayed in **Figure 3**. The percentage of phosphate removal was high in all treatments (between 91 and 99%), which revealed the good potential of crop and soil system in phosphate removal. In **Table 9**, the averages of phosphate in drained water in different treatments during growing season are displayed. The effects of soil and irrigation water on transfer of phosphor to root zone are described below:

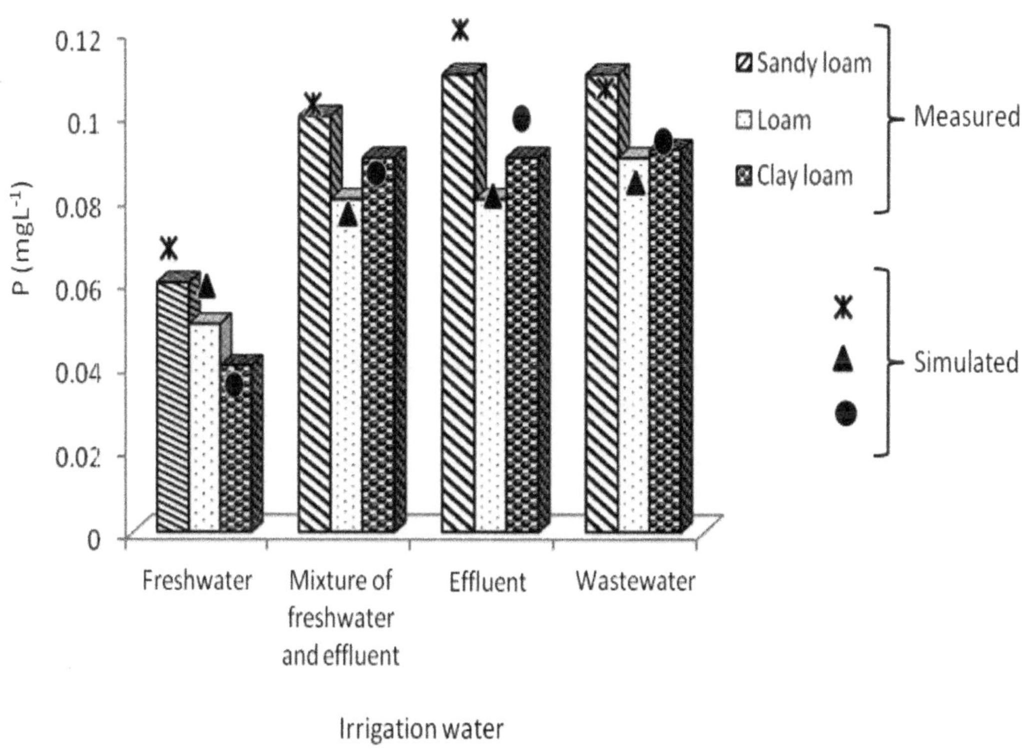

Figure 3. Mean phosphate leaching during the growing season.

Irrigation water*	Soil sample**	4.7.2004				18.7.2004				26.7.2004			
		(1)	(2)	(3)	(4)	(1)	(2)	(3)	(4)	(1)	(2)	(3)	(4)
W_1	S	0.13	0.07	0.075	50.5	0.13	0.06	0.068	43.8	0.13	0.05	0.058	37.7
W_1	L	0.13	0.052	0.059	40.2	0.13	0.05	0.055	37.1	0.13	0.04	0.046	33.6
W_1	C	0.13	0.046	0.057	35.5	0.13	0.04	0.046	32	0.13	0.04	0.047	30.0
W_2	S	2	0.10	0.14	5	1.32	0.11	0.12	8.1	1.32	0.10	0.107	5.11
W_2	L	2	0.07	0.075	3.8	1.32	0.08	0.085	6. 2	1.32	0.08	0.088	3.9
W_2	C	2	0.08	0.087	4.1	1.32	0.09	0.098	6.7	1.32	0.09	0.096	4.2
W_3	S	3.8	0.10	0.12	2.7	2.52	0.11	0.117	4.4	2.52	0.11	0.117	2.7
W_3	L	3.8	0.079	0.084	2	2.52	0.09	0.096	3.4	2.52	0.08	0.088	2.1
W_3	C	3.8	0.082	0.087	2.1	2.52	0.09	0.097	3.5	2.52	0.09	0.097	2.2
W_4	S	4.5	0.11	0.12	2.3	4. 5	0.11	0.118	2. 5	4. 5	0.11	0.118	2.2
W_4	L	4.5	0.083	0.087	1.8	4. 5	0.09	0.098	1. 9	4. 5	0.09	0.098	1
W_4	C	4.5	0.085	0.087	1.6	4. 5	0.09	0.097	2	4. 5	0.09	0.097	1

*W_1: freshwater, W_2: mixture of and effluent, W_3: effluent, W_4: wastewater.
(1)Total phosphorus inputs in terms of milligrams per liter, from irrigation water.
(2)Total phosphorus output in milligrams per liter, measured in lysimeter drainage water.
(3)Total phosphorus output in milligrams per liter, simulated in lysimeter drainage water.
(4)Percent transfer, represents the amount of total phosphorus observed in drainage water drains compared with the input values of irrigation water at each sampling time.
**S: sandy loam, L: loam, C: clay loam.

Table 9. Mean phosphate input, output, and transfers percentage.

The effect of soil: Types of soil had significant effect ($p < 0.05$) on phosphate concentration in lysimeters drained water. LSD test showed that the amount of phosphate transferred to root zone in sandy loam lysimeters was significantly higher than in loam lysimeters. Also, the amount of phosphate transferred to root zone in loam lysimeters was lower (except in control treatment) than clay lysimeters. One possible reason for this difference is considerable growth of crop in loam soil and also different permeability of different soil types. Low permeability of clay soil and phosphate absorption by soil particle are the factors influencing less transfer of phosphate to the depth. Of the loam lysimeters irrigated by effluent, wastewater, and mixture of freshwater and effluent, only about 0.97–6.2% of influent phosphor was drained. Also, in clay and sandy loam lysimeters about 1–6.7% and 1.2–8.1% of influent phosphor was drained, respectively. Since in sandy loam soil the amount of phosphor uptake by crop was not high (because of nonconsiderable growth of crop), the removal of more than 90% of phosphor in sandy loam soil suggested the ability of soil in the removal of phosphor available in wastewater and effluent. The findings are consistent with Kardos and Hook [16], who reported in their study that in loam and clay loam, the amount of phosphor leaching in the depth of 120 cm were 1 and 0.1% lower than influent phosphor, respectively. About 97–99% of phosphor removal in crop and soil system was reported by Hasan Oghli et al. [17].

The effect of irrigation water: Simulation results showed that the effect of type of irrigation water on phosphate concentration in drainage water of lysimetrs was significant at $p < 0.05$.

There was no significant difference among the amount of phosphate in drained water of lysimeters irrigated with wastewater, effluent, and mixture of freshwater and effluent. However, there were significant differences between the amount of phosphate in drained water of freshwater treatments and the other treatments. According to the findings, we can say that the amount of phosphate output from lysimeters was dependent on the growth of crop and type of soil compared to type of irrigation water.

The effect of sampling time: The findings showed that sampling time had no significant effect on the amount of transferred phosphate; however, in the middle of growing season, the amount of transferred phosphate to the depth was at the maximum level.

Once the discharge of drainage water from underground drains to surface water and groundwater is considered, the amount of phosphate phosphor should not be more than the determined standards. In our research, in the worst situations, the amount of phosphate in lysimeters drained water did not exceed 0.11 mg l^{-1}, which was lower than the standard level [10].

4. Conclusion

Inappropriate management practices in the use of wastewater in phosphorus deteriorate surface and ground water quality, mainly by causing nitrate pollution. The HYDRUS-1D model was calibrated and then validated with different datasets from a lysimeter experiment, and then used to simulate phosphorus leaching through soil under different irrigation treatment (wastewater, effluent, mixture of freshwater and effluent, and freshwater) on three types of soil (sandy loam, loam, and clay loam) to explore and develop better and safer wastewater land application strategies.

Phosphate transferred to the depths was insignificant and it was between 1.6 and 6% of inflow phosphate, which was lower than the maximum standard value of phosphate discharge to surface and groundwater.

Soil and plant systems showed high potential in filtration and removal of nitrate and phosphate, so that the concentration of nitrate and phosphate in drained treatments in all cases was lower than the limit of discharge to surface water and groundwater. It can be confirmed that through proper management and research, in addition to maintaining surface water and groundwater, the effluent, as an available and cheap source, can be used in agricultural irrigation. As there was no significant difference on nitrate leaching between treatments mixture of freshwater and effluent, and freshwater, this demonstrates that it can dilute wastewater as a suitable management strategy for reducing the leaching of impurities in the wastewater and also reduce the effects of probable hazards on soil properties. Simulation study on the process of nitrate leaching to root zone during growing season showed more matches the needs of the plant. Thus, at the time of minimum plant nutrient requirement, we can take suitable management solution such as wastewater dilution to lower leaching of elements to root zone.

Author details

Ali Erfani Agah[1,2*], Patrick Meire[1] and Eric de Deckere[1]

*Address all correspondence to: Ali.Erfaniagah@student.uantwerpen.be; Ali.Erfani68@gmail.com

1 Department of Biology, University of Antwerp, Antwerp, Belgium

2 Department of Irrigation and Drainage, Shahrood University of Technology, Shahrood, Iran

References

[1] Chen, Z., Ma, W., and Han, M. 2008. Biosorption of nickel and copper onto treated alga (*Undaria pinnatifida*): Application of isothermal and kinetic models. Journal of Hazardous Materials. 155, pp. 327–333.

[2] Erfani, A., Haghnia, G.H., and Alizadeh, A. 2001. Effect of irrigation by treated wastewater on the yield and quality of tomato. Journal of Agricultural Science and Technology, 15(1), 65–67.

[3] Erfani, A. 1996. Environmental aspects use of treated municipal wastewater in irrigation. Published and presented in Iranian National Committee on Irrigation and Drainage (IRNCID).

[4] Erfani, A., Haghnia, GH. & Alizadeh, A. 2002. Yield and Chemical Composition of Lettuce and Some Soil Characteristics as Affected by Irrigation with Wastewater. *Journal of Science and Technology of Agriculture and Natural Resources*, 6(1), 71–92.

[5] Šimůnek, J., Šejna, M., Saito, H., Sakai, M., and van Genuchten, MY. 2009. The HYDRUS-1D software package for simulating the movement of water, heat, and multiple solutes in variably saturated media. Version 4.08. HYDRUS Software Series 3. Department of Environmental Sciences, University of California, Riverside, USA, p. 330.

[6] APHA 1998. *Standard methods for the examination of water and wastewater analysis*. 20th ed. Publisher: Washington D.C.

[7] Bremner, JM., and Mulvaney, CS. 1982. Nitrogen-total. Part II. Chemical and Microbiological Properties. In: "Methods of Soil Analysis", Page, A. L., Miller, R. H. and Keeney, D. R. (Eds.). Second edition, American Society of Agronomy, Madison, WI, USA, pp. 595–624.

[8] van Genuchten, M. Th. 1980. A closed-form equation for predicting the hydraulic conductivity of unsaturated soils. Soil Science Society of American Journal, 44: 892–898.

[9] Feddes, RA., Kowalik, PJ., and Zaradny, H. 1978. Simulation of field water use and crop yield. Pudoc, Centre Agric. Publ. Doc., Wageningen.

[10] Pescod, MB. Food and Agriculture Organization. 1992. Wastewater treatment and use in agriculture. Irrigation and Drainage Paper 47. FAO, Rome, Italy.

[11] EPA Iran. 1998. Environmental regulations and standards. EPA Publication Iran.

[12] Perfect, E., Sukop, MC., and Haszler, GR. 2002. Prediction of dispersivity for undisturbed soil columns from water retention parameters. Soil Science Society of America Journal, 66, 696–701.

[13] Mojid, MA., Rose, DA., and Wyseure, GCL. 2004. A transfer-function method for analysing breakthrough data in the time domain of the transport process. European Journal of Soil Science, 55, 699–711. DOI: 10.1111/j.1365-2389.2004.00636.x.

[14] Abou Nohra, JS., Madramootoo, CA., and Hendershot, WH. 2007. Modeling phosphate adsorption to the soil: application of the non-ideal competitive adsorption model. Environmental Pollution, 149, 1–9.

[15] Sparks, DL. 2003. Environmental soil chemistry. Academic Press, San Diego, CA, p. 352.

[16] Kardos, LT. and Hook, JE.1976. Phosphorus balance in sewage effluent treated soils. Journal of Environmental Quality, 5:87–90.

[17] Hasan Oghli, A., Liaghat, A., Mirabzadeh, M., Vosoghi, M., and Fardad, H. 2002. Research on the effects of irrigation with domestic effluent on material transfer in soil and effluent quality of laisimeter. The 11th Iran National Irrigation and Drainage Convention, pp: 317–334.

8

Dual Soil Decontamination Procedures

Yoshiharu Mitoma, Alina M. Simion and
Cristian Simion

Additional information is available at the end of the chapter

Abstract

Pollutants actually existing in various types of soil, ranging from rural, agricultural soils to urban or factory soils, belong to a wide range of chemical compounds, both organic and inorganic. The modern decontamination methods were each specifically designed for a particular pollutant. Reagents and procedure conditions targeted only one particular contaminant, more rarely several pollutants, all usually belonging to the same family (e.g., several heavy metals or polychloro-*p*-dibenzodioxins and polychloro-*p*-dibenzofurans). Most reviews on the subject presented soil decontamination processes under the same auspices: specific process with specific reagent for a specific pollutant. Unfortunately, soils are often cross-contaminated with various types of pollutants, which make the decontamination procedure much more complicated: indeed, for each contaminant, a certain procedure must be carried out. This transforms the whole decontamination process in a multi-step procedure, enhancing the costs. Therefore, any method that could realize a simultaneous decontamination for at least two different types of pollutants would be extremely advantageous. In the recent years, such methods made an interesting appearance in the environmental science and engineering literature. We wish to review these dual decontamination methodologies that deal simultaneously with at least one organic and one inorganic contaminant in the same soil matrix.

Keywords: simultaneous decontamination, heavy metals, pesticides, dioxins

1. Introduction

Over two centuries of various anthropogenic emissions have caused soil contamination to be a globally widespread problem, involving not only industrialized countries but even remote areas of less developed countries [1]. From decision makers to scientists and even to individual citizens, all generally accept and understand that air and water pollution can have negative impacts on human health [2–4]. However, the same statement concerning soil pollution is harder to understand [5]. In the vast majority of papers or books that deals with the problem of soil pollution, the traditional approach is to isolate a single variable, such as a specific contaminant, and then investigate that variable as its source, fate, distribution and/or health effect [6]. Nonetheless, it is difficult to accept that, after 200 years of industrialization and intensive agriculture, rural and urban soil are only mono-contaminated. It is more than probable that in any contaminated soil horizon, more than one pollutant is present [7]. Among the most common harmful contaminants are heavy metals (37%) and mineral oils (33%), along with a large variety of persistent organic pollutants (known as POPs). Scientists and members of the medical professions are acknowledging that environmental and health effects due to soil multiple contamination present complicated issues due to synergistic relationships [8].

Almost all manuscripts that presented soil decontamination techniques have a similar approach: identification of a single pollutant and subsequent treatment. Moreover, the vast majority of these studies generally use a clean soil spiked with the chosen compound. The general reason given for not using genuine contaminated soil is often the same: since it is known that many remediation processes belong to either phyto- [9] or bioremediation [10] families, the authors wanted to avoid interferences with their own method. However, such procedures may be valuable in diminishing or even completely eliminate a particular pollutant, but in the probable case of multiple contaminations it would mean a multiple treatment of the polluted soil. Moreover, most of these treatment methods are still on the laboratory or on the ex situ scale [11]. This renders most of the laboratory-scale treatment inapplicable because of ultimately high costs: the same soil should be treated in various ways, according to each particular pollutant present [12, 13].

The general assessment criteria for the selection of the proper remediation technology are:

– the short-term versus long-term effectiveness in remediation;

– the reduction in mass or volume of the contaminants (preferably their complete eradication);

– the overall reduction of the toxicity of previously contaminated soil; and

– the cost-effectiveness.

Thus, in order to diminish the costs of a treatment procedure applied to contaminated soils, an obvious solution would be to have *in situ* treatment instead of an ex situ one (completely removing the cost of transportation and relocation of soil). A good number of such processes are also known nowadays [14–17].

There are various ways to consider the most appropriate way to treat a contaminated soil [11]. These are the following: (1) doing nothing (if the environmental assessment indicates that humans and the environment are not at risk, then no remediation activity is required, e.g., in the case of small-scale spills on sites where human and animal exposure is not likely), (2) introducing institutional controls to contain the contaminants in the infected area (a legal or institutional mechanism that limits the use or the access to the contaminated area, e.g., the Chernobyl or Fukushima areas) or (3) the removal of soil and/or destruction of contaminants (in some cases, the best option may be to physically remove the contaminated soil and move it to a special treatment, storage and disposal facility; in other cases, it is possible to remove the contaminant from the soil using technologies such as surfactant washing, soil washing or thermal desorption). Ultimately, the contaminants are destroyed, on condition that the by-products are not toxic.

The *in situ* technologies are categorized into three major groups based on the primary mechanism by which the treatment is achieved:

- Physical/chemical
- Biological
- Thermal

Physical/chemical treatment includes soil vapour extraction, solidification/stabilization, soil flushing, chemical oxidation and electrokinetic (EK) separation. Biological treatment uses microorganisms or vegetation to degrade, remove or immobilize pollutants in soil. Biological technologies include bioventing, phytoremediation and monitored natural attenuation. Electrical resistivity heating, steam injection and extraction, conductive heating, radio-frequency heating and vitrification are technologies summarized under thermal treatment.

The past few years saw an increase in the number of *in situ* treatment technologies that are effectively used in the field (e.g. chemical oxidation [18] or thermal treatment [19]), demonstrating thus that *in situ* technologies are a viable option for treating contaminated soils.

Yet, in order to have a truly useful as well as an economical process, the technology should be able to eliminate more than only one contaminant, simultaneously, since it is obvious that mono-pollution is an utopian case. For example, the soils of abandoned agricultural land contaminated by e-waste activities in Hong Kong listed no less than four classes of pollutants (polycyclic aromatic hydrocarbons (PAHs), polychlorinated biphenyls, polybrominated diphenyl ether compounds and heavy metals—cadmium, copper, chromium, lead and zinc) [20]. Fortunately, in the last few years, such methods made an interesting appearance in the environmental science and engineering literature. In the following, we review some of these dual decontamination methodologies that deal simultaneously with at least one organic and one inorganic contaminant in the same soil matrix.

2. Procedures for simultaneous decontamination of polluted soils

This chapter presents a review of the processes and technologies that allow the simultaneous removal/destruction/immobilization of more than one class of contaminants in soils, focusing on dual decontamination of at least two different pollutants, one being an inorganic, the second an organic compound. Relevant papers were retrieved using screening of the scientific literature using Scopus, ISI Web of Knowledge and Google Scholar.

As previously mentioned, the most important class of inorganic pollutants is represented by heavy metals [21], while among the organic pollutants various classes of compounds can be mentioned [22] (e.g. pesticides, polyaromatic hydrocarbons, polychlorinated compounds, etc.) and it is more than often that either rural or urban soils are polluted with both types (organic and inorganic). In order to realize a simultaneous cleansing of the soil, the method used must be suitable and effective for both classes of compounds. Indeed, some of the methods applied separately for organic or inorganic contaminants proved to be successful when both types of compounds were present in soils. These methods can be classified as follows:

– Washing using a proper solvent mixture and eventually a surfactant (including flotation processes)

– Electrokinetic methods (derived from the washing techniques)

– Bioremediation (including phytoremediation)

– Combinations of the previous

– Miscellaneous (including thermal or chemical methods)

2.1. Washing processes

Chronologically speaking, elution techniques were the first used to simultaneously clean contaminated soil with both heavy metals and organic compounds, as early as the end of the twentieth century. The main problem was identifying the proper solvent mixtures.

One of these early papers in the field of dual separation techniques investigated the ability of aqueous cyclodextrin solutions to simultaneously remove heavy metals and low-polarity organic compounds from contaminated soil [23]. In that purpose, Brusseau and co-workers used three types of soil spiked with the model organic compound (phenanthrene) and the model heavy metal (cadmium). Previously, Dunn et al. had used surfactants in a micellar-enhanced ultrafiltration, which is a separation process using surfactants and membranes, removing dissolved organic solutes or multivalent ions from water with high rejections [24]. Through this procedure, mixtures of pollutants containing phenol or *o*-cresol and Zn^{2+} and/or Ni^{2+} were separated from contaminated soils, using an anionic surfactant. Micellar-enhanced ultrafiltration was subsequently used by other authors to remove either Cu and phenol [25] or chlorinated aromatic hydrocarbons, nitrate and chromate ions [26]. Modified cyclodextrin (as glycine-β-cyclodextrin) was later successfully used in the study of the desorption behaviour of phenanthrene and lead from co-contaminated soil [27]. The authors showed that glycine-β-cyclodextrin had good solubilization properties for both phenanthrene and lead carbonate (900 g/L for phenanthrene, respectively, 2945 mg/L $PbCO_3$).

A parallel process used a combination of 2.5 N sulphuric acid and isopropyl alcohol (in a 4:9 ratio), with a dilution of 5 part solution to 1 part soil, the separation being made by ultrafiltration [28]. The contaminants removed in these experiments were heavy metals (Cd, Ag and Cu), volatile organic compounds (ethyl benzene and methyl iso-butyl ketone), halogenated compounds (chloroethene and tetrachloroethylene) and pesticides, herbicides and insecticides (lindane, methoxychlor and endrin). However, the acidic treatment cannot be applied to other types of soils than sandy ones. The ultrafiltration methods are still at the laboratory-scale level and therefore are not yet suitable for *in situ* applications. Thus, researchers tended to deepen the knowledge in the field of surfactant use.

A first step was to widen the variety of surfactants used. Thus, Marshall and co-workers, after testing the efficiency of a non-ionic surfactant (polyethylene oxide (PEO) of chain length 7.5 (Triton X-114), 9.5 (Triton X-100), 30 (Triton X-305) or 40 units (Triton X-405)), combined with iodide salts [29], introduced the use of ethylenediaminetetraacetate (EDTA) to simultaneously extract heavy metals and polychlorinated biphenyl compounds from a field-contaminated soil [30]. Since both cyclodextrin and EDTA were effective, the use of their combination was a naturally occurring step, took again by Marshall and his team at the McGill University [31]. Thus, ultrasonication was used to mix field-contaminated soil with a combination of cyclodextrin solution (10%, w/v) and 2 mmol EDTA, in the same time mobilizing polychlorinated biphenyls and much of the metals (Cd, Cr, Cu, Mn, Ni, Pb and Zn). The authors observed that a combination of randomly methylated or hydroxypropyled β-cyclodextrin with EDTA did not alter the polychlorinated biphenyls extraction efficiency nor did the presence of cyclodextrin change the efficiency of mobilization of most heavy metals (Al, Cd, Cr, Fe, Mn, Ni and Zn) but it did increase the recovery of Cu and Pb. Three sonication washings with the same charge of reagents mobilized appreciable quantities of polychlorinated biphenyls (40–76%) and quantitatively extracted the labile fraction of Cd, Cu, Mn and Pb. However, due to the low degree of biodegradability in soil [32], Marshall opted for another complexing reagent (ethylenediaminedisuccinic acid, [S,S]-EDDS). Thus, the same authors evaluated the efficacy of soil washing with a non-ionic surfactant (Brij98) in combination with [S,S]-EDDS for the simultaneous mobilization of heavy metals and polycyclic aromatic hydrocarbons from a field-contaminated soil [33, 34]. Moreover, they extended their procedure to the remediation of polluted soils with arsenic, chromium, copper, pentachlorophenol (PCP), polychlorinated dibenzo-*p*-dioxins (PCDDs) and polychlorinated dibenzofurans (PCDFs) [35]. The highest level of mobilization/detoxification was achieved in three soil washes with a mixture of 0.1 M [S,S]-EDDS and 2% Brij98 at pH 9 with 20 min of ultrasonication treatment at room temperature. This combination mobilized 70% of As, 75% of Cr, 80% of Cu, 90% of pentachlorophenol and 79% of PCDDs and PCDFs.

In order to render the process even more environmentally friendly, many authors turned their attention to naturally occurring surfactants. Thus, starting from the idea that biosurfactants are potentially less toxic to soil organisms than other chemical agents, Lima and co-workers studied the efficiency of a combination of iodide salt ligands and surfactants produced by different bacterial species in the simultaneous removal of cadmium and phenanthrene in a Haplustox soil sample [36]. For their part, Zhu et al. from Zhejiang University used saponin, a plant-derived biosurfactant, for the dual removal of phenanthrene and cadmium from contaminated soils [37]. Another Chinese team successfully treated soils from a contaminated electronic waste site that contained both polybrominated diphenyl ethers and heavy metals using common sunflower oil in a mixture with carboxymethyl chitosan [38].

Another Canadian team, led by Blais from the Université du Quebec, chose to combine alkaline-washing process with flotation in acidic solutions (in the presence of a surfactant such as cocamidopropyl betaine). They succeeded in decontaminating various types of soils from mixtures of heavy metals, pentachlorophenol and polychlorodibenzo-*p*-dioxins and furans (PCDD/F) [39]. Blais' process is mainly based on physical techniques, such as crushing, gravimetric separation and attrition. In another such study, Blais' team demonstrated that it is possible to attain removal efficiencies of 49–73% for Cu and from 43 to 63% for Zn, whereas a removal yield of 92% was measured for total PAHs [40, 41]. The results were improved replacing cocamidopropyl betaine with cocamidopropyl hydroxysultaine (up to 90% PAHs and Pb removal, in three different polluted soil types) [42]. By carefully choosing the acidic species (hydrochloric, nitric, sulphuric and lactic acids and ethanol) for leaching metals from soil in combination with non-ionic, ionic and amphoteric surfactants, Blais and co-worker studied the simultaneous removal of heavy metals and pentachlorophenol by flotation [43]. Thus, removal yields of 82–93, 30–80, 79–90 and 36–78% were obtained from As, Cr, Cu and PCP, respectively.

A different approach was undertaken by a team from the Hebrew University of Jerusalem, under the form of a sediments remediation phase transition extraction [44]. This process is based on using partially miscible solvent mixtures in which specific organic soluble chelating agents are dissolved. Extraction efficiency is improved by a phase transition cycle induced by temperature variation. With this technology, up to 90% of cadmium ions were removed within approximately 15 min, as well as practically all the organic matter, including PAHs.

2.2. Electrokinetic processes

A second family of technologies that allow concurring decontamination of co-contaminated soils are the electrokinetics processes (*aka* electrokinetics). Electrokinetics are a group of emerging techniques that are intended to separate and extract heavy metals, radionuclides and organic contaminants from various types of soils, sludges, sediments and even groundwater [45]. The goal of electrokinetic remediation is to realize the migration of subsurface contaminants in an imposed electric field via electro-osmosis, electromigration and/or electrophoresis. These phenomena occur when the soil is electrically charged with a low-voltage current. The fundamental configuration for all three processes involves the application of an electrical potential between electrode pairs that have been implanted in the ground on each side of a contaminated soil mass. There are even some emerging *in situ* electrokinetic soil remediation technologies, such as Lasagna™, Elektro-Klean™ or Electrobioremediation.

Even from the first attempts to simultaneously eliminate both heavy metals (including lead, zinc, manganese, copper and arsenic) and organic pollutants (PAHs, benzene, toluene, ethylbenzene and xylene) from co-contaminated soils, it was clear that migration occurred on the straight line between electrodes and that the process is a lengthy one, from 23 to 112 days, using a current density of 3.72 A/m^2 [46].

Pioneering studies on electrokinetics were performed at the University of Illinois in Chicago by Professor K.R. Reddy whose results demonstrated the effectiveness of this method on soils polluted either with heavy metals [47, 48] or organic contaminants [49, 50]. In an obvious continuation, electrokinetics was applied to soils co-contaminated with both heavy metals and organic pollutants [51]. As required by the theory of electrokinetic remediation technologies, the soil was acidified, using 1 M citric acid, which acted also as a chelating agent, along with ethylenediaminetetraacetic acid, and some surfactants (Igepal or Tween). The best results of the sequential extraction test were obtained with citric acid 1 M combined with Igepal CA-720 (5%) or Tween (5%). However, the authors observed that the presence of surfactants tended to reduce the electric conductivity of the soil and the electro-osmotic flow compared with the stages where citric acid was used as flushing solution. The results confirmed previous studies, in which the ability of surfactants, cosolvents, cyclodextrins, chelating agents and organic acids to remove Ni and phenanthrene from kaolin soil was tested [52, 53]. EDTA was confirmed to produce complications during electrokinetic experiments [54].

Surfactants, cosolvents and cyclodextrins (same as in washing technique) generally yielded better results for phenanthrene removal, whereas chelating agents and organic acids yielded better removal for Ni [55].

To some extent, the EK process can be applied to sediments of harbour waterways, for the rapid elimination of heavy metals (Cd, Cr, Cu, Zn and Pb) and PAHs [56]. Beside citric acid, nitric acid (which is not recommended in case of *in situ* remediation) was also tested to avoid the formation of an alkaline front into the sediment and favour the metals removal. As surfactants, sodium dodecyl sulphate (as an anionic surfactant) and Tween 20 (as a non-ionic surfactant) were used to solubilize and mobilize PAHs. However, for achieving an almost complete removal of heavy metals and PAHs, the process needs ca 10 days. The EK process was extended to real-life polluted soil, deriving from a dismissed industrial site, contaminated with several metals: Hg, Ni, Co, Zn, Pb, Cu, Cr, As and organic substances [57]. Using a Ti/Pt-Ir anode and a stainless steel cathode, the procedure allowed a fair to good removal of most of heavy metals and PAHs over an interval of 10–15 days. An addition to the previous procedure was the presence of an oxidizing leaching agent, electrochemically produced. A similar approach, by using an oxidizing agent such as H_2O_2 (hydrogen peroxide was also present in a study by Reddy and co-workers [58]), NaClO, $KMnO_4$ or $Na_2S_2O_8$ in a controlled pH (3.5 or 10), was applied in a so-called enhanced-EK remediation technology to decontaminate a heavy metal-organic compound co-contaminated soil [59]. Over ca 14 days of applying 1.0 V/cm, the results showed that there was significant migration of pyrene (favoured by the presence of oxidizing reagent such as $KMnO_4$ or $Na_2S_2O_8$) and Cu from the soil (favoured by low pH), and that the removal percentage of soil pyrene and Cu varied in the range of 30–52 and 8–94%, respectively.

Shorter elution times were obtained using vertical electrokinetic cell (just 6 days, using smaller diameter or shorter-height cells and 0.01 M HNO_3 solution as cathode chamber flow) [60], but it is hard to expect that such procedure could be adapted for *in situ* applications. Nonetheless, removal efficiencies of phenanthrene, *p*-xylene, Cu and Pb were 67, 93, 62 and 35%, respectively.

An interesting twist to the technique was brought up by Ma and co-workers, who used bamboo charcoal as adsorbent, in a bench-scale experiment conducted to investigate the simultaneous removal of 2,4-dichlorophenol and Cd from a sandy loam (artificially spiked), at different periodic polarity reversals [61]. After ca 11 days of operating, about 75% of Cd and 55% of the phenol were removed at intervals of 24 h (about half for intervals of 12 h), at soil pH values ranging from 7.2 to 7.4.

The idea of combining EK process with adsorption was investigated also by Lukman from the King Fahd University of Petroleum and Minerals, who used a locally produced granular-activated carbon from date palmpits in the treatment zones. Natural saline-sodic soil, spiked with contaminant mixture (kerosene, phenol, Cr, Cd, Cu, Zn, Pb and Hg), was submitted to a 21-day period of continuous electrokinetics-adsorption experimental run, the efficiency for the removal of Zn, Pb, Cu, Cd, Cr, Hg, phenol and kerosene being 26.8, 55.8, 41.0, 34.4, 75.9, 92.49, 100.0 and 49.8%, respectively [62]. Lukman also demonstrated the importance of the processing fluids (anolytes and catholytes), which are rapidly degrading depending on the applied voltage gradient, ultimately leading to an eventual rise in the cost of operating the remediation process [63].

A significant improvement was brought to EK process by combining it with ultrasonication. This led to an enhancement of the remediation rate of soils co-contaminated with Pb and phenanthrene [64]. The migration of water and contaminants in the porous soil media is permitted through the actions of electro-osmotic flow and electromigration by electric power and acoustic flow by ultrasonic waves. The accumulated outflow and contaminant-removal rate were higher by the addition of vibration, cavitation and sonication effects. However, if efficiency seemed to improve (the removal rates of Pb and phenanthrene were average 88 and 85% for electrokinetic test and average 91 and 90% for electrokinetic and ultrasonic test, respectively), the duration was not reduced—it still needed ca 15 days of treatment, which means that by using both EK and ultrasonication the costs are higher.

A comprehensive review on the status of *in situ* applicable electrokinetic processes (Electro-Klean™ Electrical Separation, Electrokinetic Bioremediation, Electrochemical GeoOxidation (ECGO), Electrochemical Oxidative Remediation of Groundwater, Electrochemical Ion Exchange (EIX), Electrosorb™ and Lasagna™ process) was presented by E.M. Morales, from PGATech [65].

2.3. Bioremediation processes

Bioremediation generally uses living organisms (usually microbial metabolism), in the presence of optimum environmental conditions and sufficient nutrients, to break down soil organic and inorganic contaminants. Since it represents an attractive method due to the ease of *in situ* applications, bioremediation methods have been reviewed over the past few years, mostly depending on the nature of the polluting agent (heavy metals, PAHs, polychlorinated compounds, pesticides, etc.) or on the contaminated matrix (soil, sediments, groundwater, etc.) [66–70]. Since the method proved effective for both types of contaminants, attempts for simultaneous decontamination did not take long to appear.

A first approach was the so-called 'bioaugmentation': metal-contaminated soils were enriched with metal-detoxifying microorganisms while organic-contaminated soils were supplemented with organic-degrading microorganisms [71]. In such of the first examples, a co-contaminated soil with both Cd and 2,4-dichlorophenoxiacetic acid (2,4-D), the degradation of both contaminants was realized by introducing specific microorganisms for each contaminant (*Ralstonia eutropha* JMP134 for 2,4-D and *Arthrobacter*, *Bacillus* and *Pseudomonas* species for Cd) [72]. Contaminated soil with sulphide ore ashes and aromatic hydrocarbons from a historical industrial site underwent sequential leaching by 0.5 M citrate and microbial treatments [73]. The acidic-leached soil was bioaugmented with *Allescheriella, Stachybotrys, Phlebia, Pleurotus pulmonarius* and *Botryosphaeria rhodina,* which proved to be the most effective, leading to a significant depletion of the most abundant contaminants, including 7-H-benz[*d,e*]anthracene-7-one, 9,10-anthracene dione and dichloroaniline isomers. Simultaneously, the overall metal content was sensibly diminished under the action of *P. pulmonarius*.

The discovery of the dissimilatory metal reduction [74] under the action of heavy metals reducing bacteria provided the idea for the next step. It was soon afterwards that it became clear that the reduction can occur only in the presence of a hydrogen donor, usually water or an organic compound. Thus, Cr(VI)-reducing bacteria may utilize a variety of organic compounds as electron donors for Cr(VI) reduction, though the organic compounds are generally limited to natural aliphatics, mainly low-molecular-weight carbohydrates, amino acids and fatty acids [75]. Thus, why not have the organic pollutant as hydrogen donor, resulting thus in its oxidation? A good hydrogen donor, and also a well-known organic pollutant, is phenol. Indeed, soils co-polluted with a heavy metal such as Cr(VI) and phenol can be decontaminated in the same time by using strains of *P. aeruginosa* [76]. Another strain of *Pseudomonas, P. fluorescens,* allowed a drastic reduction of the concentration of heavy metals (Cu^{2+}, Cd^{2+}, Ni^{2+} and Pb^{2+}) but also of phenols and various pesticides (hexachlorobenzene, mancozeb or 2,4-D) in water [77]. Not only *Pseudomonas* strains were able to perform dual Cr(VI) reduction/phenol degradation, but also *Stenotrophomonas* species [78].

Similar examples of metal-reducing bacteria used in dual decontamination procedures are summarized in the following. *Geobacter metallireducens* was successfully used for the biodegradation of toluene and bioleaching of As, and in the same time for an accelerated degradation rate of toluene with reductive dissolution of Fe and co-dissolution of As [79]. In another paper, *Actinobacteria,* in a mixture with some *Streptomyces* spp. *Amycolatopsis tucumanensis,* were used to remediate soil co-contaminated with Cr(VI) and lindane [80]. It is interesting that the incubation period was of only 14 days.

At a certain level, phytoremediation can be considered as a bioremediation [81, 82]. According to the same rationale, if the phytoremediation process can be applied separately to heavy metals [83] and organic pollutants [84, 85], then it can be applied to co-contaminated soils with both species [86]. However, due to its limitations (phytoremediation is applicable only to root-deep soil horizons), there are very few examples of co-decontamination.

One such example is the use of willows for a dual decontamination Cd—oil [87] and another is the use of maize for soil polluted with Cd and pyrene [88].

It is interesting to mention the fact that in a recently published review on the applicability of phytoremediation of soils with mixed organic and heavy metal contaminants, Reddy (*vide supra*) and co-workers identified the most suitable species that proved to be effective separately for heavy metals and organic contaminants, but that were never investigated in simultaneous phytoremediation.

An alternate route was taken by an Italian team led by Baldi, from the Ca Foscari University of Venice, consisting of a first step of bioprecipitation followed by fungal degradation of organic pollutants from contaminated soils [89]. Thus, the contaminated soil was leached with 0.5 M citric acid leading to a good removal of metals and a low removal of organic contaminants (12%). The leachate was then incubated with a metal-resistant *Klebsiella oxytoca* strain, capable of using residual citrate to produce an iron gel that co-precipitated metals. In the same time, the leached solid waste was bioaugmented with a fungus strain of *Allescheriella* to complete the degradation of several organic contaminants, including trichlorobenzene, naphthalene, dichloroaniline and pentachloroaniline.

2.4. Miscellaneous processes

In this area, there are two possibilities: processes that combine aspects of the above procedures (washing, EK and bioremediation) or processes that are completely different, belonging to the immobilization/sorption techniques or to the purely chemical or thermal types of technologies.

Thus, either washing or electrokinetic process can be improved if realized with the augmentation of the naturally occurring microbial activity. In a recent report, PAHs and heavy metal-polluted soil from an abandoned coking plant was in a first step cleaned up by using a methyl-β-cyclodextrin solution to enhance ex situ extraction of PAHs and metals simultaneously, followed by the addition of PAH-degrading bacteria (*Paracoccus* sp. strain) and supplemental nutrients to treat the residual soil-bound PAHs [90]. Elevated temperature (50°C) in combination with ultrasonication was also needed. In the second case, the authors studied the benefits of integrating electrokinetic remediation with biodegradation for decontaminating soil co-contaminated with crude oil and Pb, in laboratory-scale experiments lasting for 30 days [91].

Immobilization techniques imply the adsorption of both heavy metals and organic contaminants on a solid support, usually biochar [92]. Cao et al. demonstrated that incubated biochar (prepared from dairy manure) for 210 days was effective for immobilization of both atrazine and Pb (its effectiveness was enhanced by increasing incubation time and quantity) [93]. After treatment, soils to which ca 5.0% biochar was added showed more than 57 and 66% reduction in Pb and atrazine concentrations, respectively.

On the chemical side of dual degradation processes, there are mentions of some techniques that use a photochemical activation. The problem encountered by photocatalytic processes is that they need the *a priori* formation of a solution. Thus, in an aqueous solution, it was possible to simultaneously reduce Cr(VI) and oxidize benzoic acid, in a suspension of N-F-co-doped TiO_2 [94]. Therefore, photocatalytic process must follow a washing step [95].

In a purely chemical process, Mitoma and co-workers used a nano-size mixture of metallic Ca and CaO that played a double role: in combination with a hydrogen donor (naturally occurring moisture) can hydrodechlorinate harmful dioxin compounds and during mixing can immobilize heavy metals in a cement-like matrix [96]. Thus, a soil contaminated with both heavy metals and dioxins (the most common type of polluted soil from reclaimed factories) can be safely treated.

The thermal type of co-decontamination process is illustrated by only one example, in which PCDD/Fs, pentachlorophenol and mercury are simultaneously removed [97]. This is understandable since other than Hg, all the other heavy metals are thermally quite stable (Hg has a boiling point of 356.73°C).

3. Conclusion

The multiplication and diversification of the methods for simultaneous decontamination of soils co-contaminated with both heavy metals and organic pollutants represent a certainty for a more rapid cleansing of polluted soils, associated with lower costs and more environmentally friendly procedures. Thus, physical/chemical, biological and thermal methods are combined in order to offer a wide variety of procedures that allow an effective removal or immobilization of various classes of pollutants. Therefore, after profiling the pollutants composition of a particularly heavy polluted soil, one has now a large choice of methods for treatment in order to simultaneously remove two or even several pollutants in a single batch. These methods are ranging from simple washing to electrokinetic or biological procedures; others include combinations of the previous or even thermal or purely chemical methods. Choosing one or another of these methods will depend on the type of pollutants and the ratio costs/effectiveness. Nevertheless, any of these methods can allow an effective treatment with lower cost and duration than those necessary in effective but separate treatments for each pollutant.

Author details

Yoshiharu Mitoma[1*], Alina M. Simion[2] and Cristian Simion[2]

*Address all correspondence to: mitomay@pu-hiroshima.ac.jp

1 Department of Environmental Sciences, Faculty of Life and Environmental Sciences, Prefectural University of Hiroshima, Shobara City, Hiroshima, Japan

2 Department of Organic Chemistry, Faculty of Applied Chemistry and Material Sciences, Politehnica University of Bucharest, Bucharest, Romania

References

[1] J.L. Schnoor in "Environmental modeling: Fate and transport of pollutants in water, air and soil", Wiley, New York, 1996, ISBN: 978-0-471-12436-8

[2] M. Kampa, E. Castanas, Human health effects of air pollution, Environmental Pollution, 2008, 151, 362–367.

[3] R.P. Schwarzenbach, T. Egli, T.B. Hofstetter, U. von Gunten, B. Wehrli, Global water pollution and human health, Annual Review of Environment and Resources, 2010, 35, 109–136.

[4] D. Briggs, Environmental pollution and the global burden of disease, British Medical Bulletin, 2003, 68, 1–24.

[5] Science Communication Unit, University of the West of England, Bristol (2013). Science for environment policy in-depth report. Soil contamination: impacts on human health. Report produced for the European Commission DG Environment, September 2013. Available at: http://ec.europa.eu/science-environment-policy.

[6] E.C. Brevik, T.J. Sauer, The past, present, and future of soils and human health studies, Soil, 2015, 1, 35–46.

[7] M. van der Perk, ch. 16, Patterns in the soil and in the vadose zone, in "Soil and Water Contamination", 2nd Edition, CRC Press, Boca Raton, London, New York, Leiden, 2013, ISBN 978-0-415-89343-5

[8] E.C. Brevik, L.C. Burgess in "Soils and human health", CRC Press, Boca Raton, London, New York, 2012, ISBN 978-1-439-84454-0

[9] K.E. Gerhardt, X.-D. Huang, B.R. Glick, B.M. Greenberg, Phytoremediation and rhizoremediation of organic soil contaminants: potential and challenges, Plant Science, 2009, 176, 20–30.

[10] R. Margesin, F. Schinner in "Manual for soil analysis - monitoring and assessing soil bioremediation", Series: Soil Biology, Vol. 5, Springer, Verlag Berlin, Heidelberg, 2005, ISBN 978-3-642-06467-8

[11] L.V. Pavel, M. Gavrilescu, Overview of *ex situ* decontamination techniques for soil cleanup, Environmental Engineering and Management Journal, 2008, 7(6), 815–834.

[12] E.A. Vik, P. Bardos, J. Brogan, D. Edwards, F. Gondi, T. Henrysson, B.K. Jensen, C. Jorge, C. Mariotti, P. Nathanail, P. Papassiopi, The criteria for selecting remediation technologies or treatments, Land Contamination & Reclamation, 2001, 9, 119–127.

[13] A. Lodolo, E. Gonzalez-Valencia, S. Miertus, Overview of remediation technologies for persistent toxic substances, Archives of Industrial Hygiene and Toxicology, 2001, 52(2), 253–280.

[14] B. Karn, T. Kuiken, M. Otto, Nanotechnology and in situ remediation: a review of the benefits and potential risks, Environmental Health Perspectives, 2009, 117(12), 1823–1831.

[15] K.R. Reddy, Technical challenges to in-situ remediation of polluted sites, Geotechnical and Geological Engineering, 2010, 28(3), 211–221.

[16] L. Wessels Perelo, Review: in situ and bioremediation of organic pollutants in aquatic sediments, Journal of Hazardous Materials, 2010, 177, 81–89.

[17] EPA 542/F-06/013, In Situ Treatment Technologies for Contaminated Soil, Solid Waste and Emergency Response, 2006, Available at: https://clu-in.org/download/remed/542f06013.pdf.

[18] A. Tsitonaki, B. Petri, M. Crimi, H. Mosbæk, R.L. Siegrist, P.L. Bjerg, In situ chemical oxidation of contaminated soil and groundwater using persulfate: a review, Critical Reviews in Environmental Science and Technology, 2010, 40, 55–91.

[19] J.L. Triplett Kingston, P.R. Dahlen, P.C. Johnson, State-of-the-practice review of in situ thermal technologies, Ground Water Monitoring & Remediation, 2010, 30(4), 64–72.

[20] B.N. Lopez, Y.B. Man, Y.G. Zhao, J.S. Zheng, A.O.W. Leung. J. Yao, M.H. Wong, Major pollutants in soils of abandoned agricultural land contaminated by e-waste activities in Hong Kong, Archives of Environmental Contamination and Toxicology, 2011, 61, 101–114.

[21] L.K. Wang, J.P. Chen, Y.-T. Hung, N.K. Shammas in "Heavy Metals in the Environment", CRC Press, Boca Raton, London, New York, 2009, ISBN: 9-781-420-07316-4

[22] G. Chilom, J.A Rice, Organic Pollutants in the Environment. eMagRes., 2013, 2(4), 587–596

[23] M.L. Brusseau, X. Wang, W.Z. Wang, Simultaneous elution of heavy metals and organic compounds from soil by cyclodextrin, Environmental Science & Technology, 1997, 31(4), 1087–1092.

[24] R.O. Dunn, J.F. Scamehorn, S.D. Christian, Simultaneous removal of dissolved organic and divalent metal cations from water using micellar-enhanced ultrafiltration, Colloids and Surfaces, 1989, A35, 49–56.

[25] C.C. Tung, Y.-M. Yang, C.-H. Chang, J.-R. Maa, Removal of copper ions and dissolved phenol from water using micellar enhanced ultrafiltration with mixed surfactants, Waste Management, 2002, 22, 695–701.

[26] K. Baek, J.-W. Yang, Simultaneous removal of chlorinated aromatic hydrocarbons, nitrate, and chromate using micellar-enhanced ultrafiltration, Chemosphere, 2004, 57, 1091–1097.

[27] G. Wang, Y, Zhou, X. Wang, X. Chai, L. Huang, N. Deng, Simultaneous removal of phenanthrene and lead from artificially contaminated soils with glycine-β-cyclodextrin, Journal of Hazardous Materials, 2010, 184(1–3), 690–695.

[28] R. Semer, K.R. Reddy, Evaluation of soil washing process to remove mixed contaminants from a sandy loam, Journal of Hazardous Materials, 1996, 45, 45–57.

[29] M. Shin, S.F. Barrington, W.D. Marshall, J.-W. Kim, Simultaneous soil Cd and PCB decontamination using a surfactant/ligand solution, Journal of Environmental Science and Health, Part A: Toxic/Hazardous Substances and Environmental Engineering, 2004, 39(11–12), 2783–2798.

[30] S. Ehsan, S.O. Prasher, W.D. Marshall, A washing procedure to mobilize mixed contaminants from soils: II. Heavy metals. Journal of Environmental Quality, 2006, 35, 2084–2091.

[31] S. Ehsan, S.O. Prasher, W.D. Marshall, Simultaneous mobilization of heavy metals and polychlorinated biphenyl (PCB) compounds from soil with cyclodextrin and EDTA in admixture, Chemosphere, 2007, 68(1), 150–158.

[32] P.K.A. Hong, C. Li, S.K. Banerji, T. Regmi, Extraction, recovery, and biostability of EDTA for remediation of heavy metal contaminated soil, Journal of Soil Contamination, 1999, 8, 81–103.

[33] Y. Wen, W.D. Marshall, Simultaneous mobilization of trace elements and polycyclic aromatic hydrocarbon (PAH) compounds from soil with a nonionic surfactant and [S,S]-EDDS in admixture: metals, Journal of Hazardous Materials, 2011, 197, 361–368.

[34] Y. Wen, S. Ehsan, W.D. Marshall, Simultaneous mobilization of macro- and trace elements (MTEs) and polycyclic aromatic hydrocarbon (PAH) compounds from soil with a nonionic surfactant and [*S,S*]-ethylenediaminedisuccinic acid (EDDS) in admixture: PAH compounds, Journal of Hazardous Materials, 2012, 199–200, 240–246.

[35] M. Rivero-Huguet, W.D. Marshall, Scaling up a treatment to simultaneously remove persistent organic pollutants and heavy metals from contaminated soils, Chemosphere, 2011, 83(5), 668–673.

[36] T.M.S. Lima, L.C. Procopio, F.D. Brandao, A.M.X. Carvalho, M.R. Totola, A.C. Borges, Simultaneous phenanthrene and cadmium removal from contaminated soil by a ligand/biosurfactant solution, Biodegradation, 2011, 22, 1007–1015.

[37] S. Song, L. Zhu, W. Zhou, Simultaneous removal of phenanthrene and cadmium from contaminated soils by saponin, a plant-derived biosurfactant, Environmental Pollution, 2008, 156(3), 1368–1370.

[38] M. Ye, M. Sun, J. Wan, G. Fang, H. Li, F. Hu, X. Jiang, F.O. Kengara, Enhanced soil washing process for the remediation of PBDEs/Pb/Cd-contaminated electronic waste site with carboxymethyl chitosan in a sunflower oil-water solvent system and microbial augmentation, Environmental Science and Pollution Research, 2015, 22, 2687–2698.

[39] N Reynier, J.F. Blais, G. Mercier, S. Besner, Decontamination of metals, pentachlorophenol, and polychlorined dibenzo-*p*-dioxins and dibenzofurans polluted soil in alkaline conditions using an amphoteric biosurfactant, Environmental Technology, 2014, 35(2), 177–186.

[40] S. Bisone, G. Mercier, J.F. Blais, Decontamination of metals and polycyclic aromatic hydrocarbonsfromslag-pollutedsoil,EnvironmentalTechnology,2013,34(18),2633–2648.

[41] J. Mouton, G. Mercier, J.F. Blais, Laboratory-scale flotation process for treatment of soils contaminated with both PAH and lead, Journal of Environmental Engineering, 2010, 136(10), 1063–1074.

[42] J. Mouton, G. Mercier, P. Drogui, J.F. Blais, Experimental assessment of an innovative process for simultaneous PAHs and Pb removal from polluted soils, Science of The Total Environment, 2009, 407(20), 5402–5410.

[43] N. Reynier, J.F. Blais, G. Mercier, S. Besner, Treatment of arsenic-, chromium-, copper- and pentachlorophenol-polluted soil using flotation, Water, Air, & Soil Pollution, 2013, 224, 1514–1526.

[44] Z. Ludmer, T. Golan, E. Ermolenko, N. Brauner, A. Ullmann, Simultaneous removal of heavy metals and organic pollutants from contaminated sediments and sludges by a novel technology, sediments remediation phase transition extraction, Environmental Engineering Science, 2009, 26(2), 419–430.

[45] J. Virkutyte, M. Sillanpaa, P. Latostenmaa, Electrokinetic soil remediation – critical overview, Science of the Total Environment, 2002, 289(1–3), 97–121.

[46] G. Maini, A.K. Sharman, C.J. Knowles, G. Sunderland, S.A. Jackman, Electrokinetic remediation of metals and organics from historically contaminated soil, Journal of Chemical Technology and Biotechnology, 2000, 75, 657–664.

[47] K.R. Reddy, S. Chinthamreddy, Sequentially enhanced electrokinetic remediation of heavy metals in low buffering clayey soils, Journal of Geotechnical and Geoenvironmental Engineering, 2003, 129(3), 263–277.

[48] K.R. Reddy, S. Chinthamreddy, Enhanced electrokinetic remediation of heavy metals in glacial till soils using different electrolyte solutions, Journal of Environmental Engineering, 2004, 130(4), 442–455.

[49] K.R. Reddy, R.E. Saichek, Effect of soil type on electrokinetic removal of phenanthrene using surfactants and cosolvents, Journal of Environmental Engineering, 2003, 129, 336–346.

[50] K.R. Reddy, R.E. Saichek, Enhanced electrokinetic removal of phenanthrene from clay soil by periodic electric potential application, Journal of Environmental Science and Health, Part A: Toxic/Hazardous Substances Environmental Engineering, 2004, 39(5), 1189–1212.

[51] K.R. Reddy, K. Maturi, C. Cameselle, Sequential electrokinetic remediation of mixed contaminants in low permeability soils, Journal of Environmental Engineering, 2009, 989–998.

[52] A.P. Khodadoust, K.R. Reddy, K. Maturi, Removal of nickel and phenanthrene from kaolin soil using different extractants, Journal of Environmental Engineering Science, 2004, 21(6), 691–704.

[53] A.P. Khodadoust, K.R. Reddy, K. Maturi, Effect of different extraction agents on metal and organic contaminant removal from a field soil, Journal of Hazardous Materials, 2005, 117(1), 15–24.

[54] K.R. Reddy, S. Danda, R.E. Saichek, Complicated factors of using ethylenediamine tetraacetic acid to enhance electrokinetic remediation of multiple heavy metals in clayey soils, Journal of Environmental Engineering, 2004, 130(11), 1357–1366.

[55] K. Maturi, K.R. Reddy, Simultaneous removal of organic compounds and heavy metals from soils by electrokinetic remediation with a modified cyclodextrin, Chemosphere, 2006, 63, 1022–1031.

[56] M.T. Ammani, A. Benamar, H. Wang, C. Bailleul, M. Legras, F. Le Derf, F. Portet-Koltalo, Simultaneous electrokinetic removal of polycyclic aromatic hydrocarbons and metals from a sediment using mixed enhancing agents, International Journal of Science and Technology, 2014, 11, 1801–1815.

[57] L. Szpyrkowicz, M. Radaelli, S. Bertini, S. Daniele, F. Casarin, Simultaneous removal of metals and organic compounds from a heavily polluted soil, Electrochimica Acta, 2007, 52(10), 3386–3392.

[58] K.R. Reddy, M.R. Karri, Effect of oxidant dosage on integrated electrochemical remediation of contaminant mixtures in soils, Journal of Environmental Science and Health A, 2008, 43, 881–893.

[59] L. Cang, G.-P. Fan, D.-M. Zhou, Q.-Y. Wang, Enhanced-electrokinetic remediation of copper-pyrene co-contaminated soil with different oxidants and pH control, Chemosphere, 2013, 90, 2326–2331.

[60] J.-Y. Wang, X.-J. Huang, J.C.M. Kao, O. Stabnikova, Simultaneous removal of organic contaminants and heavy metals from kaolin using an upward electrokinetic soil remediation process, Journal of Hazardous Materials, 2007, 144(1–2), 292–299.

[61] J.W. Ma, F.Y. Wang, Z.H. Huang, H. Wang, Simultaneous removal of 2,4-dichlorophenol and Cd from soils by electrokinetic remediation combined with activated bamboo charcoal, Journal of Hazardous Materials, 2010, 176(1–3), 715–720.

[62] S. Lukman, M.H. Essa, N.D. Mu'azu, A. Bukhari, Coupled electrokinetic-adsorption technique for simultaneous removal of heavy metals and organics from saline-sodic soil, The Scientific World Journal, 2013, Article ID 346910, doi.: 10.1155/2013/346910

[63] M.H. Essa, N.D. Mu'azu, S. Lukman, A. Bukhari, Integrated Electrokinetics-Adsorption Remediation of Saline-Sodic Soils: Effects of Voltage Gradient and Contaminant Concentration on Soil Electrical Conductivity, The Scientific World Journal, 2013, Article ID 618495, doi.: 10.1155/2013/618495

[64] H.I. Chung, M. Kamon, Ultrasonically enhanced electrokinetic remediation for removal of Pb and phenanthrene in contaminated soils, Engineering Geology, 2005, 77, 233–242.

[65] Available at: http://www.pgatech.com.ph/documents/A Review of the Electrokinetic Process for Soil Remediation.PDF.

[66] A. Kumar, B.S. Bisht, V.D. Joshi, T. Dhewa, Review on bioremediation of polluted environment: a management tool, International Journal of Environmental Sciences, 2011, 1(6), 1079–1093.

[67] C.S. Karigar, S.S. Rao, Role of Microbial Enzymes in the Bioremediation of Pollutants: A Review, Enzyme Research, 2011, Article ID 805187, doi:10.4061/2011/805187

[68] M. Chen, P. Xu, G. Zeng, C. Yang, D. Huang, J. Zhang, Bioremediation of soils contaminated with polycyclic aromatic hydrocarbons, petroleum, pesticides, chlorophenols and heavy metals by composting: applications, microbes and future research needs, Biotechnology Advances, 2015, 33(6, Pt 1), 745–755.

[69] G. Wu, H. Kang, X. Zhang, H. Shao, L. Chu, C. Ruan, A critical review on the bioremoval of hazardous heavy metals from contaminated soils: issues, progress, eco-environmental concerns and opportunities, Journal of Hazardous Materials, 2010, 174(1–3), 1–8.

[70] J.H. Park, D. Lamb, P. Paneersevam, G. Choppala, N. Bolan, J.W. Chung, Role of organic amendments on enhanced bioremediation of heavy metal(loid) contaminated soils, Journal of Hazardous Materials, 2011, 185(2–3), 549–574.

[71] S. El Fantroussi, S.N. Agathos, Is bioaugmentation a feasible strategy for pollutant removal and site remediation?, Current Opinion in Microbiology, 2005, 8, 268–275.

[72] T.M. Roane, K.L. Josephson, I.L. Pepper, Dual-bioaugmentation strategy to enhance remediation of cocontaminated soil, Applied and Environmental Microbiology, 2001, 67(7), 3208–3215.

[73] A. D'Annibale, V. Leonardi, E. Federici, F. Baldi, F. Zecchini, M. Petruccioli, Leaching and microbial treatment of a soil contaminated by sulphide ore ashes and aromatic hydrocarbons, Applied Microbiology and Biotechnology, 2007, 74(5), 1135–1144.

[74] D.R. Loveley, Dissimilatory metal reduction, Annual Review in Microbiology, 1993, 47, 263–290.

[75] Y.T. Wang, C.S. Xiao, Factors affecting hexavalent chromium reduction in pure cultures of bacteria, Water Research, 1995, 29, 2467–2474.

[76] H. Song, Y. Liu, W. Xu, G. Zeng, N. Aibibu, L. Xu, B. Chen, Simultaneous Cr(VI) reduction and phenol degradation in pure cultures of *Pseudomonas aeruginosa* CCTCC AB91905, Bioresource Technology, 2009, 100, 5079–5084.

[77] S. Wasi, S. Tabrez, M. Ahmad, Suitability of immobilized *Pseudomonas fluorescens* SM1 strain for remediation of phenols, heavy metals, and pesticides from water, Water, Air & Soil Pollution, 2011, 220(1), 89–99.

[78] D. Gunasundari, K. Muthukumar, Simultaneous Cr(VI) reduction and phenol degradation using *Stenotrophomonas* sp. isolated from tannery effluent contaminated soil, Environmental Science and Pollution Research, 2013, 20(9), 6563–6573.

[79] K.-Y. Lee. J. Bosch, R.U. Meckenstock, Use of metal-reducing bacteria for bioremediation of soil contaminated with mixed organic and inorganic pollutants, Environmental Geochemistry and Health, 2012, 34(1), 135–142.

[80] M.A. Polti, J.D. Aparicio, C.S. Benimeli, M.J. Amoroso, Simultaneous bioremediation of Cr(VI) and lindane in soil by actinobacteria, International Biodeterioration & Biodegradation, 2014, 88, 48–55.

[81] A. Yadav, N.G. Batra, A. Sharma, Phytoremediation and phytotechnologies, International Journal of Pure and Applied Bioscience, 2016, 4(2), 327–331.

[82] L. Buhari Muhammad, R. Sulaiman Babura, N.L. Vyas, B. Sulaiman, Y. Harisu Umar, Role of biotechnology in phytoremediation, Journal of Bioremediation and Biodegradation, 2016, 7:330. doi: 10.4172/2155-6199.1000330.

[83] Y. Ma, M. Rajkumar, C. Zhang, H. Freitas, Beneficial role of bacterial endophytes in heavy metal phytoremediation, Journal of Environmental Management, 2016, 174(1), 14–25.

[84] C. Liao, W. Xu, G. Lu, F. Deng, X. Liang, C. Guo, Z. Dang, Biosurfactant-enhanced phytoremediation of soils contaminated by crude oil using maize (*Zea mays*. L), Ecological Engineering, 2016, 92, 10–17.

[85] J.W. Kang, Removing environmental organic pollutants with bioremediation and phytoremediation, Biotechnology Letters, 2014, 36(6), 1129–1139.

[86] R.A. Chirakkara, C. Cameselle, K.R. Reddy, Assessing the applicability of phytoremediation of soils with mixed organic and heavy metal contaminants, Reviews in Environmental Science and Bio/Technology, 2016, 15(2), 299–326.

[87] Y.A. Kuzovkina, M.F. Quigley, Willows beyond wetlands: uses of *Salix* L. species for environmental projects. Water, Air Soil Pollution, 2005, 162, 183–204.

[88] H. Zhang, Z. Dang, L.C. Zheng, X.Y. Yi, Remediation of soil co-contaminated with pyrene and cadmium by growing maize (*Zea mays L.*), International Journal of Science and Technology, 2009, 6(2), 249–258.

[89] F. Baldi, V. Leonardi, A. D'Annibale, A. Piccolo, F. Zechini, M. Petruccioli, Integrated approach of metal removal and bioprecipitation followed by fungal degradation of organic pollutants from contaminated soils, European Journal of Soil Biology, 2007, 43, 380–387.

[90] M. Sun, Y. Luo, Y. Teng, Z. Jia, Z. Li, S. Deng, Remediation of polycyclic aromatic hydrocarbon and metal-contaminated soil by successive methyl-β-cyclodextrin-enhanced soil washing-microbial augmentation: A laboratory evaluation, Environmental Science and Pollution Research, 2013, 20, 976–986.

[91] Z.-Y. Dong, W.-H. Huang, D.-F. Xing, H.-F. Zhang, Remediation of soil co-contaminated with petroleum and heavy metals by the integration of electrokinetics and biostimulation, Journal of Hazardous Materials, 2013, 260, 399–408.

[92] L.L. Kong, W.T. Liu, Q.X. Zhou, Biochar: an effective amendment for remediating contaminated soil, Reviews of Environmental Contamination and Toxicology, 2013, 228, 83–89.

[93] X. Cao, L. Ma, Y. Liang, B. Gao, W. Harris, Simultaneous immobilization of lead and atrazine in contaminated soils using dairy-manure biochar, Environmental Science & Technology, 2011, 45, 4884–4889.

[94] M. Antonopoulou, A. Giannakas, I. Konstantinou, Simultaneous photocatalytic reduction of Cr(VI) and oxidation of benzoic acid in aqueous N-F-codoped TiO^2 suspensions: optimization and modeling using the response surface methodology, International Journal of Photoenergy, 2012, Article ID 520123, doi:10.1155/2012/520123

[95] S. Satyro, M. Race, R. Marotta, M. Dezotti, D. Spasino, G. Mancini, M. Fabbricino, Simulated solar photocatalytic processes for the simultaneous removal of EDDS, Cu(II), Fe(III) and Zn(II) in synthetic and real contaminated soil washing solutions, Journal of Environmental Chemical Engineering, 2014, 2(4), 1969–1979.

[96] S.R. Mallampati, Y. Mitoma, T. Okuda, S. Sakita, C. Simion, Simultaneous decontamination of cross-polluted soils with heavy metals and PCBs using a nano-metallic Ca/CaO dispersion mixture, Environmental Science and Pollution Research, 2014, 21(15), 9270–9277.

[97] P.-C. Hung, S.-H. Chang, C.-C. Ou-Yang, M.-B. Chang, Simultaneous removal of PCDD/Fs, pentachlorophenol and mercury from contaminated soil, Chemosphere, 2016, 144, 50–58.

9

Soil Salinization and Mitigation Measures in Land Reclamation Regions

Shiguo Xu, Yi Xu, Yanzhao Fu and Qi Wang

Additional information is available at the end of the chapter

Abstract

Soil salinization and underground structure erosion usually occur in land reclamation regions, especially under semi-humid climate that annual evaporation is larger than annual rainfall in Northern China. Based on investigations into the status and trends of land reclamation soil along the Bohai Rim, China, this chapter summarizes the evolution of groundwater system and soil environment and analyzes the main reasons contributing to these problems. Physical and mathematical models are established to simulate the mechanism of water-salt migration in land reclamation regions. Results show that evapotranspiration and groundwater discharge during wet seasons are the main driving forces of status of soil salinization. It was pointed out that the key to soil salinity control in the reclamation region was by utilizing rainwater and flood resources to build a long-term leaching mechanism. Meanwhile, in order to rebuild and maintain a healthy and stable ecosystem in the reclaimed areas, it is necessary to design the structure of soil layers in advance, enhance the salt leaching process and plant vegetation according to the local conditions.

Keywords: land reclamation, soil salinization, mechanism, environmental change, mitigation measures

1. Introduction

Coastal regions are identified as productive and sensitive ecosystems with abundant biodiversity. They are water bodies connected with both the land and the sea, and within which seawater mixes with inland freshwater discharge. Most of the megacities in the world are located in coastal regions, and more than 3 billion people which cover almost half of the world's population live along the coastline. The overloaded population increases the pressure on land resources. Along with the population growth, the utilization and development of coastal regions have increased in recent years and the changes in socioeconomic and environmental conditions are continuing. To cope with the expansion of urbanization, land reclamation was carried out.

Overloaded population and the needs for more agricultural land and for flood protection are the main reasons for the reclaiming. Land reclamation has expanded rapidly to adjust to economic development in coastal regions. It brings about more space, which alleviates the contradiction between supply and demand of land resources. Many coastal countries, including the developed ones such as the USA, Japan and Netherlands, have long histories of coastal reclamation. Netherlands, as an example, reclaimed about 7000 km^2 from the sea and inland lakes since the 1300s, which covers up to 21% of the total land surface of the country [1]. Currently, these new formed lands have exceeded 140,000 km^2 in the worldwide scale, and still increasing rapidly in some countries such as China [2].

However, reclamation disturbs the hydro-environment near the coast. It disrupts the water-salt movement and causes engineering, environmental and ecological problems. The quality of groundwater was affected by saline intrusion in the Netherlands [1]. Mangrove forest in China has been reported to be reduced by 53% than that in 1950s [3]. These degradations of marine habitats indicate that coastal ecosystem and hydrodynamic conditions are disturbed. The high density of salt in reclaimed regions exerts pressure on the local plants. If the salt pressure is weak, the injury to the plant could be recovered. The salinity in reclamation soil is 1–4% in 1 m^3 which is much larger than the largest salinity that the most plants could bare (0.3%). The mineralization ability in groundwater is more than 50 g/L. Only plant with shallow roots and high salt tolerance could survive in reclamation areas. Once the salt pressure exceeds the salt resistance of the plant, the life cycle of it will be destroyed and hence disturbs the whole ecosystem. Apart from this, social underground infrastructures are other victims of soil salinization. Seawater accelerates the corrosion rates of reinforced concretes and underground pipe networks which would threaten the security of coastal structures. Therefore, the understanding and mitigation of soil salinization in reclamation regions are important for coastal environmental protection.

Soil salinization is a tough problem for coastal environment and has drawn attention on a worldwide scale. Efforts have been made to study the mechanism and mitigation measures of it. Armstrong et al. [4] studied the seasonal variation in water and salt distribution in fields with both grassland and arable saline-sodic clay soils under temperate rain-fed conditions. Chen and Jiao [5] analyzed the groundwater chemistry in coastal aquifer and found that groundwater pumping was the reason for seawater intrusion. Iost et al. [6] found that reclamation influenced the local pH and carbonate content by decreasing calcium, magnesium and potassium while studying the initial pedogenesis of reclaimed saline marsh soils.

The objective of this chapter is to explore the mechanism of soil salinization in reclaimed coastal regions, especially that under semi-humid climate where evaporation is more than precipitation. In this chapter, the Bohai Rim, China, is selected as an example to study the water and salt migration in reclaimed soil. Physical model and numerical model are built for understanding, quantifying and modelling the dynamic of seawater intrusion. Promising mitigation measures are also proposed.

2. Investigation and analysis

Surveying and analyzing are basic to study the physical and chemical properties of soil in reclamation regions. This chapter is based on the field surveys that were carried out along the coast line of the Bohai Sea, China. The surveys were taken on typical sea reclamation projects with similar climatic conditions.

2.1. Study area

The study area is located in the Bohai Rim, China (**Figure 1**), which has a semi-humid temperate monsoon climate with the average annual temperatures of 8.3–12.4°C. The annual sunshine duration is 2500–2900 h, and the annual total radiation is 5000–5800 MJ/m^2. The annual precipitation is 612–640 mm, which is concentrated in summer (60–75%), and the annual evapotranspiration is 1300–1900 mm. The ratio of precipitation and evaporation maintains 0.33, which makes the water balance in the soils negative. The land reclamation projects were started during the 1980s. From 1996 to 2007, 551 km^2 new lands were reclaimed and the annual reclamation area covers 80% of the country.

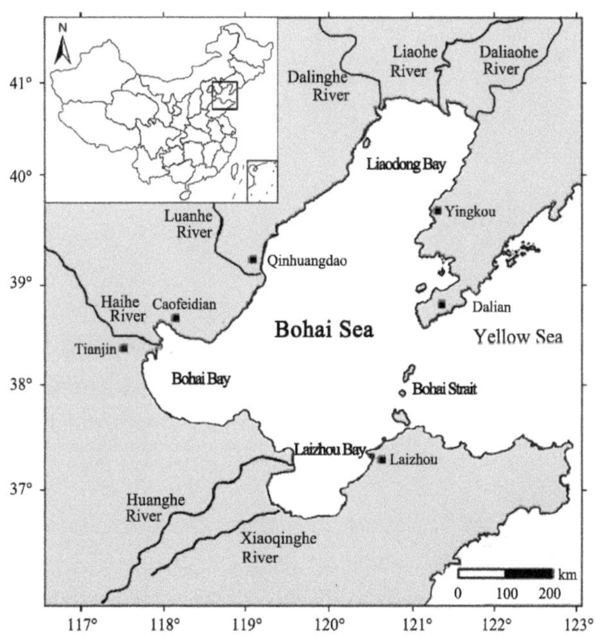

Figure 1. Location of the Bohai Rim.

The samplings of the research were carried out along the Bohai Rim. Samples of soil were collected from five main coastal cities where the reclamation projects were completed and the level of salt was stable. The reclaimed soil samples were collected using soil auger to a depth of 20 cm, and control samples were collected in the natural coast nearby. Samples were then dried at 40 °C and sieved with 1 mm plastic sieve based on the standard for classification of soils (GBJ145-90). The physical and chemical properties of the soil were then analyzed in the laboratory.

2.2. Salinization in land reclamation area

Land formation in coastal area includes natural sedimentation and artificial landfills. Every year, the Yellow River brings about 1.5 billion tons of sediment to the estuary, two-thirds of which deposit in the delta. This area has the fastest land formation speed, where the shoreline is extended with an average distance of 1.8 km and formed 21.3 km² tidelands each year. Artificial reclamation is often carried out on natural beaches using sea sand, mountain soils, minerals and construction waste, depending on the geological condition of the coast. Regions with hills and low mountains would use riprap filling method. While in river deltas, the method used is dredger filling (hydraulic filling). Although the soil structures are similar, there are differences in physical and chemical characteristics (**Table 1**).

City	Soil type	Water content (%)	Volumetric weight (kN/m³)	Specific gravity	Porosity (%)	Water-holding capacity (%)	Permeability coefficient (%)
Tianjin	Reclamation	20.2	18.8	2.38	32.8	51.1	8.33×10^{-7}
	Planting	3.9	14.2	2.22	41.1		
Caofeidian	Reclamation	3.4	13.9	2.61	45.8	22.9	2.75×10^{-5}
	Reclamation	22.9	18.2	2.63	36	39.7	5.83×10^{-6}
Laizhou	Planting	4.5	10.8	2.45	51.1		
	Reclamation	2.9	14.6	2.35	38.1	14.3	1.11×10^{-5}
Yingkou	Reclamation	10.9	13.2	2.49	51.3	29.6	3.23×10^{-5}
	Hill	26.7	16.7	2.41	44		
Dalian	Reclamation	27.2	16.6	2.39	43.8	23.6	
	Reclamation	26.1	16.9	2.43	44.2	26.8	7.19×10^{-4}
	Reclamation	19.2	17.3	2.45	35.4	21.1	

Table 1. Location of the Bohai Rim.

Our research shows that Tianjin, Caofeidian and Laizhou are dredger filling reclamation regions, and Dalian and Yingkou are riprap filling land reclamation regions. The physical properties are analyzed and are listed in **Table 1**. The grain size of dredger filling soil is smaller than that of riprap filling soil. Seventy percent of the riprap filling soil is sand, which is also more than that of the soil from the hills where the riprap filling soil comes from. The density of riprap filling soil is smaller, while the porosity and the permeability coefficients are larger, which indicate the variation in salt water migration characteristics. The soil texture of dredger filling is heavier, and the original salinity is higher, which may cause salinization. Riprap fill, on the contrary, with greater thickness and larger bottom, has better connectivity to prevent salinization. The differences between riprap land reclamation and dredger fill reclamation are listed in **Table 2**. Large areas of these projects were built directly on former salt pans, which are the extreme examples of deposited salt density in sediment. This fact aggravates the surface salinization in the backfill area. The results show that the salt contents of the reclamation soil are consistent with the surface soil of salt pans nearby, which indicate the process of salt releasing from the sediment. This consistency tends to be clearer over time. The migration of water and salt in the backfill soil is controlled by the grain composition, which reflects the aquifer permeability and adsorbing capability of the soil particle. The salt in backfilling soil is accumulated in the surface layer. The groundwater was shallow buried (1.5–2.5 m) in the sampling sites which were all within the limit depth of phreatic evaporation. Therefore, the phreatic evaporation may be the main driving force of the salt accumulation in surface layer in reclamation regions. The main types of salt are NaCl and $CaCl_2$ for dredger filling soil and $CaSO_4$ for riprap filling soil.

	Dredger filling	**Riprap filling**
Coast type	Muddy coast	Rocky coast
Geology condition	River deltas	Hills and low mountains
Chemical character of groundwater	Chloride	Bicarbonate chloride, bicarbonate chloride sulphate
Textural characteristics	Silt and clay	Brown soil
Reclamation scale	$>10\ km^2$	$<10\ km^2$
Cities	Tianjin, Caofeidian, Laizhou	Dalian, Yingkou

Table 2. Differences between dredger filling and riprap filling land reclamation.

2.3. The movement of water and salt

Rainfall, evaporation and runoff carry dissolved minerals and salts in continuous movement and form a uniform material flow. Different characteristics of salt migration occur in the different conditions of climate, soil and irrigation management, etc. In general, this process could be classified as salt leaching, salt accumulation and the release process of salts from sediment in the land reclamation regions.

2.3.1. Salt leaching process

Affected by the soil texture and structure, the impact depth of rainwater is limited, and the leaching effect in surface soil is stronger than that of the deep soil. The salinity peak declines with infiltration process until it disappears. The distribution of soil moisture in the 0–80 cm depth consists of a logarithmic curve in the absence of crops, and little effect was shown below 1.0 m. A critical mutation in salinity variety exists around 40 cm depth, and a transition region appears from 50 to 70 cm. Drizzle is not stronger enough to wash salt away, but it carries salt to the surface after rain stops. In agricultural area, large-scale irrigation contaminates freshwater, and washes minerals and nitrate away, which may cause low soil permeability and nutrient content.

2.3.2. *Salt accumulation*

There has been a history of salt and water movement under evapotranspiration conditions. Fritton et al. [7] examined the differences of water-salt distributions under various evaporation intensities, and much research was subsequently focused on salt and water transport in soil [8]. A variety of formulas are widely used to calculate evaporation. However, studies on salt transport fell behind, relatively. Groundwater has significant effects on salt accumulation, although the drainage system is blocking it with flattening landform. The higher the salinity of groundwater, the more serious the salt accumulation. In the capillary rise zone, evaporation maintains a constant generally, but it decreases rapidly at the bottom of this zone and tends to zero, while the groundwater exceeds the impact depth. Throughout this process, the cumulative evaporation has a power function in relation to the diving depth [9]. It is easier to prevent salt upward and improve the leaching efficiency with a higher hydraulic gradient when the water table is deep. However, plants cannot grow up when the root zone lacks water. Zhang and Zhang [10] considered that the groundwater should be controlled at 0.7–0.8 m during the growth period, and fell to 1.2–1.4 m without crops.

In the high salinity environment, plants play an important role in salt regulation [11]. Plants may exacerbate salinization in sea reclamation areas during evapotranspiration, and the impact depth should be the sum of root depth and capillary rise height. The movements of salt and water become stable only when the groundwater is below the limited depth of phreatic evaporation.

2.3.3. *Salt releasing from sediment*

Sediment deposited on the seabed for years has high salinity of more than 10%. When the environment changes to reclamation, salt in sediment will be gradually released. Environmental factors have obvious effects on the release of salt. For example, wind and temperature can promote this process, while the initial mineralization of groundwater will hinder it. In an acidic environment, the chemical properties maintain relative balances. Once these ions are in alkaline conditions, Mg^{2+} and Ca^{2+} would flocculate to deposition, which also accelerate the dispersion of magnesium and calcium.

Moreover, the environment of land reclamation areas is more complex than that of the natural coast. Groundwater replaces part of the saline water with the movement of fresh-salt water interface. Due to the restriction of upper soil, groundwater exchange is slow, and the accompanying removal of salt is too low. Cl^- is a conservative ion excluded by soil colloids, which can migrate within groundwater freely, and of which the concentration is determined by the salinity of groundwater. In an open system, Na^+ is released from the sediment in exchanges for a continuous supply of H^+. Meanwhile, CO_2 released by plant roots generates excess HCO_3^- and CO_3^{2-}, which may cause hydrolysis and acid erosion on rock and generate more dissolved salts. But in a closed system as reclamation region, the absence of CO_2 decreases HCO_3^- and CO_3^{2-} concentration. The bicarbonate-type groundwater transforms to chloride-type groundwater. Thus, the salinization will be intensified in land reclamation regions.

3. The driving forces of salinization

3.1. Conceptual model

The formation mechanisms of salinization are complicated for the multiple factors of water and salt movements. The sediment under the reclamation soil releases a large amount of salt, which was originally deposited in the oceanic environment over time. In a humid environment, with the upstream groundwater and rainfall supplement, the alkaline water in the reclamation soil is replaced by low-acid groundwater with higher dissolved oxygen. The fresh-salt water interface is pushed forward to the sea. However, under semi-humid climate where the evaporation rate is larger than the precipitation rate, with the effort of evapotranspiration, the salinity in reclamation soil increases, and thus, the fresh-salt water interface moves backwards to the continent (**Figure 2**).

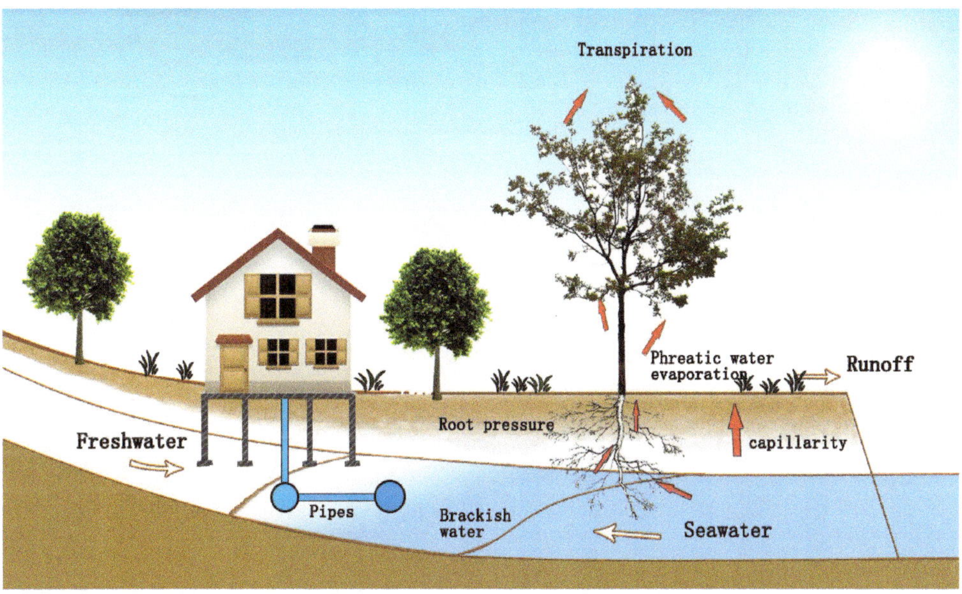

Figure 2. Concept map of water environment evolution in land reclamation regions.

3.2. Physical model

Semi-humid climate is an important incentive for coastal salinization. Taking Bohai Rim as an example, in recent years, global warming aggravates the evaporation process and salinization in sea reclamation region. Guan et al. [12] monitored the involvement of salinized coastal area around the Yellow River Estuary, and found that the saline land had expanded by 487.4 km^2 over the last 15 years. To investigate the saline formation mechanism in compacted soil, we studied the salt transport processes under the joint action of phreatic water evaporation and lateral interflow.

3.2.1. Experimental setup

The experiment was carried out outdoor using the apparatus shown in **Figure 3**. Soil collected from the reclamation site in Dalian, China, was filled into a glass tank. The tank was 0.6 m high and was divided into three sections with two perforated plates which were the simulation of coastal constructions. The diameter of the holes was 1.0 cm, and the interval between each hole was 5 cm. In total, 25 monitor holes were reserved on the side of the tank as illustrated in **Figure 3**. Seawater collected from the reclamation site was poured initially into the tank at a depth of 10 cm. The reclamation soil was then filled into the middle of the tank in layers. Each layer of the filling soil was 5 cm. The soil tank was placed outside for 9 months with light-tight cover to prevent water loss. Then, fresh groundwater was poured into the Marriotte's bottle until the water level in the left section reached to 50 cm. The purpose of this stage was to simulate the process of sideward flow in the soil. The pH value, electric conductivity and volumetric water content were measured during the experiment. Soil samples were also collected from the monitor hole at the end of each stage.

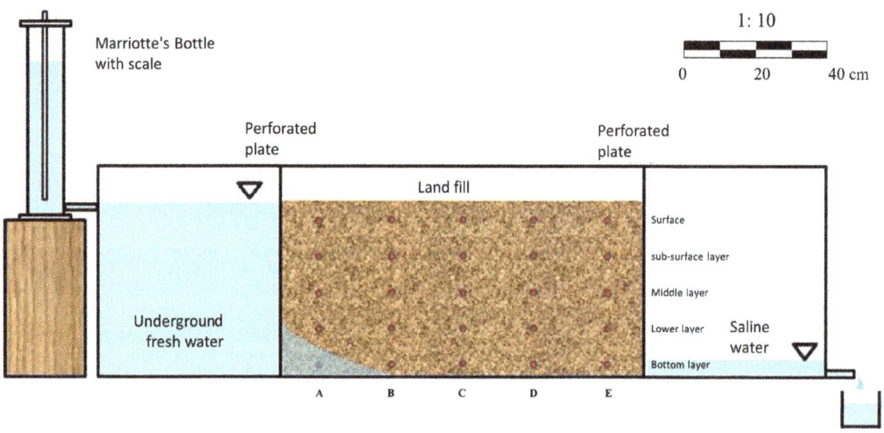

Figure 3. Testing apparatus.

3.2.2. Phreatic evaporation stage

The volumetric water content was dynamic during the experimental process (**Figure 4**). It first increased during the first month and reached a peak. Then, it decreased gradually until getting back to its initial state. In spatial scale, the water content was larger in bottom layer and smaller in surface layer. It was relatively stable in surface and subsurface layer horizontally. While under the middle layer, it was larger along the right side. These facts indicate that capillarity is the main driving force of the increase in water content. The soil matrix potential gradually decreased with the increase in water content. When the water content reached 3%, the soil matrix potential meets the bottom and was stable ever since. The actual evaporation was little at first when the water content in the surface was low. With the increase in water content, the actual evaporation raised, which leads to the total loss of water content. The higher content along the right side was supplied by the seawater.

The total dissolved salt (TDS) in the soil increased during the evaporation process. It was larger in surface layer than that at the bottom and larger along the right side than the left. This indicates the salt accumulation with plenty supplement. The initial salt type of surface soil was $CaSO_4$, and it gradually transited to $CaCl_2$ and NaCl in the salt accumulation process. The salt type in the bottom layer was NaCl; Mg^{2+} presented tendency to dissolve in seawater and stayed in the right side of the tank. The pH value decreased on the left of the tank while increased on the right.

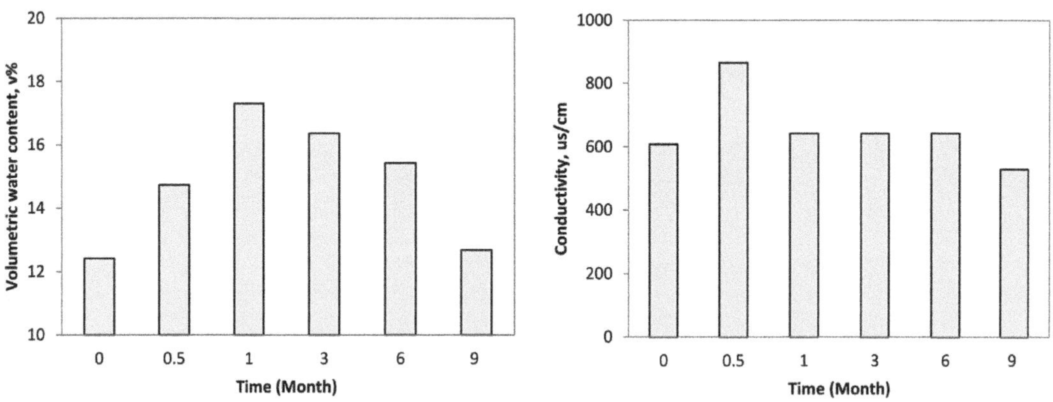

Figure 4. Variation in soil moisture and conductivity during the evaporation period.

3.2.3. Seepage stage

Under the seepage, the groundwater supply rate decreased with the increase in soil water content until 22 h later when there was water leaking out from the end of the tank (**Figure 5(a)**). The water content in the soil was stable since then. This dynamic was driven by the soil matrix potential. The potential-driven change rate (**Figure 5(d)**) decreased with the rise in the water table until the active water absorption stopped and a relatively free horizontal flow started, which leads to a stable velocity. The TDS of the soil was first increased because the crystal structure of salt in the soil was dissolved in the infilled groundwater (**Figure 5(c)**). The high TDS in the soil was carried out with the horizontal flow, and the TDS in the leaking water was high at first (**Figure 5(b)**). The time that TDS became stable was later than the flow rate indicates that the change in the salt front is later than in the wetting front.

The stable wetting front moves to the offshore, while the water table rises. Surface area above the free water is in the salt accumulation state, and the salt is mainly $CaCl_2$ and $MgSO_4$. The area below the free water surface is in the state of desalination, and the salt type is mainly

$CaCl_2$ and NaCl. Water and salt movement will change the pH value of the soil environment; the pH value will increase with the increase in salt content first decreased slightly after the rise.

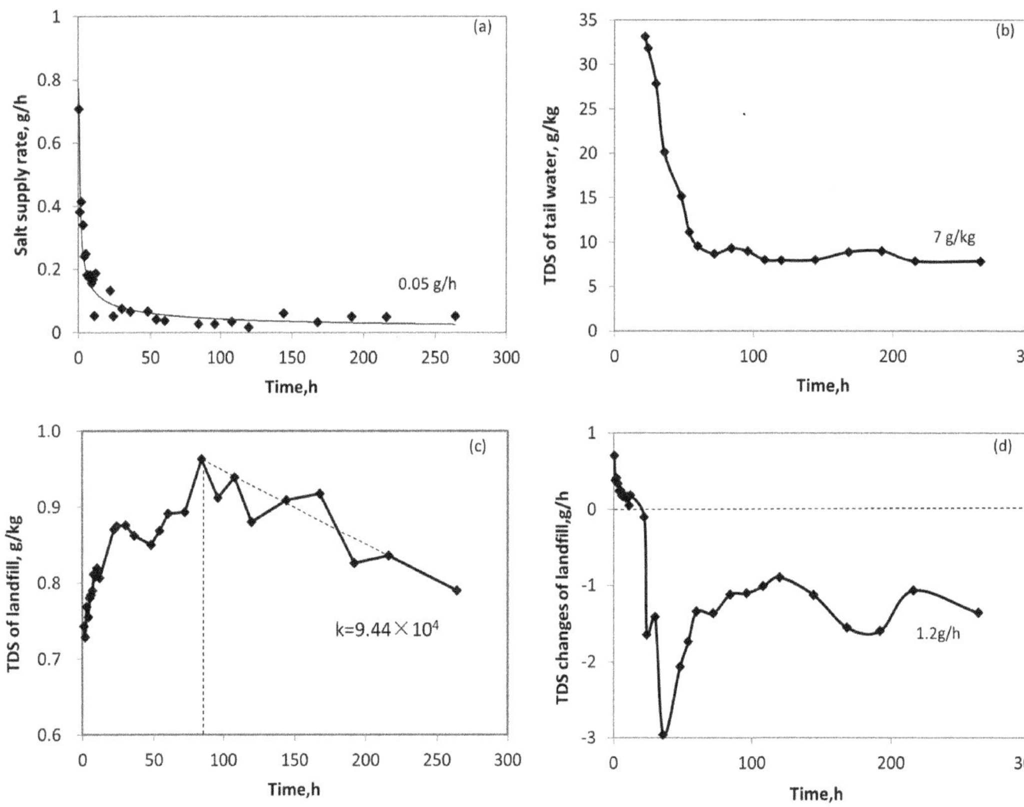

Figure 5. Variation in soil moisture, water supply and permeability.

3.2.4. *Salt migration in land reclamation regions*

In semi-humid coastal region, the water content in surface layer and water table of groundwater are the main factors of salt accumulation due to low precipitation rate but high evaporation. During the dry seasons, low groundwater supply rate from the continent pushes the freshwater and seawater interface upwards to the continent. When the water table meets the phreatic evaporation depth, Cl^- moves upwards and the $CaSO_4$ in surface layer is replaced by $CaCl_2$. With the consistent supply of Na^+ and Mg^{2+}, most of the salt in the soil is replaced by NaCl. In wet seasons, the groundwater water table rises and the wetting front moves downwards to the sea with salt front following. In reclaimed regions where there are obstacles (clay or coastal constructions), the water table would be raised. Salt in groundwater would also be raised with the water table which would then accumulate in the surface layer. $CaCl_2$ and $MgSO_4$ accumulated in the surface layer near the obstacles and NaCl accumulated at the bottom. Therefore, in terrestrial groundwater recharge conditions, land reclamation area of salinization prevention and control work should focus on underground baffle layer or foundation of a building with a relatively dense region.

3.3. Numerical model

Apart from the physical model, FEFLOW was employed to simulate the migration of water and salt within a unit of a typical land reclamation project. The combined effects of two driving factors, phreatic evaporation and rainfall infiltration, were selected to reveal the migration process in the simulation. Then, the effects of salinity suppression of different measures for rainwater utilization were analyzed.

3.3.1. Model domain description

Due to lack of geological survey data, long sequence groundwater level and solute concentration monitoring data of practical projects, the model domain was an imaginary land reclamation project which was based on Lingshui Bay land reclamation project in Dalian city, China, and several engineering examples in North China. The theoretical model (**Figure 6**) was designed to explore the mechanism of soil water and salt transport in reclamation areas in the north region of China and provide trend analysis results for practical projects.

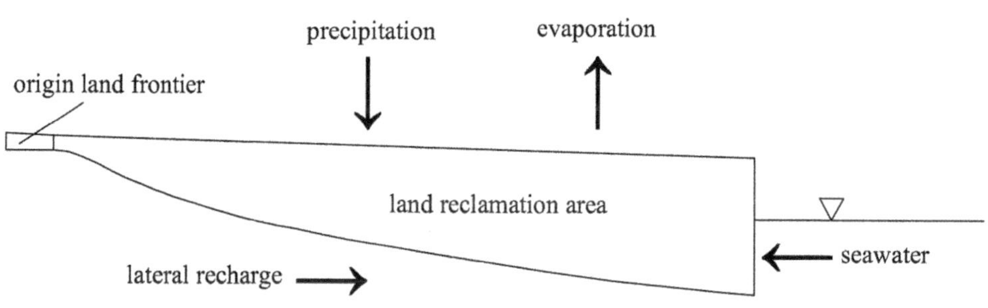

Figure 6. The schematic diagram of model domain.

The left border in **Figure 6** is the land frontier before the project implementation, and the right border is the new land frontier. Considering that the extended distance from landside to the sea was less than 1 km in general cases, the extended distance was set to be 1 km. The surface grade was 3‰, and the lateral cofferdam was 5.5 m in accordance with land reclamation engineering specifications. The structure of earth fill is designed according to Lingshui Bay land reclamation project.

3.3.2. About Feflow

FEFLOW is a finite element-based groundwater simulation system. It is considered to be a comprehensive, well-tested and reliable program for the simulation of flow, mass and heat transport processed in porous media. FEFLOW provides data interfaces for geographic information system and can generate spatial finite element grids automatically. The system is equipped with fast and accurate numeric algorithms to control and optimize the solution procedure, and advanced visual figures are embodied in output results. FEFLOW is used to compute groundwater flow dynamics in unconfined and confined aquifers and multiple free water surface(s); describe the spatial and temporal distribution of contaminants and/or temperature fields; estimate the duration and travel time of contaminants in groundwater; study saltwater intrusion and so on.

3.3.3. Mass transport model building

A two-dimensional coupled groundwater flow and mass transport model in vertical section was established (**Figure 7**). The left border was generalized as the boundary of known flow, and the groundwater flow was determined by the measured value in Lingshui Bay land reclamation project. The right border was defined as the boundary of known water level. The model contained five layers in vertical direction according to the soil layer. The top layer was planting soil layer. The second, third and fourth layers were compacted fill layer. The bottom layer was natural sediments with weak permeability. The free water surface of the unconfined aquifer was the upper boundary, and the bottom of the aquifer was impervious boundary.

Groundwater recharge in model area mainly included precipitation infiltration recharge and lateral recharge, and evaporation and runoff into the sea is the main way of groundwater discharge. Precipitation and evaporation data were referred to meteorological stations near Lingshui Bay. Spring and autumn were the evaporation periods. The precipitation was concentrated in summer, so summer was the leaching period. There was little rainfall from winter to early spring, and the evaporation was weak due to low temperature. Evaporation capacity was much higher than rainfall capacity, and the ratio was about 2.3, which meant that the driving effect by phreatic water evaporation was strong. Salt would accumulate in shallow ground in the process of migration, ultimately resulting in soil salinization. Boundary conditions including lateral runoff, sea level and chloride concentration, parameters such as groundwater chloride ion content, precipitation recharge coefficient, specific yield and porosity were assigned according to the Lingshui Bay land reclamation project. Initial groundwater level was 0.7 m, which was equal to local capillary height, and initial groundwater chloride content was equal to that in seawater. Initial dispersion coefficient values were referred to previous experience and revised repeatedly in the simulation process. The simulation period was from spring and lasted for 5 years (1825d).

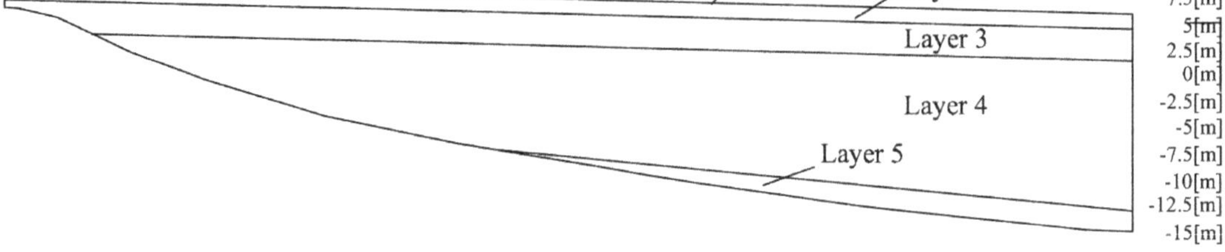

Figure 7. Soil layers of the model.

3.3.4. Simulation results and analysis

3.3.4.1. Groundwater level

After the model runs for 5 years (the fifth year after the completion of land reclamation project), the groundwater level is shown in **Figure 8**. Because of the construction of land reclamation project, the discharge outlet is cut off. Groundwater from the origin land frontier and the sea enters into the fill, causing a gradual increase in groundwater level in this area. Groundwater level in the model domain after model runs for 1, 3 and 5 years is analyzed. It indicates that groundwater level becomes stable over time. Groundwater table near the original land frontier is higher and gets closer to the limit-evaporable depth of groundwater, in which condition salt can migrate to shallow ground. In spring and summer, the evaporation is intensive, while there is no adequate supply; so groundwater table is relatively low.

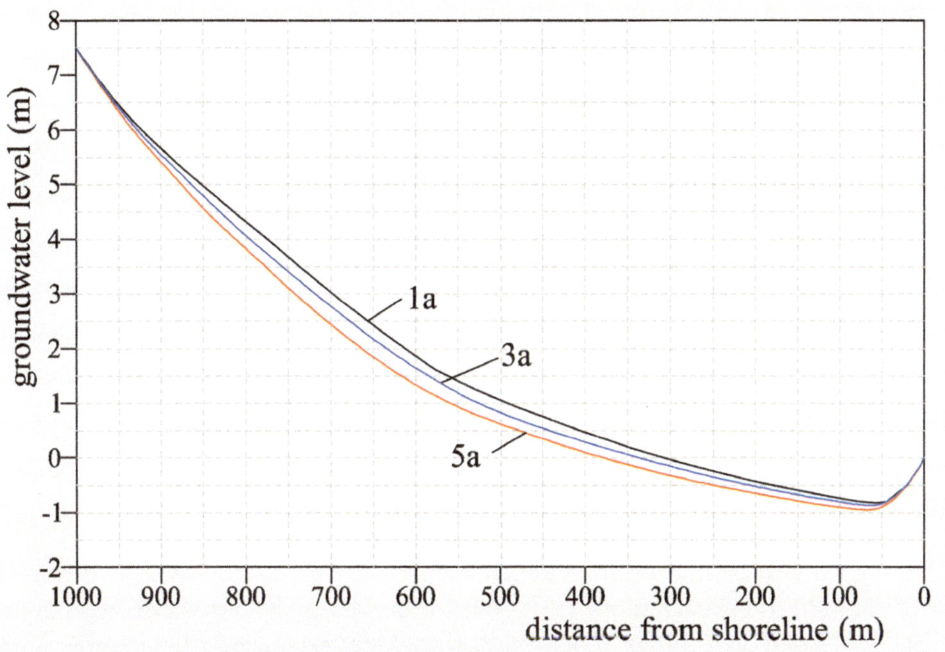

Figure 8. Groundwater level in the simulation area.

3.3.4.2. Chloride concentration

Groundwater chloride concentration after the model runs for 1, 3 and 5 years in model area is seen in **Figure 9**. As groundwater from the original land frontier enters into the reclamation area and precipitation infiltration recharge, groundwater salt within the capillary height is diluted. However, chloride concentration is still in high level. Groundwater chloride concentration near the sea is much higher. In spring and autumn, chloride concentration is over 14000 mg/L where groundwater is lower than 1 m due to insufficient lateral supply. On the whole, groundwater chloride concentration gets higher from land to the sea. Therefore, countermeasures such as recharge wells are suggested to promote salt water discharge into the sea and inhibit seawater intrusion.

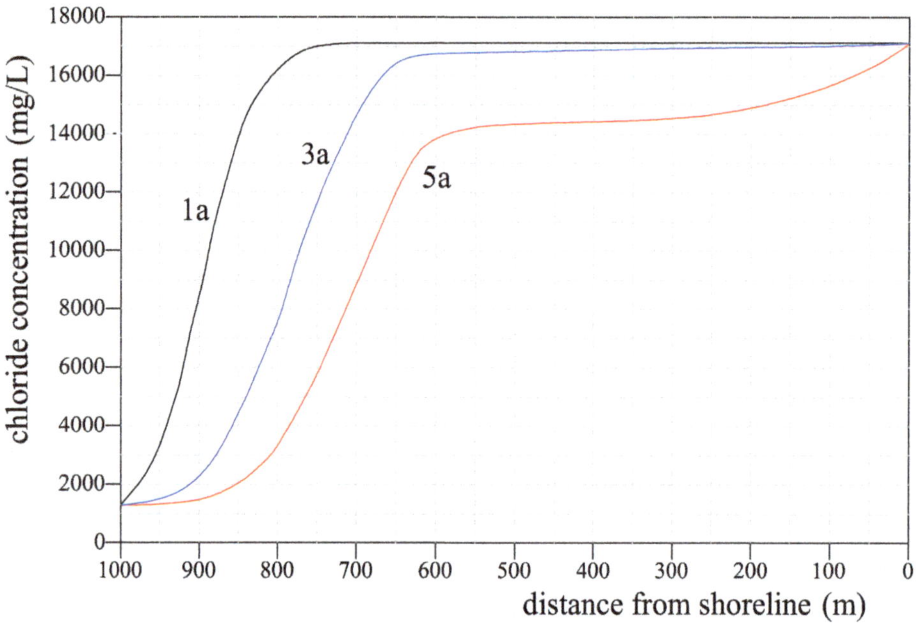

Figure 9. Groundwater chloride concentration in model area.

4. Mitigation methods

4.1. Soil structure reforms

Soil is an important medium for salt and water transport. Its structure directly determines the hydraulic conductivity. The capillary effect is the major force driving water rise, whether the water table located on the capillary rise zone has an obvious effect on the amount of evaporation. Once the capillary is destroyed if a multilayered structure is designed in the land reclamation zone, the salt accumulation process will be prevented. Therefore, it is meaningful to design multilevel backfilling technology based on the mechanism of salt movement in the reclamation area. Hornbuckle et al. [13] introduced that multilevel drainage system could provide faster leaching rate in the root zone, without increase in salt loads.

4.2. Sediment property modifications

In the reclamation regions, coastal sediment is often used to improve the physical and chemical properties of soil and construct grassy areas, because the organic matter content in offshore and river areas is high and acidic. Containing abundant microbes, sediment can increase the number of microorganisms and soil microbial population structure and improve the fertility of soil enzyme activity and content of humus. However, coastal sediment is rich in salt and heavy metal pollutants, and the soil particle size is small, which is easy to harden after dehydration. Therefore, coastal sediment needs to be modified before its application on the reclamation soil.

4.3. Rainwater utilization and desalination

Phreatic water is a sensitive element of the environment. Due to the shallow depth and high salinity of groundwater, efficient desalination system for reclaimed land is necessary. Many scholars used drainage system in salt elimination, but well-canal combined method is the main desalination technology recently [14, 15]. For coastal saline soil, desalination technology based on hydraulics was proposed. The freshwater resources in semi-humid regions are limited, and the groundwater discharge is not enough to push the fresh-salt water downwards to the sea. In the sea reclamation region, rainwater is infiltrated into soil by engineering measures, which could remedy the seasonal distribution defect of precipitation and adjust the groundwater environment. For example, rainwater is collected to construct the layer of salt leaching, and the soil drainage system is improved to control the underground water level and the salt content. Both methods above prove that it is feasible to build the long-term desalination mechanism of rain. Xu et al. [16] found that rainwater infiltration through pervious pavements can effectively resist seawater intrusion and the interface of fresh-salt water is pushed to the coastline. It is helpful to combat the salinization using rainwater infilling method in urban area.

4.4. Conservation tillage

Although the parent material under the reclaimed soil is complicated, the clay layer contains little organic matter due to lacking of supply. Soil and water loss will damage ecological resources and lead to ecological degradation. Conservation tillage has been used in the northeast of China to protect soil and water resources by preventing soil erosion. For example, straw return is an efficient strategy to enhance soil fertility, which provides a buffer for raindrops to transfer energy and promotes the infiltration rates of salt leaching. In the long term, conservation tillage will be an efficient method for saline soil restoration and ecosystem restoration.

5. Conclusion

Economic benefits could not justify the impacts of reclaiming land on coastal ecological degradation. Salinization is one of the major problems. Mitigation measures have been proposed, and some have successful achievements. However, most of these methods are based on experiences, and lack of theoretical bases. The effects of the mitigation measures on short-term and long-term monitoring are still necessary in the complicated land reclamation regions to get full understanding of the mechanism of soil salinization.

Research including field investigation, modelling and analysis was carried out in land reclamation areas. Results show that under semi-humid climate, the salinity problem in reclaimed land is severe. Driven by climate, vegetation and upstream freshwater supply, the migration of water and salt is dynamic. The processes could be classified as salt leaching, salt accumulation and the release process of salts from sediment. Measures should be taken to prevent the soil salinization. Major ways include rainwater utilization, conservation tillage, soil structure reform, desalination and sediment property modifying.

Acknowledgements

This research is supported by the project (51279022) of the National Nature Science Foundation of China.

Author details

Shiguo Xu[1*], Yi Xu[1,2], Yanzhao Fu[1] and Qi Wang[1]

*Address all correspondence to: yzfu@mail.dlut.edu.cn

1 Institute of Water and Environment Research, Dalian University of Technology, Dalian, China

2 Shenyang Academy of Environmental Sciences, Shenyang, China

References

[1] Stuyfzand, P.J. The impact of land reclamation on groundwater quality and future drinking water supply in the Netherlands. Water Science and Technology. 1995;31(8): 47-57. doi:10.1016/0273-1223(95)00356-r

[2] Xu, Y., Xu, S.G., and Xu, X.Z. Analysis of water and salt migration in sea reclamation regions under a semi-arid climate. In: XXV Iugg General Assembly; 28th June–7th July 2011; Melbourne, Australia. IAHS-AISH publication; 2011. p. 213-219.

[3] Xu, S.G. and Xu, Y. Review on water environment evolution and strategy after land reclamation. Advances in Water Science. 2013;24(1):138-145. doi:10.14042/j.cnki.32.1309.2013.01.016 (in Chinese)

[4] Armstrong, A., D. Rycroft, and T. Tanton. Seasonal movement of salts in naturally structured saline-sodic clay soils. Agricultural Water Management. 1996;32(1):15-27. doi:10.1016/s0378-3774(96)01262-0.

[5] Chen, K. and J.J. Jiao. Seawater intrusion and aquifer freshening near reclaimed coastal area of Shenzhen. Water Science and Technology: Water Supply. 2007;7(2):137-145. doi: 10.2166/ws.2007.048.

[6] Iost, S., D. Landgraf, and F. Makeschin. Chemical soil properties of reclaimed marsh soil from Zhejiang Province PR China. Geoderma. 2007;142(3):245-250. doi:10.1016/j.geoderma..08.001.

[7] Fritton, D. D., Kirkham, and R. Shaw. Soil water and chloride redistribution under various evaporation potentials. Soil Science Society of America Journal. 1967;31(5):599-603. doi:10.2136/sssaj1967.03615995003100050002x

[8] Nakagawa, K., et al. Modelling reactive solute transport from groundwater to soil surface under evaporation. Hydrological Processes. 2010;24(5):608-617. doi:10.1002/hyp.7555.

[9] Jia, Y.M. Experimental Study of Phreatic Evaporation at Different Depth of Groundwater Table. Journal of Irrigation and Drainage. 2008;27(6):021. doi:10.13522/j.cnki.ggps .2008.06.021 (in Chinese)

[10] Zhang, W.Z. and Zhang, Y.F. Examinations of balancing water and salt in irrigated districts and control of soil saline and alkaline. China Water Resources. 2003;16:24-30. doi:10.3969/j.issn.1000-1123.2003.16.005.

[11] Cui, Y.H., Zhang, R.M., et. al. Researches on the relationship of evapotanspiration and salinity. Inner Mongolia Science Technology and Economy. 2002;5:18-20. doi:10.3969/j.issn.1007-6921.2002.05.006.

[12] Guan, Y.X., Liu, G.H., Wang, J.F., Regionalization of salt-affected land for amelioration in the Yellow River Delta based on GIS. Acta Geographica Sinica. 2001;56(2):205-212. doi:10.3321/j.issn:0375-5444.2001.02.008.

[13] Hornbuckle, J.W., E.W. Christen, and R.D. Faulkner. Evaluating a multi-level subsurface drainage system for improved drainage water quality. Agricultural Water Management. 2007;89(3):208-216. doi:10.1016/j.agwat.2007.01.004.

[14] Heydari, N., A.D. Gupta, and R. Loof. Salinity and sodicity influences on infiltration during surge flow irrigation. Irrigation Science. 2001;20(4):165-173. doi:10.1007/s002710100043.

[15] Bakker, D., et al. Salinity dynamics and the potential for improvement of waterlogged and saline land in a Mediterranean climate using permanent raised beds. Soil and Tillage Research. 2010;110(1):8-24. doi:10.1016/j.still.2010.06.004.

[16] Xu, S.G., Gao, J.H., and Chang, N.B. Hydrological and environmental modeling analyses of pervious pavement impact in a coastal city. In: Book chapter, Chang, N.B. editor. Effects of Urbanization on Groundwater: An Engineering Case-based Approach for Sustainable Development. Reston: American Society of Civil Engineers (ASCE); 2010. p. 367-388. DOI:061/9780784410783.ch14.

10

Radioactive Contamination of the Soil: Assessments of Pollutants Mobility with Implication to Remediation Strategies

Ivana Smičiklas and Marija Šljivić-Ivanović

Additional information is available at the end of the chapter

Abstract

Accidental releases, nuclear weapons testing, and inadequate practices of radioactive waste disposal are the principal human activities responsible for radioactive contamination as a new and global form of soil degradation. Understanding the radionuclide distribution, mobility and bioavailability, as well as the changes caused by the variation of environmental conditions, is essential for soil rehabilitation. This chapter aims to highlight the importance of evaluating radionuclide distribution, for the selection of proper *in situ* or *ex situ* remediation strategy. Attention was focused onto remediation methods based on radioactive pollutants redistribution, for enhanced separation (chemical extraction) or containment (*in situ* immobilization). When the excavation and off-site leaching treatments are uneconomic, impractical, or unnecessary, *in situ* stabilization by the addition of appropriate reactive materials is an alternative approach. The optimization of factors in control of chemical leaching methods, selection of cost-effective immobilization agents, especially among suitable wastes and by-products, and verification of long-term effects of remediating actions are the major challenges for future investigation in this field. Furthermore, the improvement and standardization of the methods for radionuclide speciation are necessary to enable comparison between studies and monitoring of the effects achieved by the soil treatments.

Keywords: radioactive pollutants, mobility, soil remediation, extraction, immobilization

1. Introduction

Radioactivity is a phenomenon related to unstable atomic nuclei with excess of energy and/or mass, which spontaneously decompose emitting ionizing radiation in the form of electromagnetic waves (gamma rays) or streams of subatomic (alpha, beta, or neutron) particles [1]. The activity of a particular radioactive substance is characterized by the constant decay rate and the half-life ($t_{1/2}$—time taken for the activity of a given quantity of a radioactive substance to decay to half of its initial value), and it is a general rule of thumb that ten half-lives are required for each radioisotope to be eliminated [2]. Since the half-lives of various nuclei vary from seconds to billions of years [3], the time required for their total decay significantly differ as well.

Some radionuclides occur naturally in the environment, and their presence is either cosmogenic or terrestrial. The ^{3}H, $^{7,10}Be$, ^{14}C, ^{26}Al, and ^{39}Ar are the main radionuclides produced after the interaction of atmospheric gases with cosmic rays. On the other hand, the rocks, minerals, and consequently the soil, contain naturally occurring radioactive materials (NORM), characterized by a long half-life periods [3]. The most important terrestrial radionuclides are ^{238}U and ^{232}Th decay series, as well as ^{40}K. The world average values for soil activity coming from ^{226}Ra, ^{232}Th, and ^{40}K are 32 Bq/kg, 45 Bq/kg, and 420 Bq/kg, respectively [4].

The term radioactive contamination indicates the unintended or undesirable presence of radioactive substances on the surfaces or within solids, liquids, gases, or biota [5]. The origin of NORM is related to the formation of the planet; thus, their presence cannot be referred to as contamination. On the other hand, anthropogenic activities, related to the development of nuclear energy and its versatile use, have become important source of pollution. Since the middle of the last century, the radioactive contamination have appeared through the discharge of man-made radionuclides, making the ionizing radiation one of the important ecological factors, in line with other types of soil degradation (physical, chemical, and biological) [6]. Even though the radioactive contamination of the environment is relatively rare, it requires a great attention because of extreme degrading effects of ionizing radiation on living tissues. The adverse effects are in correlation with the quantity of absorbed energy, the penetrating power of the radiation, the duration of the exposure, as well as with the reproduction rate of the cells of a certain tissue [3].

In terrestrial ecosystems, soil corresponds to the major receiving pool of emitted radionuclides. Given that the nutrient cycles and the flow of energy present links between the abiotic and biotic components of the ecosystem, soils contaminated with radionuclides lose their ability to produce good quality agricultural crops and thus can be classified as degraded [6]. The issues related to the degradation of radioactively contaminated soils are being considered as an exceptional type of chemical contamination, with the additional, specific features related to the ionizing radiation.

The transport and fate of radionuclides in the soil are governed by a number of factors and the effects of their interactions; therefore, the detection and comprehension of the retention mechanisms are of great importance for the selection, development, and application of appropriate remediation technologies. In this chapter, the following topics were summarized and discussed: (i) sources of soil contamination by radioactive pollutants, (ii) interactions with soils constituents, (iii) factors influencing radionuclide mobility in the soil (iv) methods for the assessment of radionuclide mobility in the soil, and (v) the remediation strategies based on the increase or decrease of pollutant mobility.

2. The key sources of soil contamination by radioactive pollutants

Contamination of the soil with the radioactive pollutants is an important origin of hazard for the environmental and health safety, as well as for the economy. Exploitation of the nuclear energy is a key source of pollution. Radiation can enter and affect the environment at any of the stages of the nuclear fuel cycle, starting with the excavation and processing of uranium ore, over production and recycling of the nuclear fuels, to the processing and disposal of radioactive wastes. The average uranium concentration in the earth crust is 2.8 mg/kg [7]. This radionuclide is contained with variable concentrations in the range of oxide, silicate, arsenate, vanadate, and phosphate minerals. Ores, processed by conventional uranium production methods, vary from reach (>20%, Canada) to very poor (0.01%, Namibia) [8]. Uranium is extracted from the ore matrix by hydrometallurgical process, and the final product, (the so-called yellowcake), used in the following steps of the nuclear fuel production typically contain 75–85% U_3O_8. Studies of the effect of uranium production process onto environmental pollution and the potential health risks have revealed elevated activities at cites around ore processing facilities and around old mines, in particular [9, 10]. Nowadays, almost half of world-wide uranium mining, and most of the mining in the USA, Kazakhstan, and Uzbekistan, was conducted by *in situ* recovery (ISR) method [11]. This process is based on uranium leaching from the ore matrix, within the deposit. ISR is the most economically efficient method of uranium extraction; however, the associated risks include contamination of drinking-water aquifer with uranium or other heavy metals [12]. At present, approximately 60.000 tonnes of uranium ore are mined annually to supply fuel for more than 430 nuclear reactors around the world, which provide approximately one-eighth of the world's electricity [11].

Any material that is radioactive itself or is contaminated by radioactivity at levels greater than the quantities established by the competent authorities, and which cannot be of further use, is characterized as—radioactive waste. Within civil society, this kind of waste arises mainly from nuclear power production, but also from a variety of industries, medicine, agriculture, research, and education and other activities in which radioisotopes are used [13]. The radioactive wastes are being classified based on the level of radioactivity (low, medium, and high) and the half-lives of the isotopes with predominant activity [14]. In the short-lived waste, predominant activity is defined by radionuclides with $t_{1//2} < 30$ years, whereas the long-lived wastes are characterized by isotopes with $t_{1/2} > 30$ years.

Processing of radioactive waste may result in an accidental release of the radionuclides during characterization, segregation, transportation, treatment, and disposal. By the review of the inventory of fission products important in the case of accidental releases, it can be concluded that ^{89}Sr, $^{90}Sr/^{90}Y$, ^{91}Sr, ^{92}Sr, ^{95}Zr, ^{97}Zr, $^{103}Ru/^{103m}Rh$, ^{105}Rh, $^{129m}Te/^{129}Te$, $^{131m}Te/^{131}Te$, ^{132}Te, $^{131-135}J$, $^{140}Ba/$

^{140}La, ^{134}Ce, $^{144}Ce/^{144}Pr$ are important pollutants at the reactor stage; ^{90}Sr, $^{125m}Te/^{129}Te$, ^{131}I, ^{134}Cs, ^{137}Cs may be released during fuel element transport; ^{90}Sr, $^{95}Zr/^{95}Nb$, ^{106}Ru, ^{131}I, ^{137}Cs, $^{144}Ce/^{144}Pr$, and actinides are important at the fuel reprocessing stage; ^{90}Sr, ^{106}Ru, ^{137}Cs, and $^{144}Ce/^{144}Pr$ contamination may occur during fission product solidification, whereas leaching from the final disposal may result in soil contamination with ^{90}Sr, ^{137}Cs, and actinides [15]. In addition to fission products, several corrosion products may become significant soil pollutants. Namely, during nuclear reactor operation, most metallic surfaces oxidize and form a layer of corrosion film rich in oxides of structural elements. This layer is exposed to high pressures and temperatures, where radionuclides are generated under the neutron activation [16]. Depending on the composition of the reactor materials and their trace elements, reactor type and design, thermal power, years of irradiation and shutdown period, the corrosion products and their relative proportions are different. The products of steel corrosion are ^{55}Fe, ^{59}Ni, ^{63}Ni, ^{94}Nb, ^{60}Co, ^{39}Ar, ^{54}Mn, with the ^{60}Co and ^{55}Fe being the most important in the first 10 years following the closure of a reactor, and ^{63}Ni, ^{94}Nb, ^{108}Ag in the next 50 years. Reinforced concrete's corrosion products are ^{3}H, ^{14}C, ^{41}Ca, ^{55}Fe, ^{60}Co, $^{152,154}Eu$, whereas ^{3}H, ^{14}C, $^{152,154}Eu$ originates from graphite. Considering these two groups of materials, ^{3}H becomes the most prominent after 10 years, and ^{14}C, ^{41}Ca, $^{152,154}Eu$ after 50 years from the reactor shut-down. Taking into account both fission and corrosion products, 10–20 years after the reactor shutdown the most abundant radionuclides in contamination residues generally include ^{3}H, ^{60}Co,^{55}Fe, and ^{137}Cs, whereas in the period 20–30 years, ^{63}Ni, ^{137}Cs, ^{60}Co, and ^{90}Sr generally prevail [16].

Another key source of soil contamination with radionuclides is nuclear weapons tests, particularly atmospheric, which have started in 1945 in the USA [17]. In the period 1945–1980, the power of USA atmospheric tests (428 megatons) was approximately equivalent of the size of 29,000 Hiroshima bombs [17]. Finally, in 1990, thanks to the moratorium signed by SSSR, UK and USA, nuclear testing was stopped. Atmospheric detonations produce radioactive debris of different particle size, which are partitioned in the tropo- and stratosphere and my precipitate over a period of a few minutes to 1 year, or longer [18]. The concern is especially focused onto released Pu isotopes, due to the high biological toxicity and long half-lives of its relevant isotopes (e.g., 24.2×10^3, 373×10^3, 81×10^6 years, respectively, for ^{239}Pu, ^{242}Pu, and ^{244}Pu) [19]. Furthermore, ^{137}Cs, ^{90}Sr, ^{241}Am, and ^{131}I are the released radioactive isotopes with major impact on the environment and irradiation of the human body [20]. The mentioned isotopes were predominantly found in most of the nuclear test sites worldwide, especially in western US soil [21, 22].

Nuclear accident are the events that led to significant consequences to people, the environment or the facility, such as the ones in Chernobyl (Ukraine, 1986) and Fukushima (Japan, 2011). These two events caused global contamination of the environment, including air, water, soil, and living organisms. Huge amounts of radioactive elements especially ^{131}I, ^{137}Cs, ^{90}Sr and the sum activity of ^{239}Pu and ^{240}Pu were dispersed into environment [23]. Some 40% of Europe has been exposed to Chernobyl's ^{137}Cs at a level 4–40 kBq/m^2 [24]. The size of the disaster can be illustrated by the fact that the maximum radioactive contamination in the soil in the 1993 was found to be 3500 times higher than the level before Chernobyl accident.

Apart from uranium mining and related fuel cycle activities, the industrial sectors which generate technologically enhanced naturally occurring radioactive materials (TENORM) include the following: mining and combustion of coal, the oil and gas production, metal mining and smelting, production of mineral sands (rare earth minerals, titanium, and zirconium), phosphate fertilizer industry, building industry, and recycling [25–27]. The dose of radiation coming from primordial radionuclides (^{40}K, ^{232}Th, ^{235}U, ^{238}U, and the members of decay series), which are normally found in natural minerals and ores (uranium ore, coal, phosphate rock, monazite, bauxite, etc.), can be elevated in their by-products and wastes such as phosphogypsum, fly ash, and red mud. Consequently, the releases from non-nuclear industries represent a continuous source of soil contamination with natural radioactive elements, by spreading of dust from rock and solid wastes dump, as well as by the overflow of wastewater from treatment ponds. Furthermore, years of application of phosphate fertilizers enriched with TENORM may become a source of soil contamination. Depending on the contamination level, restriction of land use or the remediation measures may be necessary. Finally, soil contamination may also arise from less common sources such as incidents during use of radioisotopes in medicine, industry, and agriculture [28].

At 160 U.S. Department of Energy (DOE) sites with radioactive contamination, ^{137}Cs, ^{226}Ra, ^{238}U, $^{238-242}$Pu, ^{60}Co, ^{232}Th, and ^{90}Sr were detected as the key artificial and natural radionuclides [29].

3. The interactions of radioactive contaminants with soil matrix and the methods of their identification

3.1. The nature of radionuclide interactions with soil components

Interactions of contaminants with soil matrix, and their variation with environmental parameters, are essential for radionuclide transport and fate, as well as for the risks to the living organisms and the environment. The uptake of radionuclides by soil can occur through diverse modes of interactions, while at the same time, other mechanisms are responsible for their elimination from the soil matrix (**Figure 1**). Due to the dynamic nature, heterogeneity and the overall complexity of the soil as a system, studying, understanding, and predicting the radionuclides behavior are the major challenges.

Factors influencing radionuclide distribution in the soil include the source term and the release conditions, transport and dispersion mechanisms, and the properties of the ecosystem [30]. Source term (ions, colloids, particles, oxidation states, etc.) influences mobility properties of radionuclides, since the transfer of mobile species in the ecosystem is faster in respect to the transfer of particles. Furthermore, the properties of the particular radionuclide, its chemical form and the reactivity, control the nature of its retention in the soil and the affinity to certain soil constituents.

Soil properties are primarily grouped into physical (texture, structure, porosity, water, air, and heat regimen), chemical (chemical and mineralogical composition, pH, microelements, micronutrients, salinity (EC), cation exchange capacity (CEC), organic matter, etc.) and biological (macroflora, macrofauna (rodents, insects, woodlice, mite, snails, millipedes, spiders, worms), microflora (bacteria, actinomycetes, fungi, and algae), and microfauna (nematodes and the protozoa)) [31, 32]. All five basic components of the soil, that is, minerals, water, organic matter, gasses, and the microorganisms, affect the binding and retention of the pollutants to a greater or lesser extent, depending on the pollutant type.

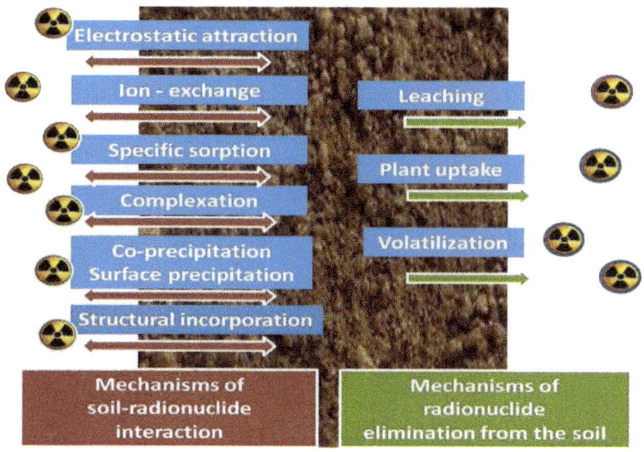

Figure 1. The mechanisms of radionuclide binding and elimination from soil matrix.

The interactions between radionuclide and the soil include physical (reversible) sorption governed by the uncompensated charges on the surface of the soil particles, and the chemical (principally irreversible) sorption through high affinity, specific interactions, and establishment of covalent bonds [33, 34]. The primary minerals in soil, mainly quarts and feldspar, are derived from the parent rock and make up most of the sand and silt fraction. Due to the relatively low specific surface area, their role in contaminant interaction is the smallest, and the attachment occurs through reversible sorption [35]. Secondary minerals, such as clay, result from physical, chemical, and biological weathering processes. Because of the unbalanced charges of structural ions, they are the carriers of permanent surface charge, which in combination with small particle size and large specific surface area make them important matrices for contaminant retention. Furthermore, oxides and (oxy)hydroxides of Fe and Al are abundant in amorphous form, with pH-dependent surface charge. Soil organic matter consists of chains of carbon atoms, containing polar and/or ionized surface functional groups, such as OH– and COOH–. Consequently, clay minerals, Fe, Al-oxides, and organic matter undergo a variety of interactions with contaminants.

3.2. Assessment of the radionuclide mobility in the soil

The bonds established between the particular radionuclide and the particular soil type can be assessed by different analytical approaches. Chemical reagents of various composition, strength, and selectivity are the most widely used, in the single stage or sequential extraction protocols [30, 36–41]. The aim of such tests was assessment of the transport mechanism in a soil profile and the potential toxicity, with implications to the risks to the biota and the ground water reservoirs. In general, weaker bonds between the pollutant and the soil components signify higher mobility of radionuclide, its increased possibility to reach the plants and soil organisms and to enter into the food chain. However, the mobility and bioavailability of closely related, they cannot be equalized in the interpretation. Bioavailability processes are defined as the physical, chemical, and biological interactions that determine the exposure of plants and animals to chemicals associated with soils and sediments, they incorporate a number of steps and represent the amount of a contaminant that is absorbed following skin contact, ingestion, or inhalation [42]. On the other hand, the bioaccessibility of the contaminant is defined as its fraction soluble in the gastrointestinal tract and available for absorption.

Review of the literature shows that a wide spectrum of single-stage extraction methods for soil analysis is in use [36, 42]. Basic groups of reagents include acids, chelating agents, and salts; moreover, reagent concentrations and other experimental conditions are considerably different (**Table 1**). In contrast to the well-established methods for the determination of soil major nutrients and fertility, the procedures for the extraction of pollutants are not standardized.

The most common chemical reagents		Common concentration ranges (mol/L)
Acidic solutions	HNO_3, HCl, CH_3COOH, H_2SO_4	0.01–2
Chelating agents	EDTA*, DTPA**	0.005–0.01
Salt solutions	$CaCl_2$, $NaNO_3$, NH_4NO_3, $AlCl_3$, $BaCl_2$	0.01–0.1
Buffered salt solutions	NH_4CH_3COO/CH_3COOH (pH 4.8, pH 7)	1

*Ethylenediaminetetraacetic acid;
**Diethylenetriaminepentaacetic acid.

Table 1. The common leaching solutions in single stage soil extraction analysis [36, 42].

In addition to acidic and salt-containing solutions, the chelating agents are applied, due to their efficiency in extracting potentially bioavailable soluble complexes of radionuclides with organic matter. The results of leaching tests represent a rough measure of mobility, as the actual mobility in the field depends also on moisture, leaching, root uptake of nutrients, activity of microorganism, and many additional factors. Furthermore, the agreement between chemically extracted and fractions available to biota should be confirmed empirically, for wide variety of contaminated samples [43].

Speciation analysis is conducted for the identification and determination of the different chemical and physical forms of elements in the soil matrix [44]. The distribution of radionuclides is related to their affinity towards certain soil components; thus, they can exist as a free ions or in the form of soluble complex ions in interstitial solution; as exchangeable ions attached to the soil surface, they can be associated with soil organic fractions, occluded, or co-precipitated with metal oxides, carbonates, phosphates, or other secondary minerals, and incorporated inside the crystal lattices of primary minerals.

The sequential extraction protocols were primarily developed for the determination of the distribution of stable macro- and micro-constituents of the soil. Identification of the mobility and availability of trace elements, both the essential ones and the pollutants, is particularly important for the improvement and protection of the plant development and growth, and for the health of the ecosystem as a whole. Different sequential extraction methods have been proposed to separate the fractions of elements from various pools. The so-called Tessier method [44] and the method proposed by the European Community Bureau of Reference—the BCR method [45], are the two commonly used protocols, while many others are based on their modifications. Additionally, a modified version of Tessier's method was proposed at the Speciation Workshop organized by the National Institute of Standards and Technology (NIST), in order to optimize the protocol of soil extraction and select operationally defined fractions which can be separated by appropriate chemical reagents [46].

Evaluation of element distribution in soils by the sequential extraction is based on the assumption that mobility decrease with each extraction step (**Figure 2**), implying that under natural conditions elements in water soluble and exchangeable fractions are the most mobile and bioavailable, whereas those in residual fractions are the most tightly bound.

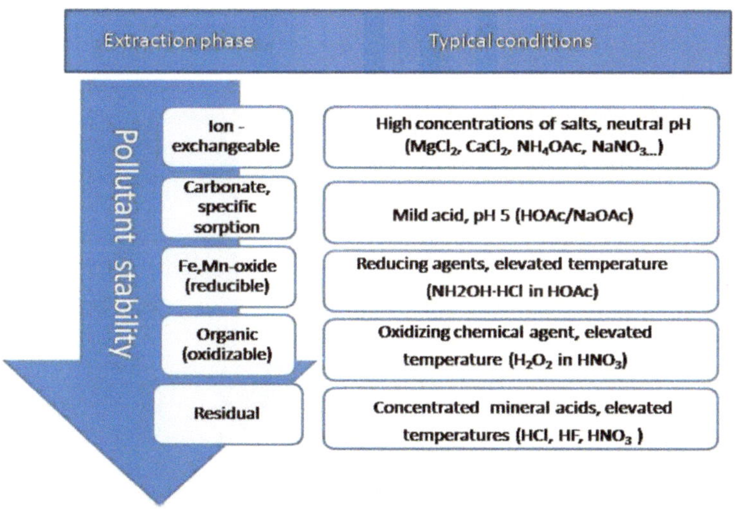

Figure 2. Common phases in sequential extractions based on Tessier's protocol [44].

The lack of the standardized procedure for the determination of pollutant mobility makes the interpretation and the comparability of the results difficult. In addition, the effect of the reagents may be questionable. For example, extracting solution having pH 5, used for the dissolution of carbonate phase, may also sequester ions specifically sorbed onto surface of other soil constituents [47]. The fractions of pollutants are defined only operationally; thus, instead of being associated with the terms *mobility* and *bioavailability*, they should actually be related to the extracting solution or the applied protocol [48]. Nevertheless, in the scientific and the technical literature, free ions, water-soluble complexes of radionuclides and the species associated by reversible, physical sorption, are commonly considered as the *mobile fraction* [49].

On the other hand, the term *inert species* refers to fraction of colloids and particles deposited in soils, together with the fraction of radionuclides irreversibly bound to or incorporated into the mineral lattices. The results of sequential extractions can be used for the calculation if the mobility factors of radionuclides (MF) [49]:

$$MF = \frac{\text{Mobile species}\,(\text{Bqm}^{-2})}{\text{Total deposition}\,(\text{Bqm}^{-2})} \times 100(\%) \tag{1}$$

where the mobile species include the fraction such as H_2O and CH_3COONH_4 extractable and that taken up by vegetation from the same site.

Although none of the methods can provide the absolute quantities associated to the specific component of the soil, such analyses represent valuable tool in elements mobility and availability assessment, and for tracking the effectiveness of soil remediation actions. Experiences achieved by practicing single and sequential chemical extractions reveal advantages of these methods but also a need for further research and developments due to increasing soil contamination which requires fast, reliable, and cost-effective assessment.

In addition to extraction methods, studies of the radionuclide retention mechanism can be complemented by the determination of the type of the surface complexes, identification of the radionuclide incorporation in the crystal lattice of existing minerals, or the formation of new solid phases, etc., for which instrumental techniques are applied (X-ray powder diffraction (XRD), Fourier transform infrared spectroscopy (FTIR), X-ray photoelectron spectroscopy (XPS), X-ray absorption spectroscopy (XAS), scanning electron microscopy with energy dispersive X-ray spectroscopy (*SEM/EDS*), etc.) [50, 51].

Furthermore, bioassay tests involving plants, animals, and microorganisms, are valuable for the analysis of radionuclide mobility and bioavailability [52]. Soil-to-plant transfer factors (TF) have been widely used in radioecology, in order to quantify the availability of soil radionuclides for plant uptake [36]:

$$TF = \frac{\text{Plant activity concentration}\,(\text{Bq kg}^{-1})}{\text{Total soil activity concentration}\,(\text{Bq kg}^{-1})} \tag{2}$$

As soil-to-plant transfer considerably differs between different plant species and the seasons, this method also gives crude estimations of potential radionuclide bioavailability. In spite of limitations, transfer factors are currently accepted as the most practical way of describing plant uptake. Also, several in vitro methods have been developed for the prediction of the relative bioavailablity of the contaminants, using physiologically based fractionation schemes [42, 53]. These methodologies mimic key processes that take place in vivo, such as contaminant dissolution, and after establishing a strong correlation between the in vivo and in vitro results, these methods have a potential to overcome the time and expense limitations of in vivo studies.

3.3. Factors influencing radionuclide mobility in the soil

A capacity of the soil itself to immobilize radionuclide is the main factor controlling activity concentrations available to biota, and it operates in conjunction with the numerous external factors. Soil texture and structure, mineral composition, organic components, redox potential (Eh) and pH, as well as rainfall, climate changes, and soil management, are recognized as important for radionuclide mobility [54]. The pH of the soil, cation exchange capacity (CEC), and total organic carbon (TOC) are the physicochemical characteristic most often correlated with the distribution of the radionuclides [40]. Alkaline soils are characterized by the presence of carbonates and have a high saturation of base cations (K^+, Ca^{2+}, Mg^{2+}, and Na^+), whereas acidity in soils comes from H^+ and Al^{3+} ions in the soil solution and sorbed to soil surfaces. The surface charge of minerals is a major contributor to soils CEC and influences the soil's ability to retain important nutrients and the pollutants. The texture of a soil is based on the relative content of sand (0.05–2.00 mm), silt (0.002–0.05 mm), and clay (<0.002 mm) fraction. Due to the finest granulation, clays minerals exhibit the largest surface area, important for soil chemistry and CEC, but also for water-holding capacity important for transporting nutrients and pollutants to soil organisms and plants. In addition, soil organic matter significantly contributes to the soil CEC and to the water-holding capacity.

Based on the literature data, the influence of soil properties and other condition on the mobility of some important pollutants is given in **Table 2**.

	Radionuclide			
	Cs	**Sr, Ra**	**U, Pu**	**I**
Chemical form	Cs^+	Sr^{2+}	PuO_2^{2+}, $Pu(NO_3)^{3+}$	I_2, I^-, IO_3^- CH_3I
		Mobility		
pH decrease	Increase	Increase	Increase	
Clay content decrease	Increase	Increase	Increase	
Sand content decrease	Decrease	Decrease	Decrease	Increase
Humus content low	Not clear	Decrease	Decrease	
CEC decrease	Increase	Increase	Increase	
Aging	Decrease	Weak effect	Decrease	

Table 2. The effect of soil physicochemical properties and aging on the mobility of radionuclides [55, 56].

Apart from soil type, different sources of variability may influence the fractionation patterns and cause the shift from less available to more available fractions, or vice versa. Generally, the increase of contaminant concentration not only increases the overall activity in the soil but also leads to redistribution from the less to the more available fractions [57]. Radioactive contamination introduces new elements into the ecosystem and, in distinction from the transport of stable elements and NORM, transfer of contaminants through the trophic chains occurs under non-equilibrium conditions. Consequently, ageing affects a decrease in the chemical mobility and biological availability of most of the radioactive pollutants [58]. The ageing process actually involves a set of reactions related to the enhancement of radionuclide sorption and fixation by the soil solid phase (i.e., the precipitation or penetration into the crystalline lattices of different mineral constituents. Aging exhibits a different effect on different ions. Increased contact times (months to years) were found to affect gradual reduction of Co^{2+} ions mobility [57, 59]. Time-dependent studies on the variation in Cs^{+} bioavailability have revealed that over years, a decrease in the labile fraction of ^{137}Cs in soils was correlated with a decrease in soil-to-plant transfer [60]. In contrast, due to low sorption affinity of ^{90}Sr towards soil constituents, impact of aging is very weak considering ^{90}Sr speciation [40]. The behavior of Sr^{2+} and its uptake by living organisms are controlled by its similarity to calcium; thus, regardless of the soil type, contamination level, and aging time, it was largely found in water-soluble and ion-exchangeable fractions of soil. Seasonal effects may also cause variations in radionuclide mobility, and these effects can be controlled by appropriate sampling plan [57].

4. Increase/decrease of radionuclide mobility as essential soil remediation strategy

As the environmental conditions change, the distribution of pollutant also changes, causing the increase or the decrease in mobility. Knowledge of such dependencies represents the theoretical background for the development of mobilization/immobilization remediation methods. Furthermore, exploration and development of suitable solid and liquid media are fundamental in support of these technologies. Mobilization techniques imply weakening of bonds with the soil constituents provoking desorption, dissolution, and chalation of the pollutant [61, 62]. On the other hand, the general idea of the radionuclide immobilization (stabilization) is to induce chemical reactions, precipitation, and other processes which cause redistribution of the contaminants from more labile to more stable forms [61, 63]. Both principles exhibit certain benefits and drawbacks. Stabilization techniques are usually less expensive and easier to perform in comparison with the alternative processes; however, the total activity concentrations remain in the soil, posing a constraint for the future uses. Otherwise, the techniques based on the pollutant exclusion from the soil matrix represent a permanent solution for the contaminated site. However, transportation, consumption of the chemicals and the energy, and further management of the resulting liquid phase with the extracted pollutants, make these techniques complicated and costly. Remediation activities may also result in some negative effects on the soil properties, including fertility; thus, evaluation of suitable strategies and decision-making process require detailed knowledge of all these aspects.

4.1. Extraction of radioactive contaminants from the soil matrix

Chemical extraction is the technique that stimulates the redistribution of contaminants from the solid phase to the solution, in order to selectively remove the contamination, or to enhance its physical separation [61, 64]. The contaminated soil is excavated and treated off-site. After the treatment, the soil is returned to its original location, while the activity remains concentrated in the extraction medium. The extract is subsequently treated to precipitate the activity and return the leaching reagents to the process. Otherwise, the extracting solutions can be implemented *in situ*, to increase the radionuclide mobility in the soil and enhance their subsequent uptake by plants (combination with phytoextraction) [65].

Radionuclides in the soil can be re-mobilized by four principal means [66]: (1) changes in the acidity, (2) changes in the ionic strength of the solution, (3) changes in the soil redox potential, and (4) formation of soluble complexes. To extract the pollutants, acids operate on the ion-exchange principle, and by dissolution of soluble soil components. Highly concentrated solutions of inorganic salts displace the radionuclides from ion-exchangeable sites by mass action, and if implemented at low pH this effect is combined with the effects of acid leaching. Chelating agents solubilize metals through complexation, while redox manipulation aims to enhance solubilization by the change of valence and thus chemical properties. The most common chemical agents are inorganic salts ($CaCl_2$, NaCl), mineral acids (HCl, H_2SO4, HNO_3), and complexing agents (EDTA, DTPA, oxalate, citrate, etc.) [61, 65, 67].

Selection of the proper chemical extracting reagent is influenced primarily by the radionuclide type, its speciation pattern and the characteristics of the soil. Pollutants that are majorly accumulated in ion-exchangeable, carbonate, and Fe, Mn oxide fractions are the most suited for the removal by chemical leaching [68]. The soils characterized by low pH, low content of clay, and humic substances are the promising candidates for such treatments [61].

In order to extract the target metal from the soil environment, the strength of the radionuclide-chelating agent complex must overcome the strength of the bonds keeping raionuclide attached to the soil surface. The efficiency of EDTA is superior, and it is usually applied at pH 4–8, as the EDTA-complexes can be re-adsorbed on soil surface sites at lower pH [69]. In addition to the high price, selectivity of EDTA towards target radionuclides, its recovery and reuse are the major drawbacks. Furthermore, its low degradability can be a persistent problem after the soil treatment. Thus provided that they enable efficient removals of pollutants, and acidic and salt-containing solutions are more acceptable due to lower environmental impact and the ease of regeneration.

In the comprehensive investigation of appropriate chelating agent for the extraction of various radionuclides, the regressive empirical predictive model was developed as a selection tool [62]. Using as the input variables, the properties of the chelators, various stability constants, radionuclide distribution, and the soil properties (mineralogical composition, pH, clay content, CEC, etc.), the following adequate chelator for target radionuclide were proposed: EDTA, DTPA, and nitrilotris(methylene)triphosphonic acid (NTTA) for Ba and Ra; 2-aminoethanethiol, EDTA, DTPA, thiobis(ethylenenitrilo)tetraaceticacid (TEDTA), and N-2-acetamidoiminodiacedicacid (ADA) for Pb and Th; whereas iminodiaceticacid (IDA), nitrilo-triaceticacid (NTA), and ethylenediiminodiacetic acid (EDDA) were suggested for the extra-ction of Pu and U.

Selective removal of ^{137}Cs and ^{90}Sr from soil poses a problem, due to the lack of suitable complexing agents [61]. Although certain crown ethers form complexes with these cations, due to the toxicity and high cost of such agents, large-scale agricultural applications are impractical. Solutions of HCl, $CaCl_2$, EDTA, tartaric, and citric acid, with different concentrations of reagents, were applied to soil artificially contaminated with Sr^{2+} and Co^{2+} ions [39]. Due to its predominant association with ion-exchangeable fraction, Sr^{2+} ions were efficiently desorbed using Ca^{2+} or acidic solutions. On the other hand, Co^{2+}, which was largely distributed between carbonate and Fe, Mn-oxide fractions, was leached most efficiently by complexing agents.

Chemical extraction processes have a large potential in the rehabilitation of the soil that have undergone radioactive contamination and their effectiveness can be additionally improved by optimizing reagent type and concentration, soil/solution ratio, pH, contact time, mixing, and other factors.

4.2. Radionuclide immobilization (stabilization) by soil amendments

Despite the fact that the main objective of the soil remediation was the removal of the maximum amount of pollution, the major obstacles for the routine application of such an approach are the processing and the disposal of the radioactive waste resulting from the soil clean-up [70]. The insufficient storage capacities, especially for waste classified as low level, long-lived, are significant and global problem. As a consequence, immobilization treatments are being rapidly developed, with main goals to reduce the risk of exposure and uptake by biota, and the risk of the spread of contamination.

The application of soil amendments is performed on site (*in situ*) which makes such technologies fast, simple, and effective. Alternatively, soil amendments can be applied in *ex situ* process, where soil is firstly physically removed from the site, pretreated, mixed with a stabilizing amendment, and then returned to its original location [71].

As the most of the radionuclides in soil exist in the cationic form, increase in pH, clay content, and CEC lead to an increase in pollutant stability (**Table 2**). Consequently, water-soluble and water-insoluble amendments are applied, with a role to modify the environmental conditions in favor of radionuclide stabilization or to directly interact with the contaminants (or both).

In order to raise pH and lower pollutants accessibility to plants, the materials traditionally applied to soil are carbonates, lime, and phosphates [72]. Other soil amendments that are currently in use or are under consideration and verification have been modeled after stabilization or encapsulation agents (such as cement) used for safe disposal of radioactive and hazardous wastes. Various forms of aluminosilicates, phosphates, carbonates, silicates, oxides, and hydroxides were largely investigated [65, 72]. In general, solid matrices that have shown superior immobilization potential towards radioactive ions in aqueous solutions are suitable for testing in the contaminated soil. Based on the numerous investigations of the sorption affinities and capacities toward variety of radioactive pollutants, the most prominent groups of materials are aluminosilicates [73–80] and phosphates [81–89]. The main operating mechanisms are quite different for these two groups: while aluminosilicate addition to soil increases the number of sorption sites, phosphate materials, mainly from the apatite group, act through several removal mechanisms (ion-exchange, formation of specific surface complexes, and structural incorporation of pollutants by co-precipitation and dissolution/precipitation processes).

Aluminosilicates, primarily clay minerals, and zeolites are inorganic ion-exchangers with high surface area, which have been conventionally used for water treatment processes, for the treatment of liquid nuclear waste, and for the protection against nuclear waste leaking [79–90]. Natural zeolites are the framework aluminosilicates, with variable porosity due to which they can selectively capture the ions having an appropriate radius. Zeolites are excellent sorbents of fission products that otherwise exhibit very low affinity for sorption on solid surfaces (such as Cs and Sr isotopes [78, 80]. Clay minerals (montmorillonite, vermiculite) are layered aluminosilicates, in which ion-exchange is typically associated with cations situated in clay mineral interlayers [72]. The stabilization of Cs^+ and Sr^{2+} contamination in the sandy soils was tested using different synthetic and natural zeolites [91]. With the addition rate of 1%, the maximum reduction of soil-to-plant transfer factor of 12.5 for Cs^+ and 24.5 for Sr^{2+}ions, was observed, as well as the significant changes in cationic composition and pH of the soil. By comparing the effect of various materials onto Sr^{2+} immobilization in the soil, zeolite has been identified as the most efficient, followed by bone char, synthetic hydroxyapatite, and phosphate rock [92]. The most of the results have been obtained on the laboratory level or out of small-scale field applications, while in solving the actual problems of soil contamination, applications are generally connected with Chernobyl and Fukushima disasters.

The other promising group of materials is the phosphate group. Among different soluble and sparingly soluble phosphate bearing materials, hydroxyapatite ($Ca_{10}(PO_4)_6(OH)_2$, HAP) exhibited superior physicochemical and sorption properties, that is, low solubility in water, high specific surface area, high buffering capacity, and the high sorption capacities towards variety of cationic and anionic pollutants [93]. HAP is by far the most selective to U and Pb, due to the removal mechanism which involve dissolution of HAP and precipitation of thermodynamically more stable Pb and U containing phases [87, 94]. In soil, apatite matrices were highly effective for U uptake; however, the increase of organic matter content influenced the decrease of amendments efficiency [95]. Furthermore, the selectivity and capacity of HAP towards Pu, Co, Ni is very high, moderate for Sr, while low considering Cs and Tc [78, 81–84, 86–88].

Comparing different apatite forms (synthetic, mineral, and biogenic), the product extracted from fish bones exhibited the best sorption properties, due to CO_3^{2-} substitutions, low trace metal concentrations, poor crystallinity, and high microporosity necessary for optimal performance in the field [87]. Giving that this sorbent is produced from the commercial fish industry waste, it is both environmental friendly and cost-effective for large-scale operations. However, the bioavailability of essential trace elements was found to decrease at high HAP addition rates (5%), while uptake of As by plants was found to increase after HA treatment [96]. These results demonstrate that HAP application for the remediation of contaminated soil must be optimized and controlled.

In addition to animal bones as the source material for apatite production, many other industrial by-products, wastes, and recycled materials are being tested as potential soil additives [65, 72]. In order to preserve natural mineral resources and reduce the costs of the immobilization treatments, application of such materials may represent a sustainable alternative. Another benefit comes from the reduction of the amount of accumulated wastes and their impact on the environment. Coal fly ash and bauxite residue (red mud) are mineral, oxide-based, residues, which exhibit high sorption potential for a range of radioactive pollutants [97–101].

Fly ash has a silt loam texture (<90% of the particles having a diameter of <0.010 mm), and it is composed mainly of aluminosilicate structures, quartz, mullite, hematite, magnetite, and calcite [102]. The pH values of fly ash vary in the wide range 4.5–12.0, depending on the content of sulfur in the parent coal. Fly ash was considered as an additive in agriculture, for improving soil properties [102], and also as an additive for stabilization of heavy metals in polluted soil, with the promising results [103–105].

Red mud is by product obtained after bauxite processing, which primarily consists of Fe, Al, Si, and Ti oxides and zeolite-like minerals [106]. Due to the nature of Al extraction process, this material exhibits extremely high pH (10–12), and it is high capacity sorbent especially for pollutants in cation form. Numerous laboratory, pot, and field studies were conducted in the past years regarding red mud utilization in remediation of heavy metal polluted soils, and its potentials (both as a liming additive, and as a sorbent) have been demonstrated [107]. However, radionuclides, as pollutants, have gained much less attention and the further research in this field is encouraged.

In general, there is a lack of the long-term studies on the overall effects of waste material additions on the soil properties. The variation in the composition of waste material and by-products adds uncertainty to their performance, and moreover, leaching of potentially hazardous substances from the waste material itself must be carefully evaluated. The activity levels of natural radinuclides can be elevated in fly ash and red mud with respect to parent coal and bauxite ore, therefore, a special attention should be paid to this aspect in order to keep activity levels in the permitted limits for soil.

5. Conclusion

The source term and a wide variety of soil and environmental parameters affect the radionuclide behavior in terrestrial systems. Weaker bonds between the pollutant and soil components implicate higher mobility of pollutant, higher potential to get into the solution and to be adopted by the biota. In addition to total concentration of the pollutant, understanding of its environmental behavior by determining distribution pattern in different fractions of the soil is of principal importance for the selection of optimal remediation technologies. Due to the large number of factors that affect the outcome of the soil rehabilitation process, selection of optimal solution must be done on a case-by-case basis. Still, some guiding principles can be derived from the research studies and the practical experience: pollutants mainly bonded in exchangeable, carbonate and reducible phase are suitable for chemical extraction, while removal of contaminants from organic and residual fraction is neither economical nor feasible. Optimization of extracting solution composition, pH, the time, and the mode of the interaction with the soil are the perspective fields of research which must include the type of the soil and the radionuclide, and the effects of the extracting solution to other important soil characteristics. Analyzing the contamination level, the size and the properties of contaminated area, *in situ* soil immobilization may prove to be more suitable solution which permanently increases sorption capacity of the soil. The use of mineral-based amendments as soil remediation additives should be as much as possible substituted by appropriate waste materials and by-products, which environmental compatibility, selectivity, and long-term effectiveness, must be verified on a variety of soil types. Immobilization technologies may be particularly useful if applied in combination with conventional *ex situ* (soil removal, chemical extraction) or *in situ* technologies (bioremediation, phytoremediation, reactive barriers, capping, monitored natural attenuation), for the stabilization of the residual activity.

Acknowledgements

This work was supported by the Ministry of Education, Science and Technological Development of the Republic of Serbia (Project III43009).

Author details

Ivana Smičiklas* and Marija Šljivić-Ivanović

*Address all correspondence to: ivanat@vinca.rs

Radiation and Environmental Protection Department, Vinča Institute of Nuclear Sciences, University of Belgrade, Belgrade, Serbia

References

[1] UNSCEAR (United Nations Scientific Committee on the Effects of Atomic Radiation). Sources and Effects of Ionizing Radiation. New York: United Nations; 2010. 20 p.

[2] Murray R. Understanding Radioactive Waste. 3rd edition. Columbus: Battelle Press; 1989. 167 p.

[3] ATSDR (Agency for Toxic Substances and Disease Registry). Toxicological Profile for Ionizing Radiation. Atlanta: ATDSR; 1999. 438 p.

[4] UNSCEAR. Sources and Effects of Ionizing Radiation – Report of the United Nations Scientific Committee on the Effects of Atomic Radiation to the General Assembly. New York: United Nations; 2000. 654 p.

[5] IAEA (International Atomic Energy Agency). IAEA Safety Glossary – Terminology Used in Nuclear Safety and Radiation Protection. Vienna: IAEA; 2007. 227 p.

[6] Aleksakhin RM. Radioactive contamination as a type of soil degradation. Eurasian Soil Science. 2009; 42: 1386–1396.

[7] Geller E, Weil J, Blumel D, Rappaport A, Wagner C, Taylor R. The Concise Encyclopedia of Chemistry. Encyclopeida of Science and Technology. Columbus: The McGraw-Hill Companies, McGraw-Hill Professional; 2004. p. 663.

[8] National Research Council. Uranium mining, processing, and reclamation. In: Uranium Mining in Virginia: Scientific, Technical, Environmental, Human Health and Safety, and Regulatory Aspects of Uranium Mining and Processing in Virginia. Washington: National Academies Press; 2012. p. 96–123.

[9] Ruedig E, Johnson TE. An evaluation of health risk to the public as a consequence of in situ uranium mining in Wyoming, USA. Journal of Environmental Radioactivity. 2015; 150: 170–178.

[10] Carvalho FP, Oliveira JM, Malta M. Radioactivity in soils and vegetables from uranium mining regions. Procedia Earth and Planetary Science; 2014; 8: 38–42

[11] OECD Nuclear Energy Agency. Uranium 2014: Resources, production and demand. 2014. Available from: http://www.oecd-nea.org/ndd/pubs/2012/7059-uranium-2011.pdf [Accessed: 2016/06/15].

[12] US EPA (Environmental Protection Agency). Technical Report on Technologically Enhanced Naturally Occurring Radioactive Materials from Uranium Mining. Volume 1: Mining and Reclamation Background. Washington: US EPA; 2008. 225 p.

[13] OECD. Radioactive waste in perspective. nuclear development. 2010. Available from: http://www.oecd-ilibrary.org/\njsessionid=1flnb9f5nsv01.x-oecd-live-03content/book/9789264092624-en [Accessed: 2016/06/15].

[14] IAEA. Classification of Radioactive Waste. IAEA Safety Standards Series No. GSG-1. Vienna: IAEA; 2009. 68 p.

[15] IAEA. Fission Product on Nuclear Data (FPND), Vol. III, In: Proceedings of a Panel on Fission Product Nuclear Data; 26–30 November 1973, Bologna. Vienna: IAEA; 1974. 388 p.

[16] IAEA. Radiological Characterization of Shut Down Nuclear Reactors for Decommissioning Purposes, Technical Reports Series No. 389. Vienna: IAEA; 1998. 184 p.

[17] Preparatory Commission for the Comprehensive Nuclear Test-Ban-Treaty. General overview of the effects of nuclear testing [Internet]. 2015. Available from: http://www.ctbto.org/nuclear-testing/the-effects-of-nuclear-testing/general-overview-of-theeffects-of-nuclear-testing/ [Accessed: 2016/05/20].

[18] Norris RS, Arkin WM. Known nuclear tests worldwide, 1945–1994. Bulletin of the Atomic Sciences. 1995; 51: 70–71.

[19] Gabrieli J, Cozzi G, Vallelonga P, Schwikowski M, Sigl M, Eickenberg J, et al. Contamination of Alpine snow and ice at Colle Gnifetti, Swiss/Italian Alps, from nuclear weapons tests. Atmospheric Environment. 2011; 45: 587–593.

[20] Prăvălie R. Nuclear weapons tests and environmental consequences: A global perspective. Ambio. 2014; 43: 729–744.

[21] Cizdziel JV, Ketterer ME, Farmer D, Faller SH, Hodge VF. $^{239, 240, 241}Pu$ fingerprinting of plutonium in western US soils using ICPMS: Solution and laser ablation measurements. Analytical and Bioanalytical Chemistry. 2008; 390: 521–530.

[22] Turner M, Rudin M, Cizdziel J, Hodge V. Excess plutonium in soil near the Nevada Test Site, USA. Environmental Pollution. 2003; 125:1 93–203.

[23] Steinhauser G, Brandl A, Johnson TE. Comparison of the Chernobyl and Fukushima nuclear accidents: A review of the environmental impacts. Science of the Total Environment; 2014; 470–471: 800–817.

[24] Yablokov AV, Nesterenko VB, Nesterenko AV. Atmospheric, water, and soil contamination after Chernobyl. Annals of New York Academy Science. 2009; 1181: 223–236.

[25] Martin A, Mead S, Wade BO. European Commission Report EUR 17625 – Materials Containing Natural Radionuclides in Enhanced Concentrations. Luxembourg: Office for Official Publications of the European Communities; 1997. 88 p.

[26] UNSCEAR. Exposure from Natural Sources of Radiation. United Nations Scientific Commettee on the Effects of Atomic Radiation. Report to General Assembly. New York: UNSCEAR; 1993. 922. p.

[27] Botezatu E, Iacob O. Radiological impact of TENORM on the environment in Romania. In: Proceedings of the 11th International Congress of IRPA; 23–28 May, Madrid. Madrid: IRPA; 2004. p. 1–10.

[28] Waltar A. The medical, agricultural, and industrial applications of nuclear technology [Internet]. 2003. Available from: http://www.laradioactivite.com/site/pages/RadioPDF/Waltar.pdf [Accessed: 2016/05/13].

[29] Hazen T, Faybishenko B. Groundwater and soil contaminants at DOE sites. Available from: http://doesbr.org/PImeetings/PI_mtg_09/present/Wednesday/Hazen DOE site Contamination.pdf [Accessed: 2016/05/17].

[30] Skipperud L, Salbu B. Sequential extraction as a tool for mobility studies of radionuclides and metals in soils and sediments. Radiochimica Acta. 2015; 103: 187–197.

[31] Brady NC, Weil RR. The Nature and Properties of Soils. New York: Macmillan Publishing Co.; 2002. 960 p.

[32] Osman KT. Soils: Principles, Properties and Management. Dordecht: Springer Netherlands; 2013. 247 p.

[33] Sparks DL. Environmental Soil Chemistry. 2nd edition. San Diego: Academic Press; 2003. 352 p.

[34] Sposito G. The Chemistry of Soils. 2nd edition. New York: Oxford University Press; 2008. 330 p.

[35] Yong RN, Mohamed AMO, Warkentin BP. Principles of Contaminant Transport in Soils. Amsterdam, New York: Elsevier Science; 1992. 327 p.

[36] Dimović S, Smičiklas I, Šljivić-Ivanović M, Dojčinović B. Speciation of ^{90}Sr and other metal cations in artificially contaminated soils: The influence of bone sorbent addition. Journal of Soils and Sediments 2013; 13: 383–393.

[37] Blanco P, Tomé FV, Lozano JC. Fractionation of natural radionuclides in soils from a uranium mineralized area in the south-west of Spain. Journal of Environmental Radioactivity. 2005; 79: 315–330.

[38] Popic J, Meland S, Salbu B, Skipperud L. Mobility of radionuclides and trace elements in soil from legacy NORM and undisturbed naturally 232Th-rich sites. Environmental Sciences: Process & Impacts. 2014; 16: 1124–1134.

[39] Smiciklas I, Dimovic S, Jovic M, Milenkovic A, Šljivic-Ivanovic M. Evaluation study of cobalt(II) and strontium(II) sorption–desorption behavior for selection of soil remediation technology. International Journal of Environmental Science and Technology. 2015; 12: 3853–3862.

[40] Smičiklas I, Jović M, Šljivić -Ivanović M, Mrvić V, čakmak D, Dimović S. Correlation of Sr2+ retention and distribution with properties of different soil types. Geoderma 2015; 253–254: 21–29.

[41] Lopez-Sanchez JF, Sahuquillo A, Rauret GG, Lachica M, Barahona E, Gomez A, Ure AM, Muntau H, Quevauviller P. Extraction procedures for soil analysis. In: Quevauviller P, editor. Methodologies in Soil and Sediment Fractionation Studies. Single and Sequential Extraction Procedures. Cambridge: The Royal Society of Chemistry; 2002. pp. 28–65.

[42] Ng JC, Juhasz A, Smith E, Naidu R. Assessing the bioavailability and bioaccessibility of metals and metalloids. Environmental Science and Pollution Research. 2015; 22: 8802–8825.

[43] Sheppard SC, Evenden WG. Bioavailability indices for uranium: Effect of concentration in eleven soils. Archives of Environmental Contamination and Toxicology. 1992; 23: 117–1124.

[44] Tessier A, Campbell PGC, Bisson M. Sequential extraction procedure for the speciation of particulate trace metals. Analytical Chemistry. 1979; 51: 844–851.

[45] Ure AM, Quevauviller P, Muntau H, Griepink B. Speciation of heavy metals in soils and sediments. An account of the improvement and harmonization of extraction techniques undertaken under the auspices of the BCR of the Commission of the European Communities. International Journal of Environmental Analytical Chemistry 1993; 51: 135–151.

[46] Schultz MK, Burnett WC, Inn KGW. Evaluation of a sequential extraction method for determining actinide fractionation in soils and sediments. Journal of Environmental Radioactivity. 1998;40:155–174.

[47] Okoro HK, Fatoki OS. A review of sequential extraction procedures for heavy metals speciation in soil and sediments. Journal of Environmental and Analytical Toxicology. 2012; 1: 181–190.

[48] Quevauviller P. SM&T activities in support of standardization of operationally defined extraction procedures for soil and sediment analyses. In: Quevauviller P, editor. Methodologies in Soil and Sediment Fractionation Studies. Single and Sequential Extraction Procedures. Cambridge: The Royal Society of Chemistry; 2002. p. 1–9.

[49] Salbu B, Krekling T. Characterisation of radioactive particles in the environment. Analyst. 1998; 123: 843–850.

[50] Kumar A, Rout S, Ghosh M, Singhal RK, Ravi PM. Thermodynamic parameters of U(VI) sorption onto soils in aquatic systems. Springerplus. 2013; 2: 1–7.

[51] Qin H, Yokoyama Y, Fan Q, Iwatani H, Tanaka K, Sakaguchi A, et al. Investigation of cesium adsorption on soil and sediment samples from Fukushima prefecture by sequential extraction and EXAFS technique. Geochemical Journal 2012; 46: 297–302.

[52] Arapis G, Goncharova N, Baveye P, editors. Ecotoxicology, Ecological Risk Assessment and Multiple Stressors. Dordrecht: Kluwer Academic Publishers; 2006. 360 p.

[53] Koch I, Reimer KJ, Bakker MI, Basta NT, Cave MR, Denys S, et al. Variability of bioaccessibility results using seventeen different methods on a standard reference material, NIST 2710. Journal of Environmental Science and Health A Toxic/Hazardous Substances and Environmental Engineering. 2013; 48: 641–655.

[54] Iurian A-R, Phaneuf MO, Mabit L. Mobility and bioavailability of radionuclides in soils. In: Walther C, Gupta KD, editors. Radionuclides in the Environment: Influence of Chemical Speciation and Plant Uptake on Radionuclide Migration. Cham: Springer International Publishing; 2015. p. 37–59.

[55] Strebl F, Ehlken S, Gerzabek MH, Kirchner G. Behaviour of radionuclides in soil/crop systems following contamination. Radioactivity in the Environment. 2007; 10: 19–42.

[56] Tamponnet C, Martin-Garin A, Gonze MA, Parekh N, Vallejo R, Sauras-Yera T, et al. An overview of BORIS: Bioavailability of radionuclides in soils. Journal of Environmental Radioactivity. 2008; 99: 820–830.

[57] Lawson LS, McComb JQ, Dong R, Han FX, Roger C, Arslan Z, et al. Binding, fractionation, and distribution of Cs, Co, and Sr in a US coastal soil under saturated and field capacity moisture regimes. Journal of Soils and Sediments. 2016; 16: 497–508.

[58] de Groot G, van der Sloot H. Determination of leaching characteristics of waste materials leading to environmental product certification. In: Gilliam TM, Willes C, editors. Stabilization and Solidification of Hazardous, Radioactive, and Mixed Wastes. Volume 2. West Conshohocken: ASTM International; 1992. p. 149–170.

[59] Degryse F, Smolders E, Parker DR. Partitioning of metals (Cd, Co, Cu, Ni, Pb, Zn) in soils: Concepts, methodologies, prediction and applications – A review. European Journal of Soil Science. 2009; 60: 590–612.

[60] Oughton DH, Børretzen P, Salbu B, Tronstad E. Mobilisation of ^{137}Cs and ^{90}Sr from sediments: Potential sources to arctic waters. Science of the Total Environment. 1997; 202: 155–165. Oughton DH, Børretzen P, Salbu B, Tronstad E. Mobilisation of 137Cs and 90Sr from sediments: potential sources to arctic waters. Science of the Total Environment. 1997; 202: 155–65.

[61] Burns PA, Clarck DE, Claxton D, Collard G, Ettenhuber E, Fazlullin M, et al. Technologies for Remediation of Radioactively Contaminated Sites. IAEA Tecdoc-1086. Vienna: IAEA; 1999. 101 p.

[62] Chao JC, Hong A, Okey RW, Peters RW. Selection of chelating agents for remediation of radionuclide-contaminated soil. In: Proceedings of the 1998 Conference on Hazardous Waste Research; 18–21 May, Snowbird, Utah; 1998. p. 142–156.

[63] Porter SK, Scheckel KG, Impellitteri C, Ryan J. Toxic metals in the environment: Thermodynamic considerations for possible immobilization strategies for Pb, Cd, As, and Hg. Critical Reviews in Environmental Science and Technology. 2004; 34: 495–604.

[64] US EPA. Selecting Remediation Techniques for Contaminated Sediment, EPA-823-B93-001. Washington, Cincinatty: US EPA; 1993. 242 p.

[65] Bolan N, Kunhikrishnan A, Thangarajan R, Kumpiene J, Park J, Makino T, et al. Remediation of heavy metal(loid)s contaminated soils – To mobilize or to immobilize? Journal of Hazardous Materials. 2014; 266: 141–166.

[66] Pickering WF. Metal ion speciation — soils and sediments (a review). Ore Geology Reviews. 1986; 1: 83–146.

[67] Gavrilescu M, Pavel LV, Cretescu I. Characterization and remediation of soils contaminated with uranium. Journal of Hazardous Materials. 2009; 163: 475–510.

[68] Peters RW. Chelant extraction of heavy metals from contaminated soils. Journal of Hazardous Materials. 1999; 66: 151–210.

[69] Dermont G, Bergeron M, Mercier G, Richer-Laflèche M. Soil washing for metal removal: A review of physical/chemical technologies and field applications. Journal of Hazardous Materials. 2008; 152: 1–31.

[70] Odile P-L, Laurence R, Chapalain E. Management of sites and soils polluted by radioactivity. Controle Review. 2012; 195: 18–20.

[71] USEPA. Technology Performance Review: Selecting and Using Solidification/Stabilization Treatment for Site Remediation. Cincinnaty: National Service Center for Environmental Publications (NSCEP); 2009. p. 21.

[72] O'Day PA, Vlassopoulos D. Mineral-based amendments for remediation. Elements 2010; 6: 375–381.

[73] Kim CK, Kong JY, Chun BS, Park JW. Radioactive removal by adsorption on Yesan clay and zeolite. Environmental Earth Sciences. 2013; 68: 2393–2398.

[74] Torstenfelt B, Andersson K, Allard B. Sorption of strontium and cesium on rocks and minerals. Chemical Geology. 1982; 36: 123–137.

[75] Li D, Kaplan DI, Knox AS, Crapse KP, Diprete DP. Aqueous ^{99}Tc, ^{129}I and ^{137}Cs removal from contaminated groundwater and sediments using highly effective low-cost sorbents. Journal of Environmental Radioactivity. 2014; 136: 56–63. Li D, Kaplan DI, Knox AS, Crapse KP, Diprete DP. Aqueous 99Tc, 129I and 137Cs removal from contaminated groundwater and sediments using highly effective low-cost sorbents. Journal of Environmental Radioactivity. 2014; 136: 56–63.

[76] Osmanlioglu AE. Treatment of radioactive liquid waste by sorption on natural zeolite in Turkey. Journal of Hazardous Materials. 2006; 137: 332–335.

[77] Hodgson JF. Cobalt reactions with montmorillonite. Soil Science Society of America Journal. 1960; 24: 165–168.

[78] Šljivić-Ivanović M, Smičiklas I, Dimović S, Jović M, Dojčinović B. Study of simultaneous radionuclide sorption by mixture design methodology. Industrial and Engineering Chemistry Research. 2015; 54: 11212–11221.

[79] Plećaš I, Dimović S, Smičiklas I. Influence of bentonite and zeolite in cementation of dry radioactive evaporator concentrates. Applied Clay Science. 2009; 43: 9–12.

[80] Smiciklas I, Dimovic S, Plecaš I. Removal of Cs1+, Sr2+ and Co2+ from aqueous solutions by adsorption on natural clinoptilolite. Applied Clay Science. 2007; 35: 139–144.

[81] Smičiklas I, Dimović S, Plecaš I, Mitrić M. Removal of Co^{2+} from aqueous solutions by hydroxyapatite. Water Research. 2006; 40: 2267–2274.

[82] Šljivić-Ivanović M, Milenković A, Jović M, Dimović S, Mraković A, Smičiklas I. Ni(II) immobilization by bio-apatite materials: Appraisal of chemical, thermal and combined treatments. Chemical Industry and Chemical Engineering Quarterly. 2016; 22:117–126.

[83] Thomson B, Smith C, Busch R, Siegel M, Baldwin C. Removal of metals and radionuclides using apatite and other natural sorbents. Journal of Environmental Engineering. 2003; 129: 492–499.

[84] Sasaki K, Goto T. Immobilization of Sr^{2+} on naturally derived hydroxyapatite by calcination of different species of fish bones and influence of calcination on ion-exchange efficiency. Ceramics International. 2014; 40: 11649–11656.

[85] Smiciklas I, Šljivić-Ivanović M. Evaluation of factors influencing Co^{2+} removal by calcinated bone sorbent using experimental design methodology. Journal of Environmental Science and Health – Part A Toxic/Hazardous Substances and Environmental Engineering. 2012; 47: 896–908.

[86] Smičiklas I, Dimović S, Šljivić M, Plećaš I, Lončar B, Mitrić M. Resource recovery of animal bones: Study on sorptive properties and mechanism for Sr^{2+} ions. Journal of Nuclear Materials. 2010; 400: 15–24.

[87] Wright J, Nw PDP, Conca JL, Lanl PD, Parsons KRR, Murphy B. PIMS using apatite II™: How it works to remediate soil & water. Environ Res. 2004; http://www.pimsnw.com/files/1howapatiteiiworks30.pdf. [Accessed: 2016/05/10].

[88] Dimović S, Smičiklas I, Plećaš I, Antonović D, Mitrić M. Comparative study of differently treated animal bones for Co^{2+} removal. Journal of Hazardous Materials 2009; 164: 279–287.

[89] Fuller CC, Bargar JR, Davis JA. Molecular-scale characterization of uranium sorption by bone apatite materials for a permeable reactive barrier demonstration. Environmental Science and Technology. 2003; 37: 4642–4649.

[90] IAEA. Inorganic Ion Exchangers and Adsorbents for Chemical Processing in the Nuclear Fuel Cycle – IAEA-TEC DOC –337. Vienna: IAEA; 1985. 282 p.

[91] Valcke E, Engels B, Cremers A. The use of zeolites as amendments in radiocaesium- and radiostrontium-contaminated soils: A soil-chemical approach. Part II: Sr-Ca exchange in clinoptilolite, mordenite, and zeolite A. Zeolites. 1997; 18: 212–217.

[92] Moody TE, Petersen SW, Torne EG, Vlcakova J, Higginbotham JF. Laboratory scale stabilization of N-springs groundwater strontium-90 using phosphatic materials. 1996. Available from: http://www.dtic.mil/cgi-bin/GetTRDoc?AD=ADA349254. [Accessed: 2016/05/11]

[93] Monteil-Rivera F, Fedoroff M. Sorption of inorganic species on apatites from aqueous solutions. In: Somasundaran P, editor. Encyclopedia of Surface and Colloid Science. 2nd edition. New York: Marcel Dekker; 2002. p. 1–26.

[94] Smičiklas I, Onjia A, Raičević S, Janać ković Đ, Mitrić M. Factors influencing the removal of divalent cations by hydroxyapatite. Journal of Hazardous Materials. 2008; 152: 876–884.

[95] Arey JS, Seaman JC, Bertsch PM. Immobilization of uranium in contaminated sediments by hydroxyapatite addition. Environmental Science and Technology. 1999; 33: 337–342.

[96] Boisson J, Ruttens A, Mench M, Vangronsveld J. Evaluation of hydroxyapatite as a metal immobilizing soil additive for the remediation of polluted soils. Part 1. Influence of hydroxyapatite on metal exchangeability in soil, plant growth and plant metal accumulation. Environmental Pollution. 1999; 104: 225–233.

[97] Li SY, Xie SB, Zhao C, Zhang YP, Liu JX, Cai T. Efficiency of adsorption of wastewater containing uranium by fly ash. Advanced Materials Research. 2013; 639–640: 1295–1299.

[98] Apak R, Atun G, Güçlü K, Tütem E. Sorptive removal of cesium-137 and strontium-90 from water by unconventional sorbents. II. Usage of coal fly ash. Journal of Nuclear Science and Technology. 1996; 33: 396–402.

[99] Apak R, Atun G, Güçlü K, Tütem E, Keskin G. Sorptive removal of cesium-137 and strontium-90 from water by unconventional sorbents. I. Usage of bauxite Wastes (Red Muds). Journal of Nuclear Science and Technology. 1995; 32: 1008–1017.

[100] Clark MW, Payne TE, Harrison JJ, Comarmond MJ, Collins RN, Reichelt-Brushett AJ. Reversibility of uranium and thorium binding on a modified bauxite refinery residue: The effects of aging temperature. Applied Geochemistry. 2015; 53: 79–90.

[101] Milenković A, Smičiklas I, Marković J, Vukelić N. Immobilization of ^{60}Co and ^{90}Sr ions using red mud from aluminum industry. Nuclear Technology and Radiation Protection. 2014; 29: 79–87.

[102] Basu M, Pande M, Bhadoria PBS, Mahapatra SC. Potential fly-ash utilization in agriculture: A global review. Progress in Natural Science. 2009; 19: 1173–1186.

[103] Lima AT, Ottosen LM, Ribeiro AB. Assessing fly ash treatment: Remediation and stabilization of heavy metals. Journal of Environmental Management. 2012; 95(Suppl.): S110–S115.

[104] Sitarz-Palczak E. Study of remediation of soil contamined with heavy metals by coal fly ash. Journal of Environmental Protection. 2012; 3: 1373–1383.

[105] Gu HH, Li FP, Yu Q, Xu YL, Zhang JY. Remediation effects of steel slag and fly ash on heavy metal contaminated acidic soil. In: Advanced Materials Research; International Conference on Biomaterial and Bioengineering, ICBB 2012; 19–20 December 2012; Hong Kong. Pfaffikon: Trans Tech Publications; 2013. p. 726–731.

[106] Wang S, Ang HM, Tadé MO. Novel applications of red mud as coagulant, adsorbent and catalyst for environmentally benign processes. Chemosphere. 2008; 106: 1621–1635.

[107] Cappuyns V. Use of red mud in soil remediation: review of applications and challenges. In: Bauxite Residue Valorisation and Best Practices Conference, 5–7 October, 2015,. Leuven; 2015. p. 81–7.

11

Mycoremediation of Atrazine in a Contaminated Clay-Loam Soil and its Adsorption-Desorption Kinetic Parameters

Wilberth Chan Cupul and
Refugio Rodríguez Vázquez

Additional information is available at the end of the chapter

Abstract

Clean-up of contaminated soils with atrazine is an ecological responsibility. The objectives of this study are to evaluate atrazine degradation in a clay-loam soil microcosm using fungal enzyme extracts from *Trametes maxima* and its co-culture with *Paecilomyces carneus* and to determine the kinetic parameters of the adsorption-desorption of atrazine in soil. Fungal co-culture extract (*T. maxima-P. carneus*) and monoculture (*T. maxima*) were able to degrade 100% of atrazine. However, we observed variation in atrazine degradation over the course of the evaluated time period, which suggests that an adsorption-desorption process is occurring in the soil. Adsorption-desorption kinetic parameters of the Freundlich model revealed that the studied soil has a significant capacity to adsorb atrazine (K_F = 8.2148; r^2 = 0.992 and P-value < 0.0001), while according to the desorption parameters (K_F = 5.4992; r^2 = 0.245 and P-value = 0.036) and hysteresis index (H = 0.573), the soil does not desorb atrazine at the same rate. Fungal enzyme extracts from a monoculture and co culture of *T. maxima* were able to degrade atrazine in a short time period (< 12 h). The ability of the contaminated soils to adsorb and desorb atrazine should be taken into account in mycoremediation systems.

Keywords: bioremediation, fungal enzyme extract, laccase, soil organic matter

1. Introduction

Atrazine (2-chloro-4-ethylamino-6-isopropylamino-1,3,5-triazine) is the most widely used herbicide around the world. In agricultural soils, approximately 29–34 million kg of atrazine are applied per year [1]. In Mexico, more than 45% of pesticides are categorized as herbicides, and atrazine is one of the most commonly used herbicides in Mexican agriculture [2]. Since 1975, atrazine has been applied to control broadleaf and grass weeds in agricultural crops, including corn, sorghum and sugar cane. Atrazine kills susceptible plants by binding to the quinone protein in photosystem II and inhibiting photosynthetic electron transport [3].

Atrazine is a pre-emergent herbicide and is considered to have low persistence in soil (<12 months). However, its low mobility in soil and its physical and chemical properties contribute towards the contamination of ground and surface waters, which represents a risk to the environment and to human health [4]. In Mexico, atrazine levels in water often exceed the maximum permissible levels for drinking water (0.1 $\mu g\ L^{-1}$) as established by Europe and by the health advisory board of the United States Environmental Protection Agency (EPA) [5, 6]. At a molecular level, atrazine has distinct fates in the environment and may be found in soil, water, biomass (plants) or air. In soil, atrazine is adsorbed by clay particles; however, other adsorption-desorption processes may be involved in its translocation in plants, movement in soil and mobility in aqueous systems, as well as its eventual abiotic or biological degradation [7].

The clean-up of soils contaminated with atrazine is an ecological responsibility, and the discovery of a safe and economical method is a major priority for land management agencies [8]. One such possibility is mycoremediation, or the use of fungal organisms and their enzymes to degrade or transform environmental pollutants [9]. This strategy has been used to degrade pesticides [10], aromatic and polycyclic hydrocarbons [11] and endocrine disruptors [12]. The degradation of environmental pollutants by fungi, specifically by white-rot fungi, is due to their ability to synthesize ligninolytic enzymes, such as laccase, manganese peroxidase and lignin peroxidase, as well as their production of hydrogen peroxide [13, 14].

However, mycoremediation faces several challenges in order to improve the feasibility of this strategy. The following issues, for example, should be addressed: (i) the competition/proliferation of native soil microorganisms (actinomycetes and bacteria) may inhibit the growth of bioremedial fungi; (ii) bioremedial fungi have a limited capacity to produce ligninolytic enzymes. Enzyme production varies depending on the strain and species and is mainly influenced by the content and availability of nutrients (carbon, nitrogen, metal ions, etc.), which stimulate fungal growth and the synthesis of ligninolytic enzymes; finally, (iii) edaphic and environmental factors may adversely affect the establishment and growth of bioremedial fungi [9].

The use of fungal extracts with a proven high activity of ligninolytic enzymes is one means of improving the degradation of pollutants in soil, which may also address some of the aforementioned challenges. Ligninolytic enzymes in white-rot fungi, for example, may be enhanced through the use of fungal co-cultures, although the mechanism by which increased enzyme activity occurs has not yet been described [15]. Given this context, the objectives of the study were: (i) to evaluate the degradation of atrazine in soil microcosms by a white-rot fungus (*Trametes maxima*) and its co-culture with a soil-borne micromycete (*Paecilomyces carneus*) and ii) to determine the absorption-desorption kinetics of atrazine in a clay-loam soil.

2. Materials and methods

2.1. Fungal source and molecular identification

The white-rot fungi *T. maxima* was isolated from a carpophore collected in a rain forest (19°32′21.23″ N, 97°00′47.29″ W) near Vega del Pixquiac, San Ándres Tlalnelhuayocan, Veracruz, Mexico. To obtain the isolate, 0.5–1 cm fragments of the carpophore were cut and removed; these were washed in ethanol (70%) for 1 min, in sodium hypochlorite (50%) for 3 min and finally, in sterile, distilled water. The washed and disinfected fragments were placed on potato-dextrose agar plates (Bioxon®, Mexico) and supplemented with chloramphenicol (20 mg/L; Sigma-Aldrich, St. Louis, MO, USA) to prevent bacterial contamination and benomyl (3 mg/L; Biesterfeld Co., Mexico) to inhibit mold growth.

The soil-borne micromycete *P. carneus* Duché & R. Heim (Trichocomaceae: Ascomycota) was donated by the Micromycetes Laboratory of the Institute of Ecology (INECOL A.C.) located in Xalapa, Mexico. This strain was isolated from an andic acrisol soil (texture: loam-silt loam) from a coffee plantation in Huatusco, Veracruz, Mexico (location: 19°12′57″ N, 96°53′7″ W). The carpophores of *T. maxima* (Mont.) A. David & Rajchenb (Polyporaceae: Basidiomycota) and the *P. carneus* strain are stored in the herbarium (XAL) and Micromycetes Culture Collection of INECOL. Both strains were maintained and subcultured in potato dextrose agar.

2.2. Soil sampling and characterization

Soil samples were collected from the first horizon of <20 cm profundity at a sugar cane plantation in Mahuixtlan, Veracruz, Mexico (location: 19°23′21.3″ N, 96°53′34.9″ W). Plant residues and rocks were removed manually. Soil was sieved in 2 mm mesh in the laboratory and dried at 20°C for 5 days prior to use. The physical and chemical characteristics of the soil were determined using standard methods to establish texture (clay loam soil), soil organic matter (4.35%), pH (4.86), NH_4-N (5.8 mg kg^{-1}), soluble salts (5.38 S m^{-1}), acidity (0.053 meq 100 g^{-1}), cation exchange capacity (16.41 meq 100 g^{-1}), water holding capacity (WHC) (53.6%) and electrical conductivity (53.75 μS cm^{-1}).

2.3. Production of ligninolytic enzymes through fungal co-culture

Modified Sivakumar culture medium [16] was used to produce laccase, MnP and H_2O_2 for the monoculture of *T. maxima* and the co-culture of both *T. maxima* and *P. carneus*. To establish the co-culture, four agar plugs of *T. maxima* (7 days old) were deposited in a 250 mL Erlenmeyer flask with 120 mL of modified Sivakumar culture medium. After 3 days, four agar plugs of *P.*

carneus (9 days old) were added. Monocultures of both fungi were established at the same time. Fungal cultures were incubated at 25°C and 120 rpm for 6 days. After this step, the fungal enzyme extracts (FEEs) were centrifuged at 7000 rpm during 10 min. The supernatant was filtered with a 0.2 mm nylon filter; this process allows a cell-free extract to be obtained, which was used to determine laccase and MnP activity and H_2O_2 content.

2.4. Ligninolytic enzyme activity and H_2O_2 quantification

2.4.1. Laccase determination

Laccase activity was determined according to More et al. [17] by measuring the oxidation of ABTS [2,2′-azino-bis (3-ethylbenzthiazoline-6-sulfonic acid)] in a reaction mixture (1 mL) containing 100 µL of ABTS (0.5 mM, Sigma, St. Louis, MO, USA), 800 µL of acetate buffer (100 mM, pH 4.5) and 100 µL of enzyme extract. Absorbance changes in the presence of the enzyme were monitored during 5 min at 420 nm ($\varepsilon = 3.6 \times 10^4$ M^{-1} cm^{-1}). One unit of laccase activity was defined as the amount of enzyme required to oxidize 1 µmol ABTS per minute per milligram of protein under the assay conditions.

2.4.2. Manganese peroxidase assay

MnP activity was determined at 610 nm ($\varepsilon = 4460$ M^{-1} cm^{-1}), following the method described by Kuwahara et al. [18]. The reaction mixture contained the following: 700 µL of enzyme extract, 50 µL of phenol red (0.2%), 50 µL of sodium lactate (0.5 mM), 50 µL of egg albumin (0.1%), 50 µL of manganese sulfate (2 mM) and 50 µL of H_2O_2 (2 mM). The reaction was carried out in 50 µL of sodium succinate buffer (20 mM) at pH 4.5. After 5 minutes, 50 µL of NaOH (2N) was added to stop the reaction. One enzyme unit was defined as 1 µmol of the product formed per minute per milligram of protein under the assay conditions.

2.4.3. Hydrogen peroxide content

H_2O_2 content of the fungal enzyme extracts (FEEs) was determined using the iodide/iodate method, according to Klassen et al. [19]. Three milliliters of the FEEs were mixed with 3 mL of a solution containing KI (33 g), NaOH (1 g) and $(NH_4)_6Mo_7O_{24} \times 4H_2O$ (0.1 g) in 500 mL of distilled, deionized water, in addition to 3 mL of a solution containing $C_8H_4KO_4$ (10 g) in 500 mL of distilled, deionized water. The absorbance of the resulting solution was measured at 351 nm in a 3 cm^3 cuvette. The blank absorbance was determined by substituting the FEEs with a sterile Sivakumar culture medium in the reaction mixture. Hydrogen peroxide content was calculated by substituting with H_2O_2 reagent (30%, J.T. Baker™) according to the standard curve of the absorbance of known concentrations (0.0, 0.5, 1.0, 1.5, 2.0, 2.5, 3.0 and 3.5 mg L^{-1}).

2.5. Biodegradation studies

Biodegradation of atrazine was evaluated in sterile soil microcosm conditions. First, 20 g of air dried soil was placed in serological flasks (100 cm^3). Then, the sterile soil was contaminated with atrazine (Sigma-Aldrich Co., USA) at the field application rate of 5mg/kg [8], and 20 mL of methanol were added (analytical grade, Honeywell Burdick & Jackson, Muskegon, MI, USA). Soil-methanol-atrazine was mixed using a sterile spatula until the complete evaporation of methanol under a laminar flow hood.

Three treatments were evaluated: *T. maxima* extract, *P. carneus* extract and their co-culture extract (*T. maxima-P. carneus*). Soil microcosms were adjusted to a water holding capacity (WHC) of 40% using 0.215 mL of fungal extract per gram of soil. Four experimental units (serological flasks) were used per treatment. Atrazine degradation was evaluated at 1, 3, 6 and 12 h using high-pressure liquid chromatography (HPLC) analysis. In addition, abiotic (sterile soil) and biotic (nonsterilized soil) controls were used.

2.6. Adsorption-desorption studies

Experiments were conducted using six sorbate concentrations of atrazine (0.5, 1, 5, 10, 20 and 30 mg/kg). Two grams of all soil samples were added to a polypropylene bottle (20 mL), and immediately 5 mL of a methanol solution with the sufficient amount of atrazine was added to obtain the established concentration. Bottles were shaken vigorously (24 h) and placed on a flat rotator shaker (120 rpm) at room temperature (27 ± 1°C) [20]. Four replicates were used for each initial concentration of atrazine. After an equilibration period (24 h), samples were centrifuged in cold (5°C) at 7000 rpm during 20 min. Then, 0.2 mL of supernatant was filtered through a 0.22 μm nylon syringe. The filtrate was used to analyze the atrazine adsorbed using HPLC.

Desorbed atrazine was determined by examining the solid phase of the centrifuged samples; 5 mL of methanol was added in each bottle and shaken during 24 h at 120 rpm in a flat rotatory shaker. After the agitation period, the bottles were centrifuged and filtered as mentioned above for further atrazine analysis.

2.7. Atrazine analysis

The analysis of atrazine degradation and its desorption-adsorption was performed using a Thermo-Scientific HPLC system coupled to a diode array detector (SpectraSystem UV8000), a sampling injector (SpectraSystem AS3000) and a pump (SpectraSystem P4000) equipped with a Restek ultra C18 column (5 mm × 150 mm × 4.6 mm). The column was operated at 25°C with a flow rate of 1.0 mL min^{-1} and an injection volume of 20 μL. An isocratic mobile phase was established using acetonitrile-water at a ratio of 70:30. The HPLC-photodiode array detector was monitored at 215 nm [8]. The HPLC method had a running time of 10 min and a retention time of 3.8 min, which enabled the detection and quantification of atrazine. The atrazine detection limit was 0.05 mg g $soil^{-1}$. The standard curve for atrazine [atrazine = (peak area – 491818)/804962] was made using a standard analytical solution (Sigma-Aldrich Co., USA) at different concentrations, and the r^2 value was >0.99. The extraction efficiency of this method was 105%, and this value was taken into account in the final quantifications.

3. Results and discussion

3.1. Enzyme characterization of fungal extracts

Laccase activity and H_2O_2 production in the fungal co-culture (laccase = 18956.0 U/mg of protein and H_2O_2 = 6.2 mg/L) were significantly higher (T = 6.19, P = 0.0004) than in the *T. maxima* monoculture (laccase = 12866.2 U/mg of protein and H_2O_2 = 4.2 mg/L). Regarding MnP activity, we did not find significant differences between the fungal co-culture and the *T. maxima* monoculture (T = 0.27, P = 0.3957). Since *P. carneus* is a soil microfungus (Hyphomycete), it did not show laccase or MnP activity; only H_2O_2 production (0.9 mg/L) was detected, which was significantly lower (F = 126.4, P = 0.00001) than in the *T. maxima* monoculture (4.2 mg/L) and fungal co-culture (6.2 mg/L, **Table 1**).

	Fungal enzyme extracts			
Variable	*P. carneus*	*T. maxima*	**Co-culture**	**Mean Comparison test**
Laccase (U/mg of protein)	ND	12866.2 ± 446.7	18956.0 ± 204.0	*t*-student [T = 6.19, P = 0.0004]
MnP (U/mg of protein)	ND	572.4 ± 31.8	542.6 ± 43.5	*t*-student [T = 0.27, P = 0.3957]
H_2O_2 (mg/L)	0.9 ± 0.07 c	4.2 ± 0.10 b	6.2 ± 0.15 a	Fisher [F = 126.4, P = 0.00001]

Laccase and MnP were compared with the *t*-student test, and H_2O_2 content was compared using an ANOVA and LSD test for mean comparison. Means with different letters are significantly different from each other (P = 0.05). ND = No detected.

Table 1. Amount of enzymes in fungal extracts.

Laccase is an important enzyme in white-rot fungi; this enzyme is a defence mechanism against saprotrophic and parasitic microfungi. This phenomenon has been reported for *Lentinula edodes* [21], *Agaricus bisporus* [22] and *Pleurotus ostreatus* [23] when infected with *Trichoderma* sp. These macromycetes have been studied due to their importance as edible mushrooms, and *Trichoderma* is their naturally antagonistic fungus, especially in production systems. In particular, recent studies have sought solutions stemming from fungal interactions to obtain relevant biotechnological solutions and products. Thus, the interaction between white-rot fungi (Basidiomycetes) and other soil-borne micromycetes (hyphomycetes) has received greater interest in recent years [24, 25].

One of the principal applications of fungal co-cultures is to increase ligninolytic enzyme activity (laccase, MnP and LiP), and these may then be applied to resolve environmental problems, such as the contamination of soil and water with pesticides or the presence of endocrine disruptors, medical drugs, hydrocarbons, dyes or other emerging contaminants in the environment. Several studies have reported that soil-borne micromycetes enhance ligninolytic enzyme activity in white-rot fungi; for example, Baldrian [25] reported that *Sphaerospermum* sp., *Acremonium* sp., *Fusarium reticulatum*, *Humicola grisea* and *Penicillium rugulosum* enhanced laccase activity in *Trametes versicolor* and *Pleurotus ostreatus* when co-cultivated. Dwivedi et al. [26] reported an increase in the laccase activity of *Pleurotus ostreatus* when co-cultured with *Penicillium oxalicum*. In addition, Chan-Cupul et al. [15] recently demonstrated that laccase and MnP activity in a specific co-culture may be increased if the culture media are optimized. In that study, a 1.8- and 2.9-fold increase in laccase and MnP activities, respectively, was recorded for the co-culture of *T. maxima* and *P. carneus*.

3.2. Biodegradation studies

Figure 1 shows atrazine degradation by fungal enzyme extracts (FEEs) from the monocultures of *T. maxima* and *P. carneus* and their co-culture. One hour after application, the co-culture enzyme extract degraded 100% of atrazine at a significantly higher rate ($F = 331.31$, $P = 0.00001$) than *T. maxima* and *P. carneus* extracts, which degraded 80.0% and 27.3% of atrazine, respectively (**Figure 1A**). At 3 h after application, the monoculture extract of *T. maxima* (84.5%) statistically achieved the same level of atrazine degradation as the co-culture extract (89.1%); however, both values were higher ($F = 320.5$, $P = 0.0001$) than atrazine degradation by the *P. carneus* enzyme extract (5.3%, **Figure 1B**).

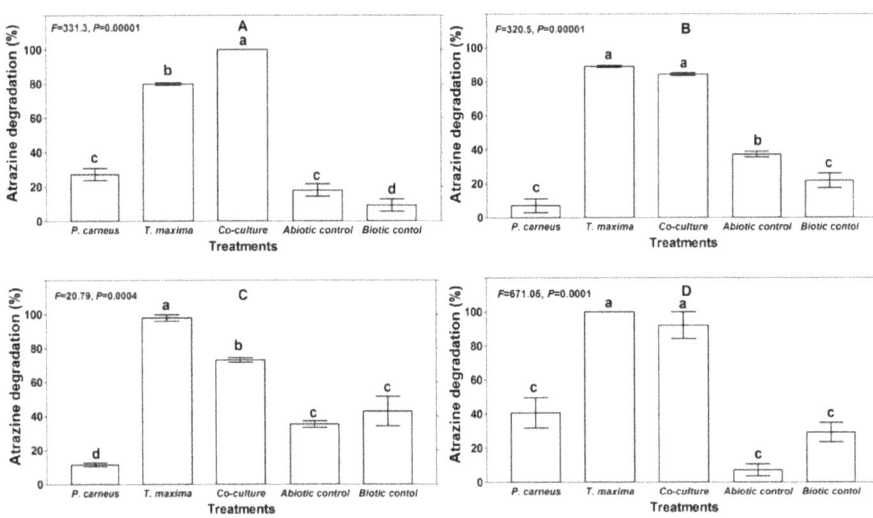

Figure 1. Atrazine degradation at 1 h (A), 3 h (B), 6 h (C) and 12 h (D) after application of fungal enzyme extracts in a clay-loam soil. Bars (mean ± standard error) with different letters are statistically different from one another (LSD test $P = 0.05$).

At 6 h after application (**Figure 1C**), the relationship of *T. maxima* and its co-culture with *P. carneus* was inverted. Atrazine degradation by the co-culture enzyme extract decreased by 23.9% in comparison to its initial rate of degradation (at 1 h). This may be attributed to the absorption of atrazine by the soil, which motivated the investigation of the kinetic absorption-desorption parameters of atrazine in the studied clay-loam soil. Meanwhile, degradation of atrazine by the fungal monoculture extract increased to 97.9% (6 h), and *P. carneus* showed the lowest percentage of degradation (8.9%; $F = 20.79$, $P = 0.0004$).

However, during evaluation the degradation of atrazine by the fungal co-culture enzyme extract increased once again (92.2% at 12 h). This may be due to a desorption effect of atrazine previously absorbed by soil particles, principally clay. Meanwhile, the *T. maxima* monoculture extract degraded 100% of atrazine by this time, and the *P. carneus* extract also reached its maximum level of atrazine degradation (40.7%). At the end of evaluation period (12 h), both the *T. maxima* extract and its co-culture with *P. carneus* degraded 100% of atrazine. However, the increase in degradation by the *P. carneus* extract was not significant and did not reach levels of greater than 25% ($F = 671.05$, $P = 0.0001$, **Figure 1D**).

During mycoremediation, a single strain is commonly used. The application of bioremedial fungi in the soil is often based on the inoculation of immobilized mycelium in organic substrates, such as pine sawdust, wood chips, peat, corn cobs, wheat straw, bark, rice grains, sugarcane bagasse, coffee pulp or sugar beet pulp [27–30]. However, this technology has several challenges to overcome, which are as follows: (i) the competition and proliferation of native soil microorganisms (microfungi, bacteria and actinomycetes) with bioremedial fungi [9]; (ii) the limited capacity of inoculated fungi in the soil to produce sufficient amounts of the ligninolytic enzymes responsible for degrading contaminants [31–33]; (iii) the adverse effects of environmental and edaphic conditions on the establishment or growth of bioremedial fungus [14] and (iv) the amount of contaminants in the soil, which in some cases may be toxic to the bioremedial fungi [14].

One alternative for overcoming these challenges is the use of fungal enzyme extracts produced in fungal co-culture systems, which may enhance the amount of ligninolytic enzymes [34]; these extracts may then be applied to soil through irrigation systems by drenching or by immobilizing ligninolytic enzymes in chitosan, alginate or nanoparticles [35]. In our study, we applied fungal enzyme extracts from a co-culture to degrade atrazine in a clay-loam soil and found efficiencies of 100% at 6 and 12 h. Other studies have reported the ability of white-rot fungi extracts to degrade atrazine. For example, *Phanerochaete chrysosporium* extract can degrade atrazine in the soil microcosm (38% at 8 days), although its volumetric enzyme activity is low (MnP = 77.6 U/L, LiP = 149 U/L), as this species has low or null laccase activity [32]. In batch studies, Pereira et al. [36] reported that 39% of atrazine was degraded using a broth culture of *Pleurotus ostreatus* INCQ40310; the rate of degradation was enhanced to 71% when the broth culture was optimized by manipulating the nutritional compounds of the culture medium.

Several additional studies have used fungal co-cultures or their products, such as ligninolytic enzymes, to degrade contaminants. Recently, Pan et al. [37] demonstrated the feasibility of the fungal co-culture extract between *Coprinopsis cinerea* and *Gongronella* sp. to decolorize indigo dye. However, the native laccase from the fungal extract did not degrade indigo dye, and it was necessary to add ABTS (2,2′-azinobis(3-ethylbenzthiazoline-6-sulfonic acid)) as a redox mediator to degrade 75% of the dye. In another study, Qian and Chen [38] reported that the crude extract from the co-culture of *T. versicolor* and *Phanerochaete chrysosporium* degraded 20% more benzo-α-pyrene than the crude extracts of the monocultures of both fungi.

3.3. Atrazine absorption-desorption in a clay-loam soil

Figure 2 shows the adsorption-desorption of atrazine in the studied clay-loam soil. Depending on the concentration of atrazine dissolved in soil, between 39% and 77% is absorbed (**Figure 2A**). More atrazine is adsorbed than desorbed, or in other words, the desorption of atrazine is slower than its adsorption given the studied the soil type and time period (24 h, **Figure 2B**). Atrazine desorption is slower when high concentrations are adsorbed by the soil; i.e., when 20 mg/L was absorbed, only 1% was desorbed at 24 h. In this sense, Davidchik et al. [39] suggest that the adsorption of atrazine may be irreversible if a high concentration is found in the soil; these authors consider that oxidative binding is the most probable mechanism of atrazine incorporation into the organic matter.

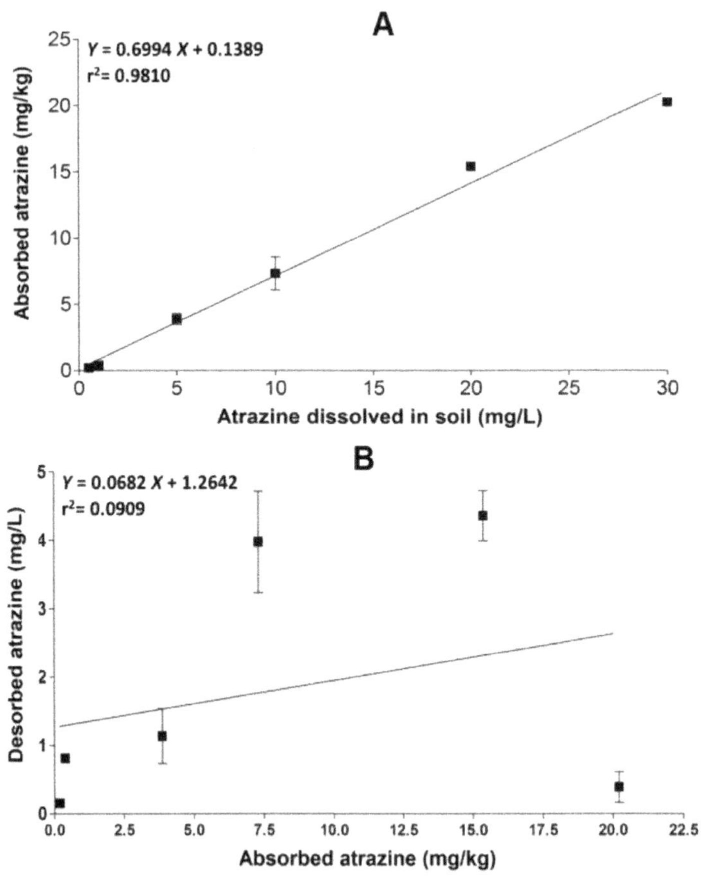

Figure 2. Adsorption (A) and desorption (B) of atrazine in a clay loam soil.

Adsorption and desorption values were linearized using the Freundlich equation (Eq. (1)), where q_e is the amount adsorbed at equilibrium (mg of atrazine/g of soil) and C_e the equilibrium concentration of atrazine in the solution (mg of atrazine/L). **Figure 3** shows the linearized Freundlich isotherms for atrazine adsorption and desorption, while **Table 2** describes the Freundlich isotherm parameters and hysteresis index.

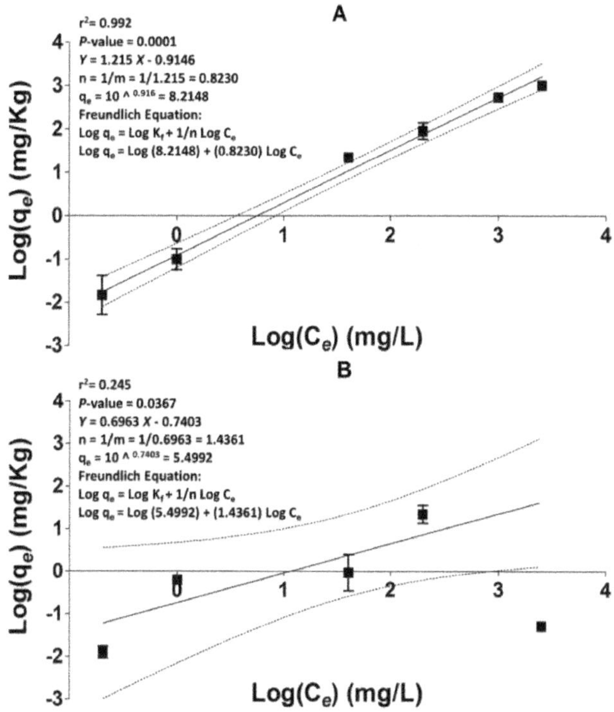

Figure 3. Linearized Freundlich isotherms for atrazine adsorption (A) and desorption (B) from a clay-loam soil.

Soil	Adsorption				Desorption				
	K_F	n	r^2	*P*-value	K_F	N	r^2	*P*-value	H
Clay-loam	8.2148	0.8230	0.992	0.0001	5.4992	1.436	0.245	0.036	0.573

K_F = Freundlich adsorption-desorption constant; n = absorbent constant; r^2 = regression coefficient; H = hysteresis index.

Table 2. Freundlich isotherm parameters and hysteresis index values for atrazine adsorption-desorption in a clay loam soil.

$$\log q_e = \log K_F + 1/n \log C_e \tag{1}$$

The Freundlich constant for adsorption of atrazine was 8.2148, which was higher than that reported by Kulikova et al. [7], who studied the absorption of atrazine to three soils with different textures (silt-loam: sod-podzolic [K_F = 4.51] and gray forest [K_F = 0.81] and clay-loam: chernozem [K_F = 5.54]). These authors suggest that clay-loam soil has high levels of organic carbon (organic matter), which leads to a high rate of atrazine absorption. In our study, the soil also possessed this characteristic, as a high organic matter content (4.35%) was detected in the soil analysis due to the incorporation of crop residues (sugarcane stalks) to the soil. In another study, Naga-Madhuri et al. [40] reported a lower K_F (=2.66) for atrazine adsorption in a silty clay-loam soil; the authors suggested that this value is high and may be due to the high electric conductivity and organic matter content of the studied soil.

On the other hand, the Freundlich desorption constant for atrazine was lower (K_F = 5.4992) than the adsorption constant (K_F = 8.2148). This was reflected in the hysteresis value (H = 0.573), which has a maximum value of 1; in this case, values near 1 indicate that almost all adsorbed atrazine is readily desorbed [7]. In this study we found that the clay-loam soil used in mycoremediation experiments does not desorb the adsorbed atrazine to a great extent, due to the high organic carbon content of the soil. Future studies will need to further examine the effect of enzyme extracts from fungal co-cultures and the adsorption-desorption phenomenon of atrazine in contaminated and bioremediated soils.

4. Conclusions

We conclude that:

1. The co-culture of *T. maxima* and *P. carneus* increases laccase activity and H_2O_2 content in the fungal enzyme extract.

2. Both the fungal enzyme extract of the monoculture of *T. maxima* and its co-culture with *P. carneus* were able to degrade atrazine in a short period of time (12 h) in a contaminated clay loam soil at a field application rate of 5 mg/kg.

3. Atrazine was highly adsorbed by the studied clay-loam soil. This was reflected by its high Freundlich coefficient for adsorption and low coefficient for desorption.

Acknowledgements

We thank Dra. Gabriela Heredia Abarca for donating the *P. carneus* isolate and the Center for Research and Advanced Studies of the National Polytechnic Institute for providing the equipment and supplies for the analysis of atrazine in the HPLC system.

Author details

Wilberth Chan Cupul[1*] and Refugio Rodríguez Vázquez[2]

*Address all correspondence to: chancupul@gmail.com

1 Biological Control and Applied Mycology Laboratory, Faculty of Biological and Agricultural Sciences, University of Colima, Mexico

2 Xenobiotics Laboratory, Center for Research and Advanced Studies of the National Polytechnic Institute, Mexico City, Mexico

References

[1] Islas-Pelcastre M, Villagómez-Ibarra JR, Madariaga-Navarrete A, Castro-Rojas J, González-Ramirez CA, Acevedo-Sandoval OA. Bioremediation perspectives using autochthonous species of *Trichoderma* sp. for degradation of atrazine in agricultural soil from the Tulancingo Valley, Hidalgo, Mexico. Tropical and Subtropical Agroecosystems. 2013;16:265–276.

[2] González MLC, Hansen AM. Adsorption and mineralization of atrazine and their relationship with soil parameters in the irrigation district 063 Guasave, Sinaloa.

[3] Sene L, Converti A, Ribeiro S, Garcia S. New aspect on atrazine biodegradation. Brazilian Archives of Biology and Technology 2010;53:487–496.

[4] Cejudo EE, Ramos VA, Esparza GF, Moreno CP, Rodríguez VR. Short-term accumulation of atrazine by three plants from a wetland model system. Archives of Environmental Contamination and Toxicology 2008;56:201–208. DOI: 10.1007/s00244-008-9193-7.

[5] Kumar GP, Philip L, Bandyopadhyay M. Management of atrazine bearing wastewater using an upflow anaerobic sludge blanket reactor-adsorption system. Practice Periodical Hazardous Toxic and Radioactive Waste Management 2005;9:112–121. DOI: 10.1061/(ASCE)1090-025X(2005)9:2(112).

[6] Farré M, Martínez E, Ramon J, Navarro A, Radjenovic J, Mauriz E, Lechuga L, Marco MP, Barcelo D. Part per trillion determination of atrazine in natural water samples by a surface plasmon resonance immunosensor. Analytical Bioanalytical Chemistry 2007;388:207–214. DOI: 10.1007/s00216-007-1214-2.

[7] Kulikova N, Davidchik VN, Stepanova EV, Koroleva O. Enhanced adsorption of atrazine in different soils in the presence of fungal laccase. In: Multiple stressors: a challenge for the future. Mothersill C, Mosse I, Seymour C (Eds.). Springer, 2007; 391–403. DOI: 10.1007/978-1-4020-6335-0_29.

[8] Bastos AC, Magan N. *Trametes versicolor*: potential for atrazine bioremediation in calcareous clay soil, under low water availability conditions. International Biodeterioration and Biodegradation 2009;63:389–394. DOI: 10.1016/j.ibiod.2008.09.010.

[9] Singh H. Fungal biodegradation and biodeterioration. In: Singh H (Ed.), Mycoremediation fungal bioremediation (1st edn.). Wiley-Interscience, New Jersey, 2006, pp. 1–28.

[10] Singh SB, Lal PS, Pant S, Kulshrestha G. Degradation of atrazine by an acclimatized soil fungal isolate. Journal of Environmental Science and Health, Part B: Pesticides, Food Contaminants, and Agricultural Wastes 2008;43:27–33. DOI: 10.1080/036012307017352 27.

[11] Field JA, Jong E, Feijoo G, De Bont JAM. Biodegradation of polycyclic aromatic hydrocarbons by new isolate of white rot fungi. Applied and Environmental Microbiology 1992;58:2219–2226.

[12] Cabana H, Jean-Louis HJ, Rozenberg R, Elisashvili V, Pennickx M, Agathos SN, Jones JP. Elimination of endocrine disrupting chemical nonylphenol and bisphenol A and personal care product ingredient triclosan using enzyme preparation from the white rot fungus *Coriolopsis polyzona*. Chemosphere 2007;67:770–778. DOI: 10.1016/j.chemosphere.2006.10.037.

[13] Izcapa-Treviño C, Loera O, Tomasini-Campocosio A, Esparza-García F, Salazar-Montoya JA, Díaz-Cervantes MD, Rodríguez-Vázquez R. Fenton (H_2O_2/Fe) reaction involved in *Penicillium* sp. culture for DDT [1,1,1-trichloro-2-2-bis(p-chlorophenyl)ethane)] degradation. Journal of Environmental Sciences and Health Part B. 2009;44:798–804. DOI: 10.1080/03601230903238368.

[14] Chan-Cupul W, Abarca HG, Rodríguez VR, Salmones D, Hernández RG, Gutiérrez EA. Response of ligninolytic macrofungi to the herbicide atrazine: dose-response bioassays. Revista Argentina de Microbiología 2014;46:348–357. DOI: 10.1016/S0325-7541(14)70094-X.

[15] Chan-Cupul W, Heredia-Abarca G, Martínez-Carrera D, Rodríguez-Vázquez R. Enhancement of ligninolytic enzyme activities in a *Trametes maxima-Paecilomyces carneus* co-culture: key factors revealed after screening using a Plackett-Burman experimental design. Electronic Journal of Biotechnology 2014;17:114–121. DOI: 10.1016/j.ejbt.2014.04.007.

[16] Sivakumar R, Rajedran R, Balakumar C, Tamilvendan M. Isolation, screening and optimization of production medium for thermostable laccase production from *Ganoderma* sp. International Journal of Engineering Science and Technology 2010;2:7133–7141.

[17] More S, Renuka P, Pruthvi K, Swetha M, Malini S, Veena S. Isolation, purification and characterization of fungal laccase from *Pleurotus* sp. Enzyme Research 2011;1:1–7. DOI: 10.4061/2011/248735.

[18] Kuwahara K, Glenn J, Morgan M, Gold M. Separation and characterization of two extracellular H_2O_2-dependent oxidases from ligninolytic cultures of *Phanerochaete chrysosporium*. FEBS Letters 1984;169:247–250. DOI: 10.1016/0014-5793(84)80327-0.

[19] Klassen NV, Marchington D, Mcgowan HCE. H_2O_2 determination by the I^{3-} method and by $KMnO_4$ titration. Analytical Chemistry 1994;66:2921–2925. DOI: 10.1021/ac00090a020.

[20] Inoue MH, Oliveira RS, Regitano JB, Tomenta CA, Constantin J, Tornisielo VL. Sorption-desorption of atrazine and diuron in soil from southern Brazil. Journal of Environmental Science and Health, Part B: Pesticides, Food Contaminants, and Agricultural Wastes 2006;41:605–621. DOI: 10.1080/03601230600701767.

[21] Hatvani N, Kredics L, Antal Z, Mécs I. Changes in activity of extracellular enzymes in dual culture of *Lentinula edodes* and mycoparasitic *Trichoderma* strain. Journal of Applied Microbiology 2002;92:415–423. DOI: 10.1046/j.1365-2672.2002.01542.x.

[22] Flores C, Vidal C, Trejo-Hernández MR, Galind E, Serrano Carreón L. Selection of *Trichoderma* strain capable of increasing laccase production by *Pleurotus ostreatus* and *Agaricus bisporus* in dual cultures. Journal of Applied Microbiology 2009;106:249–257. DOI: 10.1111/j.1365-2672.2008.03998.x.

[23] Velazquéz-Cedeño M, Farnet AM, Billete C, Mata G, Savoie JM. Interspecific interactions with *Trichoderma longibrachiatum* induce *Pleurotus ostreatus* defence reactions based on the production of laccase isozymes. Biotechnology Letters 2007;10:1583–1590. DOI: 10.1007/s10529-007-9445-z.

[24] Hiscox J, Baldrian P, Rogers HJ, Boddy L. Changes in oxidative enzyme activity during interspecific mycelial interactions involving the white-rot fungus *Trametes versicolor*. Fungal Genetic and Biology 2010;47:562–571. DOI: 10.1016/j.fgb.2010.03.007.

[25] Baldrian P. Increase of laccase activity during interspecific interactions of white-rot fungi. FEMS Microbiology and Ecology. 2004;50(3):245–253. DOI: 10.1016/j.femsec.2004.07.005.

[26] Dwivedi P, Vivekanand V, Pareek N, Sharma A, Singh RP. Co-cultivation of mutant *Penicillium oxalicum* SAUE-3.510 and *Pleurotus ostreatus* for simultaneous biosynthesis of xylanase and laccase under solid-state fermentation. New Biotechnology. 2011;28:616–626. DOI: 10.1016/j.nbt.2011.05.006.

[27] Young D, Rice J, Martin R, Lindquist E, Lipzen A, Grigoriev I, Hibbett D. Degradation of bunker C fuel oil by white-rot fungi in sawdust cultures suggests potential applications in bioremediation. Plos One. 2015;10(6):1–15. DOI: 10.1371/journal.pone.0130381.

[28] Rhodes CJ. Mycoremediation (bioremediation with fungi)-growing mushrooms to clean the earth. Chemical Speciation Bioavailability. 2014;26(3):196–198. DOI: 10.3184/095422914X14047407349335.

[29] Mohammadi A, Nasernejad B. Enzymatic degradation of anthracene by the white rot fungus *Phanerochaete chrysosporium* immobilized on sugarcane bagasse. Journal of Hazardous Materials. 2009;161(1):534–537. DOI: 10.1016/j.jhazmat.2008.03.132.

[30] Fragoeiro S, Magan N. Impact of *Trametes versicolor* and *Phanerochaete chrysosporium* on different breakdown of pesticides mixtures in soil microcosm at two water potentials and associated respiration and enzymes activity. International Biodeterioration and Biodegradation. 2008;62(4):376–383. DOI: 10.1016/j.ibiod.2008.03.003.

[31] Chirnside AEM, Ritter WF, Radosevich M. Biodegradation of aged residues of atrazine and alachlor in a mix-load site soil by fungal enzymes. Applied and Environmental Soil Science 2011;10:21. DOI:10.1155/2011/658569.

[32] Canet R, Birnstingl JG, Malcolm DG, Lopez-Real JM, Beck AJ. Biodegradation of polycyclic aromatic hydrocarbons (PAHs) by native microflora and combinations of

white-rot fungi in a coal-tar contaminated soil. Bioresource Technology. 2001;76(2):113–117. DOI: 10.1016/S0960-8524(00)00093-6.

[33] Magan N, Fragoeiro S, Bastos C. Environmental factors and bioremediation of xenobiotics using white rot fungi. Microbiology. 2010;38(4):238–248. DOI: 10.4489/MYCO.2010.38.4.238.

[34] Hu HL, Van der Brink J, Gruben BS, Wosten HAB, Gu JD, de Vries RP. Improved enzyme production by co-cultivation of *Aspergillus niger* and *Aspergillus oryzae* and with other fungi. International Biodeterioration and Biodegradation. 2011;65(1):248–252. DOI: 10.1016/j.ibiod.2010.11.008.

[35] Mohajershojaei K, Mohammad NM, Khosravi A. Immobilization of laccase enzyme onto titania nanoparticles and decolorization of dyes from single and binary systems. Biotechnology and Bioprocess Engineering. 2015;20(1):109–116. DOI: 10.1007/s12257-014-0196-0.

[36] Pereira PM, Sobral-Teixeira RS, Leal de Oliveira MA, da Silva M, Ferreira-Leitao VS. Optimized atrazine degradation by *Pleurotus ostreatus* INCQS 40310: an alternative for impact reduction of herbicides used in sugarcane crops. Microbial and Biochemical Technology. 2013;S12:006. DOI: 10.4172/1948-5948.S12-006.

[37] Pan K, Zhao N, Yin Q, Zhang T, Xu X, Fang W, Hong Y, Fang Z, Xiao Y. Induction of a laccase Lcc9 from *Coprinopsis cinerea* by fungal coculture and its application on indigo dye decolorization. Bioresource Technology 2014;62:45–52. DOI: 10.1016/j.biortech.2014.03.116.

[38] Qian L, Chen B. Enhanced oxidation of benzo[a]pyrene by crude enzyme extract produced during interspecific fungal interaction of *Trametes versicolor* and *Phanerochaete chrysosporium*. Journal of Environmental Sciences. 2012;24(9):1639–1646. DOI: 10.1016/S1001-0742(11)61056-5.

[39] Davidchik VN, Kulikova NA, Golubeva LI, Stepanova EV, Koroleva OV. *Coriolus hirsutus* laccase effect on atrazine adsorption and desorption by different types of soil. Applied Biochemistry and Microbiology. 2008;44(4):440–445. DOI: 10.1134/S0003683808040121.

[40] Naga-Madhuri KV, Chandrasekhar-Rao P, Suba-Rao M, Prathima T, Giridhar V. Adsorption-desorption of atrazine on vertisols and alfisols. Indian Journal of Weed Science. 2013;45(4):273–277.

12

Ex Situ Surfactant-Enhanced Bioremediation of NAPL-Impacted Vadose Zone

Roger Saint-Fort

Additional information is available at the end of the chapter

Abstract

This work presents a review of surfactant-enhanced bioremediation of hydrophobic organic contaminants in the soil with a focus on *ex situ* method. Conventional strategies of disposal methods in secure landfill and incineration have become cost prohibitive and environmentally risky and do not restore the contaminated soil, whereas chemical and physical methods have shown very limited success and can also be expensive.Traditional bioremediation pertaining to remedial technology of hydrophobic organic contaminants in soil has empirically demonstrated limited success due to their low aqueous solubility. Addition of single synthetic surfactant or biosurfactant, or in combination, has the potential to increase their mass transfer phase, hence their bioavailability. Surfactant-enhanced biodegradation represents a promising cost-effective alternative to complete mineralization of hydrophobic organic contaminants in soil. In this work, the potential of surfactants on the remediation of contaminated soil in an *ex situ* approach is reviewed with considerations given to the practical aspects of field components. Surfactant-enhanced biodegradation represents a promising cost-effective alternative to complete mineralization of hydrophobic organic contaminants in soil. In this work, the potential of surfactants on the remediation of contaminated soil in an *ex situ* approach is reviewed with considerations given to the practical aspects of field components.

Keywords: surfactant, NAPL, vadose zone, bioremediation, bioreactors, CMC, hydrophobic, soil

1. Introduction

In recent years, improper management of non-aqueous phase liquid (NAPL) hydrocarbons such as polyaromatic hydrocarbons, petroleum hydrocarbons, as well as other hazardous substances such as creosote and coal tar, has resulted in the formation of source zone plumes, virtually recalcitrant, in the vadose zone. The impacted vadose zone containing pooled NAPL and its residual are commonly referred to as the source zone. Generally, NAPLs are hydrophobic, low water-soluble liquids with a specific density that can be greater or less than 1. Nonetheless, NAPL chemical constituents that are soluble enough in the vadose source zone architecture may travel downward because of gravitational and capillary forces to contaminate the groundwater [1]. Many NAPL compounds are volatile and their behavior in the vadose zone may cause vapor intrusion concerns. The potential adverse impact of NAPL contamination has engendered significant concerns among the public, policymakers, environmental regulators, and scientists. Even at very low concentrations, NAPL constituents are considered highly toxic, mutagenic, and/or carcinogenic or can pose some other harm to humans and other environmental receptors [2]. Costly site-specific remediation strategies are often warranted and sometimes with limited success for the NAPL source zone and its associated plumes. In many instances, remediation strategies are designed towards partial mass removal, plumes containment, source zone stabilization, relative to a formulated acceptable risk-management objective. Surfactant-enhanced soil bioremediation has been proven as a promising technology through both empirical studies and field applications as a result of its low cost and the lack of toxic metabolites. Traditional framework of bioremediating NAPL-impacted soil is a very difficult process because of the mass transfer dissolution limit into the soil solution matrix, sorption onto the soil matrix, toxicity of constituents to soil biota, alteration in soil matrix physical properties. These factors have made the traditional bioremediation design approach at contaminated sites ineffective. Increasing dissolved mass transfer phase is a vital prerequisite towards achieving successful biodegradation of NAPL-impacted soil. Surfactants or surface active agents represent a class of chemicals that has the ability to increase the bioavailability of NAPL constituents by acting as solubilizing agents in the source zone. An *ex situ* remediation design properly strategized will allow exponential optimization of biotreatment process by enhancing the native capability of the soil microorganisms and risk mitigation. This work provides a fundamental review and approach of *ex situ* surfactant-enhanced bioremediation of NAPL-contaminated vadose zone as it pertains to an *ex situ* design program.

2. Architecture of NAPL in the vadose zone

Once accidentally released in the vadose zone, a NAPL will begin to create a dynamic source zone as it is contacting the soil matrix. A simplified conceptual site model (CSM) of a NAPL release in the vadose zone is shown in **Figure 1.** A NAPL heavier than water is defined as a dense NAPL (DNAPL), and if the NAPL has a density less than water, it is referred to as a light NAPL (LNAPL). In some cases, the source release may be single or a mixture of both types of

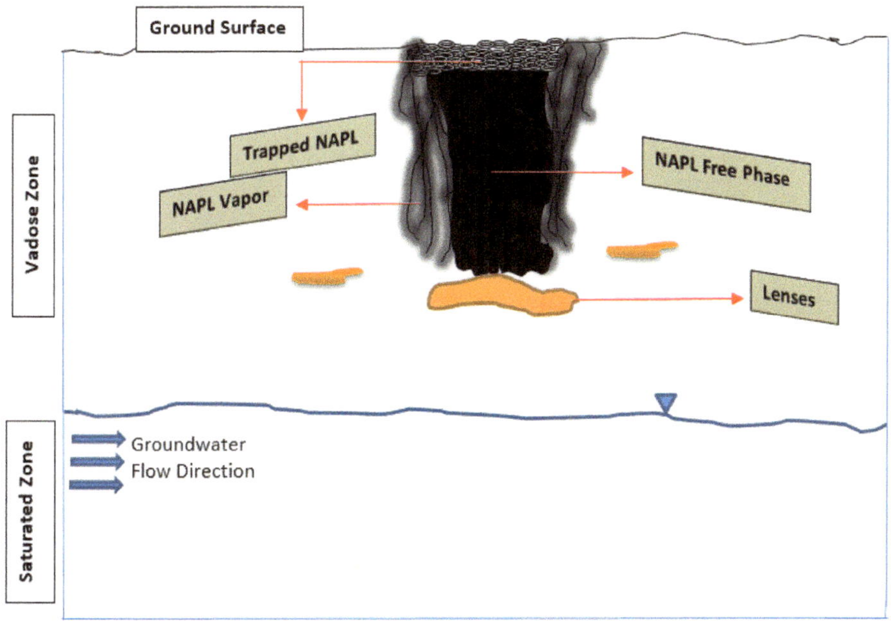

Figure 1. Simplified CSM of a NAPL release in the vadose zone.

NAPL. Irrespective, the NAPL will typically consist of multi-component of chemical compounds with varying degree of solubility. **Table 1** provides examples of characteristics for DNAPL and LNAPL compounds commonly encountered at contaminated sites. When released in significant quantity, the presence of air in soil pores in the vadose zone allows a NAPL to move downward under the force of gravity expressed in the capillary and bond number without overcoming a displacement pressure [3]. Soil NAPL saturation [4] is defined as Eq. (1):

$$S_s = \text{NAPL Volume released/Volume of Open Pore Space} \tag{1}$$

Fraction of the NAPL is held in place by capillary forces in the soil open pores space through which it flows. This immobile fraction under static conditions is termed residual saturation or globules. As a result, this creates a persistent source of contamination for groundwater. The relative fraction of a NAPL fluid immobilized and a continuous NAPL becomes discontinuous in a given volume of soil is termed residual saturation, R_s, which is expressed as Eq. (2):

$$R_s = (\text{Volume of NAPL/Volume of voids})100 \tag{2}$$

In addition, retention capacity (R_c) [5] has also been used to describe residual saturation of the non-wetting phase in the vadose zone as in Eq. (3):

$$R_c = R_s \times \text{soil porosity} \tag{3}$$

Compounds	NAPL type (D or L)*	Molecular formula	Molecular weight (g/mole)	Aqueous solubility (mg/L)	Density (kg/m³)	Vapor pressure (mmHg) @ 25°C	Viscosity (cP)
Chloroform	D	$CHCl_3$	119.38	8000	1483	160	0.58
Perchloroethylene	D	C_2Cl_4	165.83	1100	1623	14	0.89
Aroclor 1254	D	$C_{12}H_5Cl_{15}$	326.43	0.057	1540	$7.71E^{-05}$	1800
Aroclor 1242	D	$C_{12}H_6C_{14}$	261	0.200	1381	$1.00\ E^{-03}$	1350
Carbon tetrachloride	D	CCl_4	153.82	8000	1590	90	0.91
Methylene chloride	D	CH_2Cl_2	84.93	13,000	1330	435	0.44
Naphthalene	D	$C_{10}H_8$	128.17	30	1140	$9.44E^{-02}$	0.9684 @80°C
Nitrobenzene	D	$C_6H_5NO_2$	123.11	2090	1204	0.245	1.863
Anthracene	D	$C1_4H_{10}$	178.23	1.29	1250	$6.56E^{-06}$	3.00^{-01}
Nitrobenzene	D	$C_6H_5NO_2$	123.11	2090	1204	0.245	1.863
Benzene	L	C_6H_6	78.11	1840	876	95	0.75
Ethylbenzene	L	$C_6H_5CH_2CH_3$	106.17	152	866	9.998	0.669
Toluene	L	C_7H_8	92.14	520	862	21	0.59
Xylenes: 0-Xylenes m-Xylene p-Xylene	L	C_8H_{10}	106.16	178 161 162	880 860 860	7 8.29 9	0.812 0.62 0.61
MTBE	L	$C_5H_{12}O$	88.15	55,000	740	250	0.35 @15°C
Phenol	L	C_6H_6O	94.11	82,800	1060	0.40	9.7 @20 oC

Source: PubChem, Properties are typically at 20°C;
*D =Dense; *L = Light.

Table 1. Properties of select NAPL common pollutants.

Depending on the vadose zone NAPL-related characteristics and volume released, several distinct plumes may emerge. As the system strives to maintain a locale-scale equilibrium, contaminants may be transferred between phase media as environmental conditions change in accordance with equilibrium constants (**Figure 2**). In the vadose zone, the vapor and dissolved phases are significant in terms of mass transfer and transport as well as further spreading of contamination. Under most conditions in low conductivity areas into which diffusion and migration of a NAPL plume have occurred, these migration pathways can become intermittent sources of low-level contamination after the NAPL source mass has disappeared [4]. If the source zone and/or pooled NAPL is not timely and effectively risk managed, downward migration of NAPL constituents will eventually enter the phreatic zone resulting to further spreading of contamination at the site and significant additional remediation costs. The presence of moisture in the soil as well as infiltrating precipitation is required for downward movement of dissolved NAPL contaminants. The fundamental mass transport equation for the vadose zone can be applied according to Eq. (4):

$$R\,(\partial C/\partial t) = D_s\,(\partial^2 C/\partial z^2) - V\,(\partial C/\partial z) - \eta\, C + \zeta \quad (4)$$

where

1. C = solute concentration in the aqueous solution at time t
2. D_s = soil moisture diffusion coefficient
3. $\partial C/\partial z$ = concentration gradient
4. η = rate decay

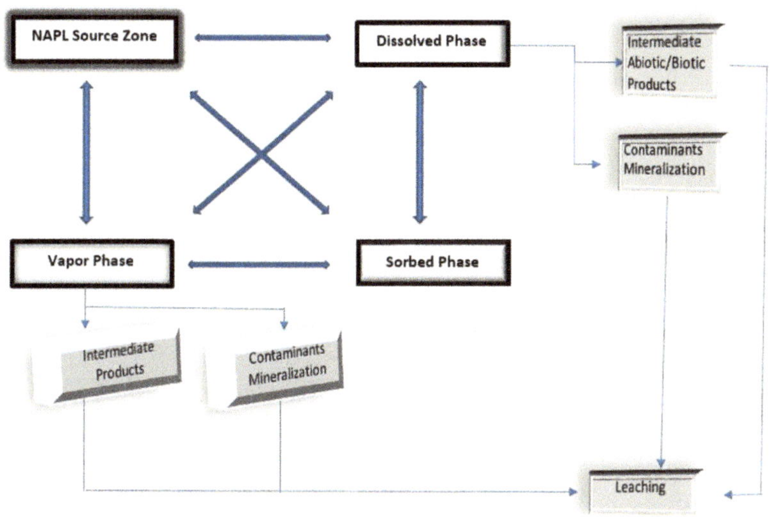

Figure 2. Dynamic of chemical phases in mass distribution of NAPL in the vadose zone.

NAPL movement once it reaches the saturated zone will be a function of its density. Evidence suggests that Darcy's equation used to describe fluid movement through a permeable bed can be equally applied. Numerical models have been used to predict movement of NAPLs in porous media [5–7]. In a one-dimensional model, hydraulic conductivity variable, K, is replaced by intrinsic permeability, κ to take into consideration the varying hydraulic characteristics pertaining to a NAPL fluid [8]. The negative sign in Eq. (5) is to indicate that flow is in the direction of decreasing head:

$$V = -(\kappa\, pg / \omega)\, dh / dL \quad (5)$$

where

V = Darcy velocity (cm/s)

κ = intrinsic permeability (1 darcy = 1×10^{-8} cm^2)

p = density of NAPL (g/cm^3)

g = force of gravity (980 cm/s^2)

ω = dynamic viscosity (cp) of NAPL

dh/dL = hydraulic gradient of NAPL mass
in Eq. (5), the hydraulic gradient is derived as described in Eq. (4), then Eq. (6) is expressed as:

$$dh/dl = (\beta + Q/pg)\, dL \quad (6)$$

where

β = reference elevation

Q = atmospheric pressure.

3. Solubilization of NAPLs by surfactants

Chemical surfactants and natural surfactants (biosurfactants) are surface active agents. The first ones are manufactured by petrochemical plants, whereas the latter are produced by biological organisms. However, the majority of surfactants produced and utilized are chemicals because of economic factors. In their common form, surfactants are amphipathic molecules constituted by both a hydrophobic moiety (chain) and a polar or ionic moiety (head) of varying length in different surfactants. The chain can be linear or branched:

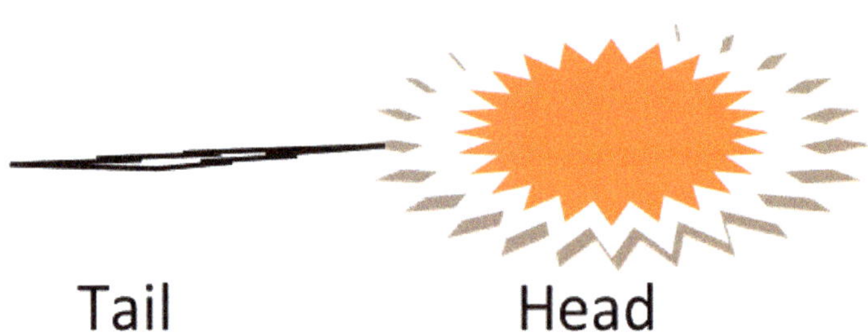

They tend to partition preferentially at the interface between fluid phases of different degrees of polarity and water bonding, consequently, making them the most versatile chemicals. Roy and Griffin [9] reported that the hydrophilic head group is the main factor responsible for the special chemistry of surfactants. Surfactants that are generated chemically are referred to as synthetic surfactants. They are generally grouped into various categories depending on the nature of the polar moiety (**Table 2**). The hydrophobic portion of these molecules are alkylbenzenes, alcohols, olefins, paraffin, or alkyl phenols, while the polar moiety will consist of either a sulfonate, sulfate, or a carboxylate group in the case of anionic surfactants. A quaternary ammonium group is found in cationic surfactants. The hydrophilic moiety of non-ionic surfactants is represented by sucrose, polypeptides, or polyoxyethylene groups. In contrast, biosurfactants are grouped according to the chemical composition of the different molecules representing the hydrophobic and hydrophilic moieties as well as microbial origin. Alternatives to petrochemicals and microbial generated surfactants are plant-based classified surfactants. As a natural solution for environmental remediation and daily common applications, plant-based surfactants offer the same very qualities and effectiveness that are found in a synthetic or biosurfactants.

Head Charge	Group Class
 Negative	Anionic
 Positive	Cationic
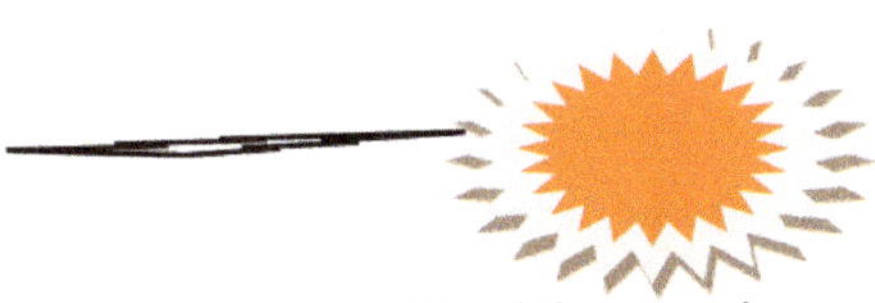 Positive and negative	Zwitterionic or Ampholytic
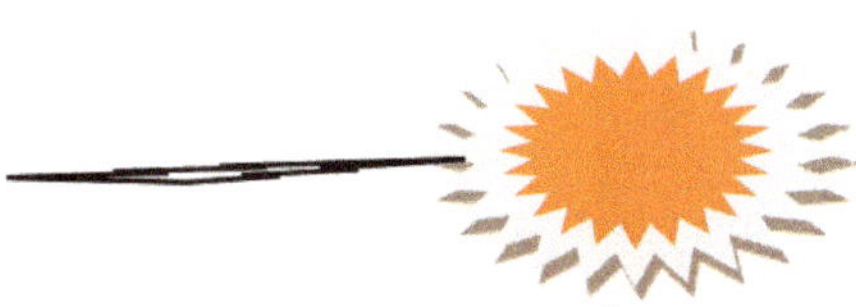 No charge	Nonionic

Table 2. Summary of chemical surfactants classification.

It has also been suggested that biosurfactants can be conveniently divided into low-molecular mass molecules or high-molecular mass polymers. An adaptation of their classification is provided in **Table 3** [10, 11].

Class type biosurfactants		
Microorganisms group		**Phytogenic group**
Low mass	**High mass**	
Glycolipids: Conjugates of fatty acids and carbohydrates. *Most common biosurfactants:* trehalopids, Sophorolipids, rhamnolipids. Burkholderia plantarii, *Producing microorganisms:* *Mycobacterrium, Arthrobacter spp, Pseudomonas aeruginosa*	**Polymeric biosurfactants:** Typically consists of three to four repeating sugars with fatty acids attached to them. *Most common biosurfactants:* emulsan, liposan, alasan *Producing microorganisms:* *acinetobacter calcoaceticus, candida lipolytica*	Saponins,lecithins, soy protein, lactonic, soybean oil, glycolipid, Sunflower seed
Lipopeptides and lipoproteins: Consist of a lipid attached to a polypeptide chain. *Most common biosurfactants:* surfactin and lichenysin *Producing microorganisms:* *Bacillus sp.*	**Particulate biosurfactants:** Can be extracellular vesicles and whole microbial cell. *Most common biosurfactants:* vesicles, whole microbial cells. *Producing microorganisms:* *acinetobacter calcoaceticus, pseudomonas marginalis, cyanobacteria*	
Phospholipids, fatty acids and neutral lipids: Length of hydrocarbon chain in their structures determines the hydrophilic and hydrophobic balance. *Most common biosurfactants:* corynomycolic acid, phosphatidylethanolamine *Producing microorganisms:* *Rhodococcus erythropolis, corynebacterium lepus*		

Table 3. Summary of biosurfactants classification (Adapted with permission from [10, 11]).

Hydrogen bonding property between water molecules is the primary factor responsible for NAPL insolubility in water. Surfactants can solubilize NAPL constituents by reducing surface and interfacial tensions of water (**Figure 3**). Reduction in the surface tension of water may range from 70 mN m^{-1} to less than 30 mN m^{-1} [12], thereby increasing the wetting ability of water. Surfactant molecule that is unable to form hydrogen bonding in an aqueous phase leads to an increase in the free energy of the system. This leads to an increase in NAPL solubilization in the water phase achieved through the formation of micelles. It has been reported that the aggregation number to form micelles is between 50 and 100 surfactant molecules [12]. Increasing surfactant concentration to above a critical micelle concentration (CMC) will lead to the formation of dynamic micelles by incorporating the hydrophobic solubilizates into the hydrophobic cores of the micelles [12]. Surfactant molecules that exist as monomers below the surfactant's CMC have minimal effects in the aqueous solubility of the system. As surfactant concentrations above the CMC threshold increase, the solubilization process of hydrophobic contaminants increases linearly with surfactant concentration. Invariably, micelle formation allows increased mobilization and partitioning of sorbed NAPL contaminants into the soil solution by lowering capillary forces. The lower the CMC value of a given surfactant in a system, the more stable will be the micelles and therefore the mass transfer process.

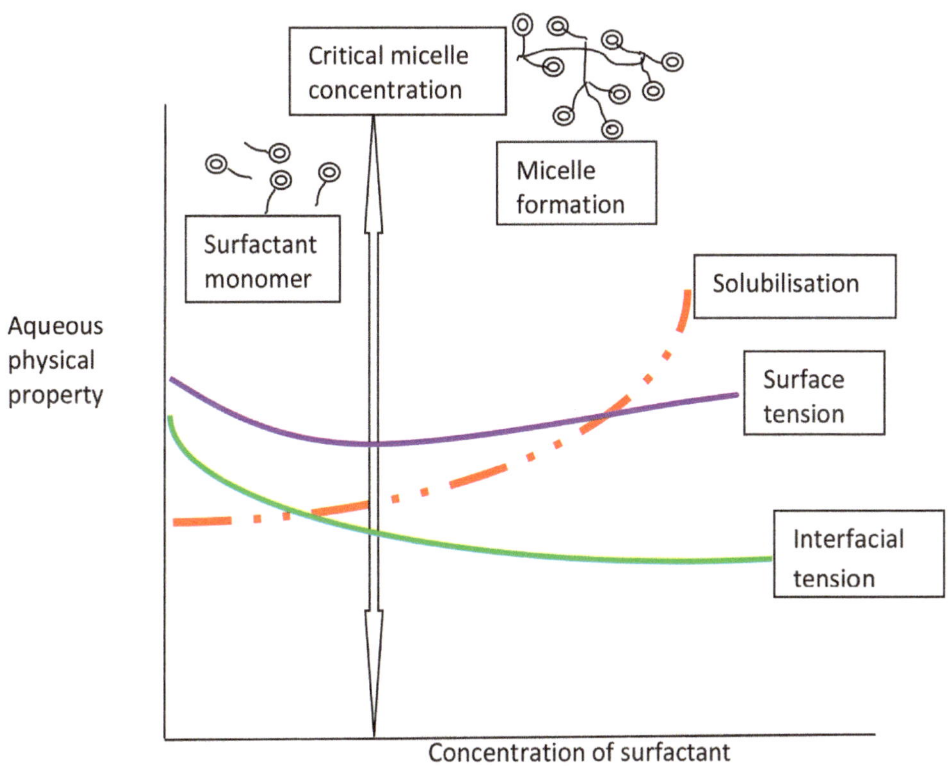

Figure 3. Interplay between hydrophobic contaminant solubility, surface tension, interfacial tension and micelle in the case of a specific surfactant at the core-water interface.

The capacity of surfactants to affect micellar solubilization of hydrophobic organic compounds is affected by the following factors:

- Temperature: CMC's typically increases with increase above a certain temperature as micelle formation is opposed by thermal agitation, termed the Krafft point. However, non-ionic surfactants do not show Krafft points. Consequently, increasing temperature tends to decrease their solubility. The temperature at which non-ionic surfactants begins to exhibit surface active properties loss is termed the cloud point.

- Salinity: presence of electrolytes tends to reduce repulsion forces between charged groups of the micelle and consequently inhibit CMC formation.
- Surfactant hydrophobic property: As the hydrophobicity portion of a surfactant increases, this results in a decrease in the formation of CMC. Above C_{18}. CMC appears constant. This is ascribed to coiling of the long hydrophobic moiety in the aqueous phase.
- Soil moisture content: Soil moisture level must be high enough to allow mass-transfer. Heavy soils relative to a coarse soil type will require a higher level of moisture in the system to enhance contaminant solubilization by a specific surfactant.
- Presence of other organic molecules: May affect water structuring such as to create a shift in CMC. Structure makers such as sugars are known to lower CMC, while structure breakers like urea and formamide typically will increase surfactant solubility. In a mixed surfactant mixture system, CMC may synergistically occur at a lower level than any of the CMC's of the single pure surfactants.
- Sorption: It reduces the concentration of surfactant monomers in the aqueous phase. Under such conditions will not aggregate to form micelles of colloidal-size until the sorption process is overcome through addition of more surfactant. CMC becomes more appreciable.
- pH: Depending on the nature of the surfactant and the degree of humification of the soil organic matter, CMC may be affected. Enhanced solubility of organic chemical may be observed at pH values at which soil humus and surfactant are found mostly ionized and at opposite charged.
- Interfacial energy: The interfacial tension of a given surfactant solution decreases with correspondingly increase in the surfactant monomers in a system. This leads to an attainment of a minimum free energy state. Enhanced micellar solubilization of hydrophobic organic compounds is favored.

The effectiveness of a particular surfactant in solubilizing a NAPL constituent may be represented by the molar solubilization ratio (MSR) [13] defined as expressed in Eq. (7):

$$MSR=(S-S_{CMC})/(C_s-CMC) \quad (7)$$

where

MSR = moles of organic contaminant solubilized per mole of surfactant added to the aqueous phase

S = apparent solubility of organic contaminant at a given surfactant concentration

C_s = apparent solubility of organic contaminant at CMC (i.e., C_s > CMC)

CMC = critical micelle concentration

By plotting solute concentration as a function of surfactant concentration, MSR can be determined from the slope of the linearly fitted regression equation. The micelle aqueous-phase partition coefficient (K_m) is often used as another approach to quantify the solubilization capacity of a single surfactant [14]. K_m can be defined according to Eq. (8):

$$K_m = X_m / X_a \tag{8}$$

where

X_m = the mole fraction of hydrophobic compounds encapsulated in the micellar phase given by {MSR/(1 + MSR)}

X_a = the mole fraction of hydrophobic compounds in the aqueous phase.

Studies on mixed surfactant systems competitive effects on hydrophobic contaminants solubilization has been investigated and reported elsewhere [15–18].

4. Mineralization of NAPL

The most widely applied soil bioremediation approach to organic contaminants involved the biostimulation of natural microbial biodegraders. Biodegradation requires a source of carbon (organic contaminant) and nutrients, as amendment. The hydrophobic organic contaminants represent the carbon source as electron donors, while nitrogen and phosphorous are essential for microbial growth for cellular metabolism. Addition of nitrogen particularly is often necessary due to heavy demands by the biodegradation process. Phosphorous is usually amended in lower concentration. Optimizing nutrient status of a contaminated soil can have direct impact on microbial activity and contaminants biodegradation. In some instances, the negative effects of high nutrients amendment with NPK on soil biodegradation especially on aromatics have been reported [19–21].

The ultimate microbial aerobic degradation process of converting bioavailable NAPL constituents in a contaminated soil matrix:

Bioavailable hydrophobic organic contaminants in bioreactors + O_2 + N + P + moisture --------------> (Acclimatized biodegraders) inorganic cpds + CO_2 + water + cell proteins

This process is commonly referred to as mineralization. The degradation process is brought about under aerobic conditions. NAPL constituents are hydrophobic organic chemicals that exhibit limited or no solubility in contaminated soils and thermodynamically tend to partition to the soil solid phase. Sorption may account for more than 95% of the total contaminant mass. As a consequence, the hydrophobic contaminant exhibits limited dissolved mass transfer phase and bioavailability, which limits its biotic degradation in the soil. Therefore, in a contaminated soil environment, biodegradation of an organic hydrophobic compound should be envisioned as a stepwise process involving contaminants bioavailability and species of biodegraders.

The use of surfactants represents a cost-effective and promising method that can enhance bioremediation of organic hydrophobic contaminants in soils. Many studies have shown that surfactants can solubilize and mobilize hydrophobic organic contaminants sorbed onto soil matrices [22–24]. Adding surfactant to a contaminated soil matrix is expected to enhance microbial degradation through mobilization or emulsification. Mobilization takes place at concentrations below CMC and the solubilization process above the surfactant CMC, whereas emulsification allows for dispersion of one phase into the other. A certain amount of surfactant in the slurry system will inevitably be sorbed onto the soil particles. Sorbed surfactant does not contribute to the solubilization and bioavailability of contaminants during treatment. The more surfactant is sorbed, the less effective will be the surfactant. Furthermore, soil hydrophobicity may increase as more surfactant becomes sorbed onto the contaminated soil matrix.

Considerations	**Remarks**
Environmental factors	
Acclimation	Proper biodegraders; enzymatic adjustment for metabolic process
Temperature	Mesophiles 15–45°C
Oxygen	Aerobes; DO > 0.30 mg/L
pH	Optimum range 5–9
Nutrients	Sufficient N, P not limiting biodegraders growth; C:N:P ratio of 100:50:1
Redox potential	Greater than 70 mV; promote aerobes
System slurry	Optimized to promote mass transfer; 50–80% of soil water intrinsic saturation
Metabolites	Non-toxic
Salinity	Low inhibition of CMC formation
Surfactant properties	
Environmental risk	Pose no risk to the environment
Toxicity	No inhibitory effects; not toxic to any receptors
Substrate source	Not a preferential growth substrate
Sorption behavior	Low sorption onto soil constituents
Effective concentration	Efficient in increasing aqueous solubility of organic compounds at low concentration
Recalcitrancy	Non-persistent; biodegradable and mineralizable
CMC	Effective below CMC; partial micelle encapsulation of contaminant; low sequestration vis-à-vis target contaminant

Table 4. Relevant environmental and surfactant considerations for *ex situ* surfactant-enhanced bioremediation.

Surfactants can enhance metabolic degradation and thereby, contaminants mineralization in the soil by two main mechanisms [25]. One mechanism involves the increase in the contaminant bioavailability for microorganisms. The second mechanism is due to interaction with cell surface resulting in the hydrophobicity increases in the cell surface allowing hydrophobic organic chemicals to interact with bacterial cells. Environmental factors and surfactant properties affecting the metabolic capability of biodegraders in the soil vis-à-vis hydrophobic organic contaminants are summarized in **Table 4**.

The role of treatability studies for *ex situ* surfactant-enhanced bioremediation of hydrophobic organic contaminants contaminated soil is vital. It will allow to derive crucial information that will serve as blueprint to optimize field operation. Typically, a treatability study will be conducted in laboratory microcosms to inform (a) on the dosage of surfactant required to optimize contaminant mass transfer, (b) on the effect of temperature on contaminant bioavailability as temperature may affect surfactant efficiency and microbial activity, (c) on optimum biostimulation through the addition of appropriate nutrient amendments such as N, P and other elements, (d) optimum moisture level as it will vary with soil type, (e) selection of appropriate surfactant, (f) modeling rate of contaminants degradation under varying environmental factors, (g) rate of oxygen and nutrients consumption under different environmental conditions, (h) implement bioaugmentation utilization by inoculation with acclimated bacteria strains, (i) the complimentary effects of combined bioaugmentation and biostimulation, (j) determine whether targeted level of cleanup is attainable, (k) formulation of an efficient and effective monitoring program for field treatment operation, (l) the engineering design, (m) potential surfactant toxicity and means to reduce it, (n) sorption behavior of a surfactant.

The two main strategies can be highlighted for assessing a bioremediation system performance. A material balance approach consists of extracting and quantifying residual parent compounds and monitoring partitioning in the headspace phase. The other strategy involves monitoring the system for CO_2 production. A direct correlation occurs between mineralization of the parent compound and CO_2 production.

The biodegradation during the treatability assessment may be modeled through either a first- or zero-order power rate model [26]. A zero-order reaction indicates the biodegradation of a parent contaminant in the microcosm occurs at a constant rate and independent of concentration and time. If the parent compound C is mineralized to CO_2, the rate of disappearance of C is given by Eq. (9):

$$dC/dt = -k \tag{9}$$

integration yields Eq. (10):

$$C_t = C_o - kt \tag{10}$$

where

C_t = parent compound present at time t

C_o = initial concentration of parent compound

k = zero-order reaction rate constant

t = corresponding sampling time.

First-order reactions have rates that depend on mass transfer of parent compound concurrent to its biodegradation, Eq. (11):

$$dC/dt = -kC \tag{11}$$

where

C = parent compound concentration

t = corresponding sampling time

k = first-order reaction rate constant
integration yields Eq. (12):

$$Ln(C_t) - Ln\ (C_o) = Ln\ (C_t/C_o) = -kt \tag{12}$$

where

C_t = parent compound present at time t

C_o = initial concentration of parent compound

k = first-order reaction rate constant ($time^{-1}$)

t = corresponding sampling time.

Solving for concentration yields Eq. (13):

$$C = C_o e^{-kt} \tag{13}$$

and parameters are as defined above.

Biosurfactants may be the strategic choice for increasing contaminant bioavailability in bioreactors while minimizing toxicity to biodegraders. An examination of the literature indicates that synthetic surfactants while effective for increasing contaminant mass transfers at the recommended concentration may show inhibitorial effects on the microorganisms in the bioreactor [27, 28]. In such case, this will inhibit cell proliferation and thus the biodegradation of organic contaminants. According to empirical evidence, surfactant toxicity was found to be primarily dependent on its molecular structure, in order of toxicity, generally non-ionic <anionic< cationic [28]. Several practical approaches may be implemented to reduce surfactant cytotoxicity in a bioreactor by considering a suitable biosurfactant as an alternative to a synthetic surfactant, adding a surfactant at concentration below CMC, using a suitable non-ionic surfactants, using a suitable combination of biosurfactant and synthetic surfactant, in some instances, strategically increasing the surfactant concentration to decrease contact of biodegraders with the contaminant, prescreening for a suitable additive such as Ca and Mg as they were found to stabilize the cell membrane, thereby decreasing surfactant toxicity [29].

5. Field implementation

First and foremost, site access should be restricted to minimize human and wildlife exposure to contamination. As a contaminated site, safety should be implemented and followed at all

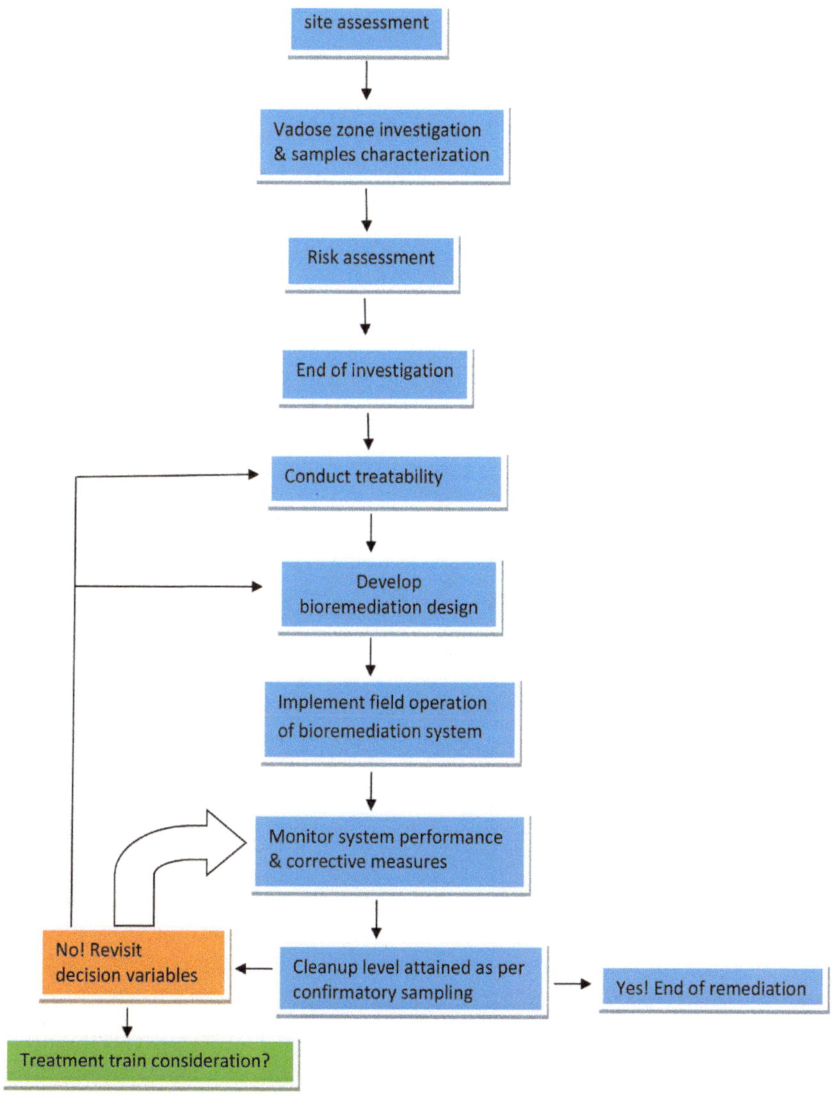

Figure 4. Approach to main components of implementing a field bioremediation program.

time. A site assessment and site characterization program should be conducted prior to excavating and bioremediating the contaminated soil. A site characterization will involve a more rigorous and field testing program (i.e., drilling and installing groundwater monitoring wells, chemical parametrization of soil samples, soil gas sampling). A good site assessment program should provide basic and qualitative information such as how much? When? What types of contaminants was released? It should also allow to generate site-specific information pertaining to soil physical, chemical, and biological properties critical in the success of the bioremediation program. **Figure 4** provides a simplified overview of environmental site assessment approach. The site characterization should be conducted in a phased approach. Each evolutionary phase should be designed to assess the CSM. As such, this will increase the investigation capacity to perform risk analysis.

Once the areal extent of vadose contamination has been delineated and staked out, excavation can safely proceed ahead. The excavation process should be managed to prevent any additional pollution and protect the environment and human health. Common equipment used to excavate and move soils around the site includes but not limited to: a bulldozer or dozer pushes soil with a hydraulically controlled blade. A backhoe uses a toothed bucket attached to a loom or dipper stick. Front-end-loaders are tractors equipped with buckets that can be used for excavation, lifting, hauling and dumping soil material, hydraulic excavator with primary function for digging, and articulated trucks are used as versatile hauling units.

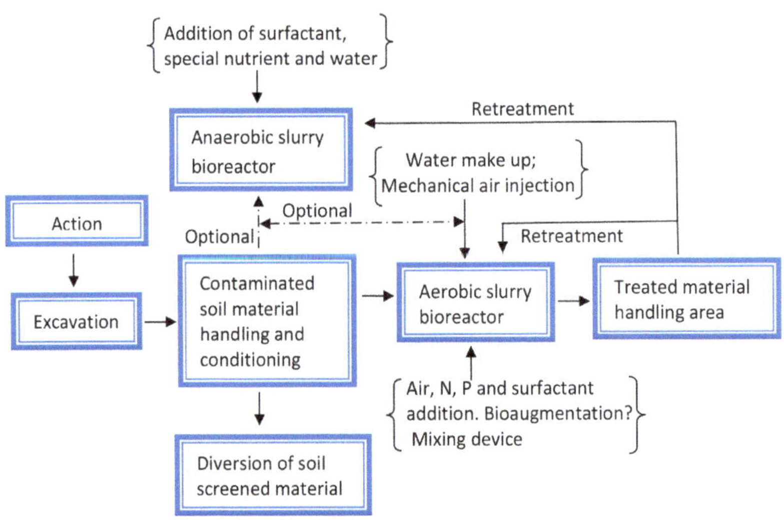

Figure 5. Flow diagram of a typical batch sequencing slurry bioreactor.

Several bioremediation option processes can be contemplated for on-site and off-site treatment of the excavated contaminated soil material. Irrespective of the system configuration and design, process fundamentals of a surfactant-enhanced-bioremediation efficiency requirements must be optimized prior and during project implementation. Aqueous slurry conditions typically ranging from 20 to 40% w/v are one of the most important types of *ex situ* technique [30–32]. A slurry bioreactor may consist of a vessel or a lined lagoon, which is typically run in a batch or semi-continuous operation mode. Sometimes, they may be operated in sequencing batch reactors to achieve a desired treatment train objective as illustrated in **Figure 5**. Dehalogenation under anaerobic conditions of chlorinated contaminants is initially necessary prior to aerobic treatment. When dehalogenation is not required, the treatment process can be carried out aerobically only. During treatment, slurry mixing may be performed with mechanical or pneumatic devices in a rather intermittent than continuous mode.

CSF key variables	Better success
Site access	Restricted
Site safety	Followed at all time
Equipment	Available on site
Season	Summer, spring, fall
Volume of soil	No restriction
Working area	Sufficient for footprint needed
Characterization	All contaminants of concern
Contaminant types	Organic hydrophobic
Contaminants	Non-toxic level
Acclimation	System time-dependent
Contaminant phase	Liquid or sorbed
Anaerobic bioreactors	Critically ≤ − 10 mV for dehalogenation
Redox	Critically ≥ + 5 mV for mineralization
C:N:P	100:50:1
Surfactant cost	Low
Remediation cost	Competitive
Surfactant sorption	Low
Timeframe	Fast
CMC	Low
Surfactant availability	Readily
Surfactant toxicity	Non-toxic
Surfactant persistence	Biodegradability balanced with effectiveness
Encapsulation effects	Minimal on bioavailability
Mixed surfactants	Synergistic effects
Soil:liquid ratio	Optimize slurry consistency as per soil type
Public perception	Positive
Regulatory perception	Positive
Surfactant metabolite	Non-toxic
Environmental compatibility	Very good

Table 5. Matrix of CSF for *ex situ* surfactant enhanced bioremediation.

Mixing will play a critical role by increasing mass transfer rates and bioavailability of contaminants as enhanced by the presence of surfactant, provide slurry homogenization, keep solid particles in suspension, and help achieving oxygen transfer in aerobic bioreactor. In its simple design, a SB construction will consist of soil handling and conditioning area, aeration device, the bioreactor (anaerobic/aerobic) itself, drying and storage area of treated material, off-gas treatment, and chemical storage area. Air quality monitoring should be conducted at and around the site.

A matrix summary of critical success factors (CSFs) for *ex situ* surfactant enhanced bioremediation has been best summarized in **Table 5**.

6. Summary

Vadose zone contamination by NAPL hydrocarbons through either natural or industrial processes represents a worldwide concern due to its potential hazard to the environment and health impact to biological receptors. Several scientific and engineering remediation strategies have been researched, developed, field-tested, and subsequently implemented to restore these contaminated sites. For a successful risk management of a contaminated vadose, the contamination must be prevented from spreading and be removed as economically as possible in a time-efficient and practical method. In these capacities, *ex situ* surfactant enhanced bioremediation has been attracting increasing attentions in recent years. Biosurfactants and chemically synthesized surfactants are relatively low-cost production industrial process. They have been playing an increasing and pivotal role in *ex situ* remediation of contaminated soil due to their unique desorption function capability, strong solubilizing power of hydrophobic organic chemicals, and considerable enhancement of contaminants bioavailability. Several critical issues have, however, to be vigorously researched. However, the data current available indicate some research gap areas. Therefore, a concerted research endeavor is currently needed to better elucidate the fate and behavior of synthetic surfactants in natural ecosystems, mechanism of soil biota toxicity and regulation, hysteresis effect on treated soil properties, metabolites production during biodegradation, soil hydrophobicity increase, synergistic properties of mixed surfactants, combined use of surfactants with additives on enhancing bioreactors performance. Furthermore, the prospects of future development and industrial production of mixed surfactant systems combined with low CMC are very promising alternatives to either biosurfactants or chemically synthesized surfactants. This new generation of surfactants will offer the possibility of removing the large-scale remediation impediments associated with current *ex situ* surfactant-based soil remediation technology.

Author details

Roger Saint-Fort

Address all correspondence to: rsaintfort@mtroyal.ca

Department of Environmental Science, Faculty of Science and Technology, Mount Royal University, Calgary, AB, Canada

References

[1] Pennell KD, Abriola LM, Weber WJ. Surfactant-enhanced solubilisation of residual dodecane in soil columns. 1. Experimental design. (1993) Environ Sci Technol 27(12): 2332–2340.

[2] Kuepper BH, Wealthall, Smith, JWN, Leharne, SA, Lerner, DN. Environmental Agency. An illustrated handbook of DNAPL transport and fate in the subsurface. R&D Technical Publication 133. (2003) Environment Agency Bristol: 1–9.

[3] Feenstra S, MacKay DM, Cherry JA. A method for assessing residual NAPL based on organic chemical concentrations in soil samples. (1991) Groundwater Monit Rev. 11(2): 128–136.

[4] Mercer JW, Cohen RM. A review of immiscible fluids in the subsurface: properties, models, characterization and remediation. (1990) J Contam Hydrol 6(2):107–163.

[5] Bradford SA, Abriola LA. Dissolution of residual tetrachloroethylene in fractional wettability porous media: incorporation of interfacial area estimates. (2001) Water Resour Res. 37(5):1183–1195.

[6] Ngien SK, Ken TL. Numerical model for NAPL migration in double-porosity subsurface systems. E-Proceedings of the 36th IAHR World Congress 28 June–3 July, 2015, The Hague, the Netherlands (2015).

[7] Pasha AY, Hu L, Meegoda JN. Numerical simulations of a light nonaqueous phase liquid (LNAPL) movement in variably saturated soils with capillary hysteresis. (2014) Can Geotech J 51(9):1046–1062.

[8] Bedient BP, Rifai HS, Newell CJ. Ground Water Contamination: Transport and Remediation, 2nd edition. (1999) Prentice Hall: Upper Saddle River, 391–439.

[9] Roy WR, Griffin RA. Surfactant-and-chelate-induced decontamination of soil materials: current status. Environmental Institute for Water Management Studies. Open File Report 21. The University of Alabama, Tuscaloosa, AL, USA, 35487 (1988).

[10] Desai JD, Banat IM. Microbial production of surfactants and their commercial potential. (1997) Microbiol Mol Biol Rev 61(1):47–641.

[11] Mulligan CN, Yong RN, Gibbs BF. Surfactant-enhanced remediation of contaminated soil: a review. (2001) Eng Geol 60(2):193–207.

[12] Christofi N, Ivshina IB. Microbial surfactants and their use in field studies of soil remediation. (2002) J Appl Microbiol 93(6):915–929.

[13] Santanu P. Surfactant-enhanced remediation of organic contaminated soil and water. (2008) Adv Colloidal Interf Sci 138(1):24–58.

[14] Edwards DA, Luthy RG, Liu Z. Solubilisation of polycyclic aromatic hydrocarbons in micellar nonionic surfactant solutions. (1991) Environ Sci Technol. 25(1): 127–133.

[15] Zhentian S, Jiajun C, Jianfei L, Ning W, Zheng S. Anionic-nonionic mixed-surfactant-enhanced remediation of PAH-contaminated soil. (2015) Environ Sci Pollut Res. 22(16): 12769–12774.

[16] Parekh P, Varade D, Parikh J, Bahadura P. Anionic-cationic mixed surfactant systems: micellar interaction of sodium dodecyl trioxyethylene sulfate with cationic Gemini surfactants. (2011) Colloids Surf A: Physicochem Eng Asp 385(1):111–120.

[17] Mohamed A, Mafoodh AM. Solubilisation of naphthalene and pyrene by sodium dodecylsulfate (SDS) and polyoxyeethylenesorbitan monooleate (Tween 80) mixed micelles. (2006) Colloids Surf A: Physicochem Eng Asp 287(1):44–50.

[18] Kimball SL. Surfactant-enhanced soil flushing: an overview of an in situ remedial technology for soils contaminated with hydrophobic hydrocarbons. In: Kostecki PT, Calabrese EJ, Bonazountas M. (Eds.), Hydrocarbon Contaminated Soils, Vol. II. Boca Raton: Lewis Publishers (1992).

[19] Oudot J, Merlin FX, Pinvidic P. Weathering rates of oil components in a bioremediation experiment in estuarine sediments. (1998) Marine Environ Res 45(2):113–125

[20] Chaineau CH, Rougeux G, Yèprèmian C, Oudot J. Effects of nutrient concentration on the biodegradation of crude oil and associated microbial populations in the soil. (2005) Soil Biol Biochem 37(8):1490–1497.

[21] Pacwa-Plociniczak M, Plaza GA, Piotroska-Seget Z, Cameotra SS. Environmental applications of biosurfactants: recent advances. (2011) Int J Mol Sci 12(1):633–654.

[22] Hongo Z, Michael DA. Surfactant-enhanced desorption and biodegradation of polycyclic aromatic hydrocarbons in contaminated soil. (2010) Environ Sci Technol. 44(19): 7260–7265.

[23] Yu H, Zhu L, Zhou W. Enhanced desorption and biodegradation of phenanthrene in soil-water systems with the presence of anionic-nonionic mixed surfactants. (2007) J. Hazard Mater. 142(1):354–361.

[24] Martienssen M, Schirmer M. Use of surfactants to improve the biological degradation of petroleum hydrocarbons in a field site study. (2007) Environ Technol. 28(5):573–582.

[25] Mulligan CN, Gibbs BF. Types, production and applications of biosurfactants. (2004) Proc Indian Nat Sci Acad. B70(1):31–35.

[26] Roger SF. Sulfolane attenuation by surface and subsurface soil matrices. (2006) J Environ Sci Health Part A. 41(6):1211–1231.

[27] Woo SH, Jean CO, Park JM. Phenanthrene biodegradation in soil slurry systems: influence of salicylate and Triton X-100. (2004) Korean J Chem Eng. 21(2):412–418.

[28] Jin DY, Jiang X, Jing X, Ou ZQ. Effects of concentration, head group, and structure of surfactants on the degradation of phenanthrene. (2007) J Hazard Mater 144(1):215–221.

[29] Willumsen PA, Karlson U. Effect of calcium on the surfactant tolerance of a fluoranthene degrading bacterium. (1998) Biodegradation 9(5):369–379.

[30] Mohan VS, Shailaja S, Krishna MR, Reddy KB, Sama PN. Bioslurry phase degradation of di-ethyl phthalate (DEP) contaminated soil in periodic discontinuous mode operation: Influence of bioaugmentation and substrate partition. (2006) Process Biochem. 41(3):644–652.

[31] Mueller JG, Lantz SE, Blattman BO, Chapman PJ. Bench-scale evaluation of alternative biological treatment processes for the remediation of pentachlorophenol and creosote contaminated materials: slurry-phase bioremediation. (1991) Environ Sci Technol. 25(6):1055–1055.

[32] Christodoulatos C, Koutsopyros A. Bioslurry reactors. In: Lewandowsky GA, Defilippi LJ. (Eds.), Biological Treatment of Hazardous Wastes. New York: Willey, 69–103 (1998).

13

The Electrokinetic Treatment of Polluted Soil by Hydrocarbon: From Laboratory to Field

Maribel Pérez-Corona, Zsuzsanna Plank and Erika Bustos

Additional information is available at the end of the chapter

Abstract

Soil contaminated with hydrocarbons (HC) in all over the world is a recurring problem arising from distribution, storage and illegal connections. A wide range of methods are used in all the world like remediation with biological and physicochemical treatments, however, for the purpose of reducing time and increasing the scope of new technologies that have proven its viability in experimental laboratory tests later tested implemented on field are necessary. One of the main advantages of electroremediation processes (ER) is the relatively short implementation time as well as its ease of removing contaminants in highly heterogeneous soils with low permeability. In this chapter, the ER process is described starting from the laboratory scale, determining the supporting electrolyte used, through the choice of material of the electrodes as well as its configuration; finally pilot-scale implementation and fieldwork.

Keywords: electrokinetic, polluted soil, hydrocarbons, resistivity

1. Introduction

Oil is the main source of energy in developed countries, its derivatives such as diesel, paraffin and liquefied gas are used for transport, heating and electricity production; in contrast, the pollution generated by the production processes required for the production, processing, transportation and distribution has generated a serious environmental problem affecting bodies of water, soil and air.

Polluted soil by hydrocarbons (HC) is importante in the environment, because the soil could be affected changing its physicochemical characteristics and lossing its ability to regenerate itself, or this regeneration takes a long time.

Because of this it is necessary to implement new techniques that allow the rehabilitation of soils in a shorter time and facilitate the process of restoration to the affected area.

There are plenty of available technologies to remediate soils contaminated with hydrocarbons, which are divided under three main headings: biological, physical, chemical and thermal treatments; however in the world most of the companies are dedicated to remediate soil using biological methods (bioremediation). Another important part of the mostly used technologies are soil washing, chemical oxidation and physical separation.

However, over the years a technique called physical chemistry electro-remediation has been developed, which has proved its viability on the laboratory level and has been successfully applied in the field in some countries of Europe and the USA.

The electro-remediation (ER) process, also called electrokinetic electrochemical treatment and/or electro-claim among others [1], is a technique within physicochemical treatments. ER has been considered to be a promising process addressing problems such as heterogeneous soils and low permeability, also can be applied *in situ* or *ex situ*, and is especially useful for remediation of inaccessible sites with minimal disruption to the surface, where other technologies fail. Besides it is also sensitive to a wide variety of contaminants.

The ER process is relatively safe, effective, easy to implement, economic and flexible from the points of applying on various types of soils and contaminants. Moreover, most of the *in situ* conventional methods present difficulties in treatment time in the case of fine-grained soils with high water content and high organic or clay content; in contrast, the ER method is suitable for these types of soils too [2].

ER is a technology to restore contaminated soil based on the generation of an electric field from imposing direct current. For the application of potential difference or direct current, the use of electrodes (anode and cathode) placed in wells previously dug into the ground is required, usually the soil is wetted with an electrolyte to improve conditions driving the electric field. The action of the electrolyte makes the pollutant transport to the wells where it will be extracted. Unlike the fluid drag, this technique allows for a directed migration preventing contaminant dispersion outside the treatment zone [3].

The main mechanisms of the electric field leading the contaminants to the electrodes are electro-migration, electro-osmosis and electrophoresis. The first two processes have the greatest influence on contamination transport. These processes are described below [1]:

Electro-migration is a phenomenon in which ions in solution and colloids having electric charge move through the electric field with a velocity that is proportional to the product of the strength of the electric field and mobility of the ion or particle.

The negatively charged ions (anions) will move toward the positively charged electrode and the positively charged ions (cations) will gravitate toward the negatively charged electrodes (cathode).This process is favored when the contaminations to be removed are metals with different oxidation states.

Electro-osmosis is the transport process describing the movement of mass of fluid through the pores of the soil under the influence of a potential gradient. When a potential gradient is generated in presence of moist soil, a movement of ions appears toward the electrode polarized in order of their electric charge, generating a migration of them by movement of the water due to the hydration of these ions, and causing movement of fluid through the soil pores. The electro-migration of species and the establishment of a double layer at the solid-liquid interface generate an electro-osmotic flow through the soil pores [4].

Electro-osmosis is the most important transport mechanism for removal of compounds uncharged or weakly charged as organic pollutants. The electro-osmotic component almost disappears in the cases of coarse sands and plastic clays wherein the electro-migration is the most important contaminant removal mechanism. It becomes as important as the electro-migration in the cases of fine sand and silt with high amounts of water and low conductivity [4].

Electrophoresis is a mechanism observed when particles, colloids or micro-surface electric charges that the contaminants bound to this material can be transported by the electric field [3].

Other mass transport phenomena occurring during ER are:

- Diffusion: the movement of the species due to concentration gradients and advection generated hydraulic gradients [4].
- The electrolysis of water: this occurs on the surface of the electrodes when applying electrical current, which creates an acidic border (with a pH value near 2) at the anode and a basic border (pH value about 12) at the cathode due to the generation of H^+ and OH^-, respectively. The reactions are:

  At the anode: $2H_2O\text{-}4e^- \rightarrow O_{2(gas)} + 4H^+_{\ (ac)}$ $E° = +1.229$ V

  At the cathode: $2H_2O + 2e^- \rightarrow H_{2(gas)} + 2OH^-$ $E° = -0.828$ V

It is noteworthy that electrolysis reactions depend on the type and arrangement of the electrodes [5] as well as the chemical species and electric potential applied during electro-remediation. Thus, protons generated at the anode move through the soil to the cathode by:

- Migration of ions due to electrical gradient.
- Advection of fluid through the pores due to electro-osmotic flow.
- Fluid flow through the soil pores due to a difference in hydraulic potential internally generated or externally applied.
- A diffusion caused by chemical gradients.

The alkaline medium developed at the cathode moves toward the anode by ion migration and diffusion of OH^-, which is transport overshadowed by the electro-osmotic flow and neutralization of H^+, ranging to the cathode where the ions can recombine to form water [4].

In the last three decades, there have been several investigations at laboratory and pilot even applying electrokinetic basis to remove a variety of contaminants. The electro-remediation has been successfully tested in the USA [1–7]. There are even companies offering it as an alternative remediation method within the portfolios of their services a large scale in soils with high clay content.

The ER method has demonstrated its ability to remove some organic contaminants in studies at laboratory, pilot or field [6], but its main application was on sites contaminated with metals in order to remove elements such as chromium, cadmium, mercury, lead, zinc, etc. [7].

In several studies, the application of the ER process has helped to achieve efficiencies close to 100% removal, particularly if the pollution is caused by a single metal (Pb). In *on-site* applications, the results depended on soil-type variables and the type of pollutant [3].

One example is the consortium formed by Monsanto, DuPont and General Electric, where the applied technology was called Lasagna™ ER *in situ* to remove trichloroethylene, achieving removal of 98% [8].

Another practical example was developed by Sandia National Laboratories, for electrochemical *in situ* remediation of soil contaminated with chromium, where electrodes of Iridium/Titanium were used with applying a power of 1572 kW/h; after 5 months of continuous treatment 64% efficiency was obtained [9].

Also, the ER was made at the Centro de Investigación y Desarrollo Tecnológico en Electroquímica, S. C. (CIDETEQ) at laboratory level in order to be able to apply it at pilot and on field level. For that reason, several investigations were developed that led to get familiar with different aspects of field application helping implementation of the technique in a petroleum industrial area. Meanwhile the Geological and Geophysical Institute of Hungary developed an analytical method for investigating the physical and chemical characteristics of soil.

2. Methodology

2.1. Selecting the type of electrodes

The activity was carried out with testing cyclic voltammetry using a potentiostat BAS Epsilon and a glass cell with three electrodes as a reference electrode Ag|AgCl saturated with KCl, wire Ti as auxiliary electrode and plaques of different materials to evaluate as working electrodes. The supporting electrolyte used in these tests was phosphate buffer solution at pH 12 (i = 0.1), because it has been reported that hydrocarbons are best removed in alkaline medium [10].

2.2. Choosing the supporting electrolyte

Solutions of KOH, NaOH, K_2HPO_4, Na_2HPO_4, KH_2PO_4 and Na_2HPO_4 were all prepared at 0.1M in water, which was used to wet soil for an electrolysis. UV-Vis spectrophotometry was used to verify the removal of HC after electrochemical treatment in the different solutions used [11].

2.3. Choosing the best treatment

The technologies described below were compared in order to find the best treatment for decontaminating soils; in all the three cases the removal of oil by Soxhlet extraction at the end of treatment was evaluated. The initial content of fats and oils of the contaminated soil was 4000 mg HC/kg of dry soil [12].

Soil washing surfactant Triton X-114: Triton X-114 (4% V/V) was passed at a flow rate of 1.5 mL/min into a tubular reactor containing 30 g soil, for a period of 5 h.

Biological treatment with solid culture: 30 g of soil was added to agro-industrial waste bagasse and filter cake with a-residue soil agro-industrial 100:2:2, together they were placed in glass containers while maintaining a temperature of 28°C, for a period of 15 days with aeration every 3 days for 20 min.

Electro-remediation of contaminated soil: a tubular reactor was used with 30 g soil and 0.1 M NaOH as supporting electrolyte with a flow rate of 1.5 mL/min, by applying a current of 2 mA for a period of 3.5 h; the working electrodes were titanium mesh (cathode) and Ti|IrO_2-Ta_2O_5 (anode).

2.4. Choosing the best configuration of electrodes

Three-electrode configurations were evaluated: (a) face to face consisting of four cathodes and eight anodes (all rectangular) placed opposite the cathodes; (b) the arrangement of alternating electrodes consisted of six cathodes and six anodes alternating rows of three; and (c) the circular configuration resided in a central cathode and six anodes around this one [13]. The sample amount was 1.9 kg for the three cases and hydrated for a period of 18 h with 800 mL of 0.1M NaOH; the current applied was 0.23 A for a period of 6 h. The used working electrodes were made of Titanium plates and IrO_2-Ta_2O_5|Ti as cathodes and anodes respectively, all at a distance of 6 cm.

The removal process was followed by Soxhlet extraction on the ground and in the solution for determination of chemical oxygen demand (COD), samples for fats and oils were obtained near the anodes and cathodes, as well as in the half-cell.

2.5. ER pilot scale *in situ* and *ex situ*

The arrangement of circular electrodes was used during ER pilot scale *in situ* and *ex situ*. The cathode was used in the center of the electrochemical cell, and the IrO_2-Ta_2O_5|Ti anodes were used around this one. All the electrodes were used during ER pilot system with dimensions of 60 cm length × 24 cm diameter, which were placed 117 cm between them. In these experiments the amount of soil type Vertisol pelic treated was 3.3 m^3 [13, 14].

Ex situ: The soil was contaminated with 1126 mg/kg by gasoline. To ER a constant current of 9 A during 4.5 h by day was applied, adding every day 60 L of 0.7 μM NaOH as supporting electrolyte.

In situ: Soil contamination by hydrocarbon was up to 58,000 mg/kg, a current of 11 A was applied for a period of 7.5 h; in this case hydrate first with water and then 135 L of the supporting electrolyte is added (0.1M NaOH) to the cathode hole.

The removal of fats and oils (F&O) were measured by Soxhlet extraction.

2.6. Application of ER in the field

Antrosol-type soil (275 m^3) contaminated with hydrocarbons was treated, a constant current of 9 A was applied for 4 h for each cell in a six-cell system mounted in series, the soil removed to insert the electrodes was treated *ex situ* and then returned to its place. The volume necessary for moisturizing the soil was 120 L of 0.1M NaOH per cell, and the solution extracted at the end of the process of ER was treated by an advanced oxidation process.

The treatment consisted of applying the electric field for 4 h to the first block of six cells, once it is completed the first block of the treatment is continued with the second block and so on until the end of treatment with a total of 14 blocks for complete 84 cells mounted on a three-week period, the *ex situ* process is followed on par with the same operating conditions.

DC resistivity measurements were carried out using a Digital Ground Resistance Tester Model 4500 AEMC® INSTRUMENTS applying a current of 2 mA, using four copper electrodes, placed at a distance of 1 m, before and after treatment.

Determination of hydrocarbon medium (NMX-AA-145-SCFI-2008) and heavy (NMX-AA-134-SCFI-2006) fractions was performed, as well as polycyclic aromatic hydrocarbons (NMX-AA-146-SCFI-2008) before and after electrochemical treatment.

3. Results and discussion

3.1. Selecting the type of electrodes

The argument for selecting the electrode material was based on selecting the material with the greatest electro-active area. **Table 1** shows the electrode materials evaluated with the corresponding electro-active areas, having been calculated using the equation Randles-Sevcik with cyclic voltammetry at different sweep speeds of 20–150 mV/s in the presence of 1 mM $Cl_3Ru(NH_3)$. It showed a reversible behavior only with the reticulated vitreous carbon (RVC) and quasi-reversible for all other electrodes in 0.1 M KCl [15].

The electrode showing the highest electro-active area was the RVC; however, its use was discarded because it has a great capacity to adsorb organic compounds from its surface. Therefore, IrO_2-Ta_2O_5|Ti anode was used during the different ER treatment, because they have an effective life of 5–10 years [16] and as a cathode of Ti.

Material	Electro-active area (cm^2)
Ti I RuO_2-SnO_2	14.362
RVC 1000ppp	30.328
Ti I IrO_2-Ta_2O_5	13.333
Ti I RuO_2-IrO_2	5.741
Ti I RuO_2	2.724
Stainless steel 316	1.414

Table 1. Electro-active area for different materials.

3.2. Choosing the supporting electrolyte

Of the solutions prepared from KOH, NaOH, K_2HPO_4, Na_2HPO_4, KH_2PO_4 and Na_2HPO_4 were chosen for the process of ER KOH and NaOH because they have the highest ionic molar conductivity, in this case for K^+ and Na^+ 73.5×10^{-4} and 50.1×10^{-4} sm^2/mol respectively [11].

NaOH was used as electrolyte for the higher removal of HC than KOH, because of its higher molar ionic conductivity.

This behavior can be attributed to the ability of adsorption of K^+ in the ground which is higher than that of Na^+ (17). Concentration of K^+ in solution decreased, causing an increase in electrical resistance in soil, and decreasing the removal efficiency of HC [17–20].

3.3. Choosing the best treatment

Table 2 shows the comparison of the three evaluated treatments. It can be observed that the electrochemical treatment shows the best removal rate with 81.9% with a period of 3.5 h [12, 21].

Treatment	Operation time (h)	Removal (%)
Soil washing, surfactant: Triton X-114	5	11.9
Biological treatment with solid culture	360	44.4
Electro-remediation with NaOH 0.1 M	3.5	81.9

Table 2. Comparison of remediation treatments.

According to these tests, the process of ER proved to be the most efficient treatment and with 3.5 h of application time, besides being a technology that can remove both organic and inorganic contaminants in soils with high clay content and low permeability. These characteristics make the electrokinetic treatment a viable process to be applied on large scale in HC-contaminated soils.

3.4. Choosing the best arrangement of electrodes

Table 3 summarizes the three proposed arrangements where the circular one shows the best results in removal of HC (47.81%) in soil and the highest amount of COD in solution (8880 mg/L) associated with the presence of organic pollutants transported into the solution. In the results reported in **Table 3** and **Figure 1**, the lowest and highest amounts of HC removed from all the sampled points were chosen at each of the arrangements [13, 22, 23].

Configuration	Removal F&O (%)		COD (mg/L)
	Minimum	Maximum	
Face to face	0.51	21.35	3830
Alternating	3.65	29.29	3080
Circular	14.97	47.81	8880

Table 3. Results of F&O in soil and COD in solution of three-electrode configurations.

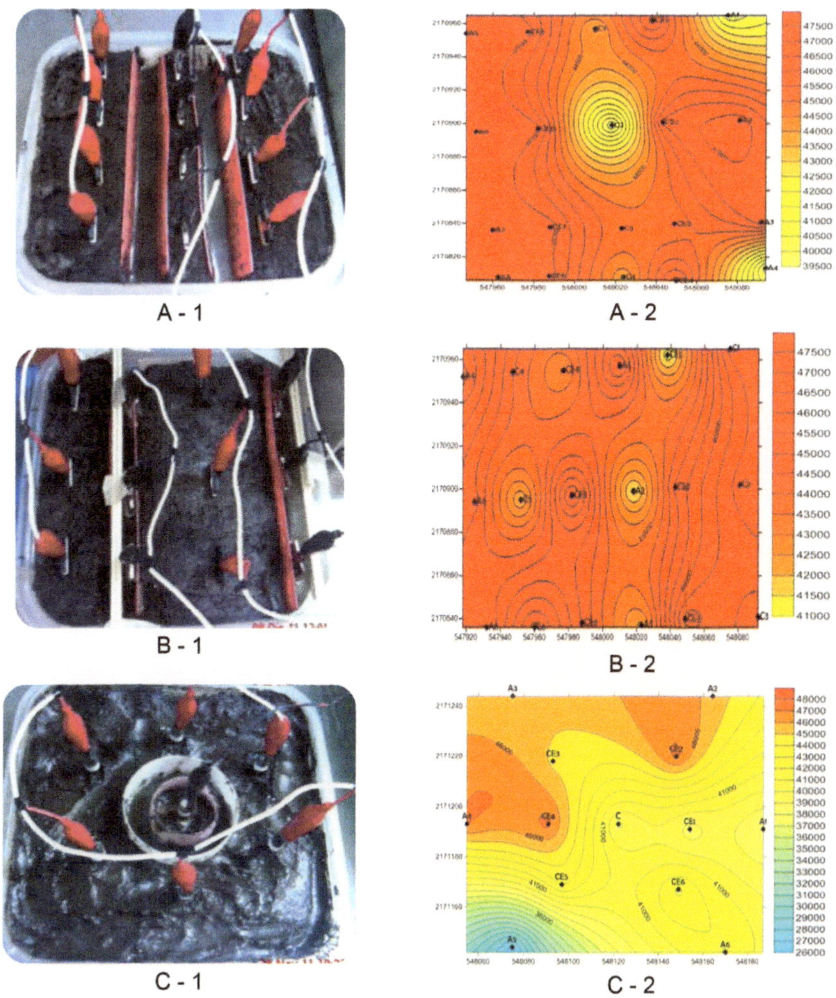

Figure 1. *Representation of the different configuration of electrodes (1): face to face (A), alternating (B) and circular (C) where the red alligators are the anodes, and black alligators are the cathodes, with their corresponding removal of HC in mg HC/kg of dry soil (2).*

Based on these results, one can be convinced that the circular is the best choice for electrode configuration to be used in fieldwork, because this arrangement allows the concentration of all pollutants to the cathode hollowed by the influence of the electric field where the power lines all converge anode to the cathode.

3.5. ER pilot scale *ex situ and in situ*

Ex situ: Samples of fats and oils have been collected for analysis as taken from different sections of the soil cell, because the soil heterogeneity represents different behaviors throughout the cell. After three weeks of electrochemical treatment, a decrease of about 84–88% was observed in the concentration of gasoline in the different sampled points (**Figure 2A**) which is due to electro-migration, electro-osmosis and electrophoresis, aided by water electrolysis. The contribution of the use of modified anodes IrO_2-Ta_2O_5|Ti is also considered, provided the chemical conditions are adequate to desorption and/or destruction of hydrocarbons present in the soil particles [14, 24].

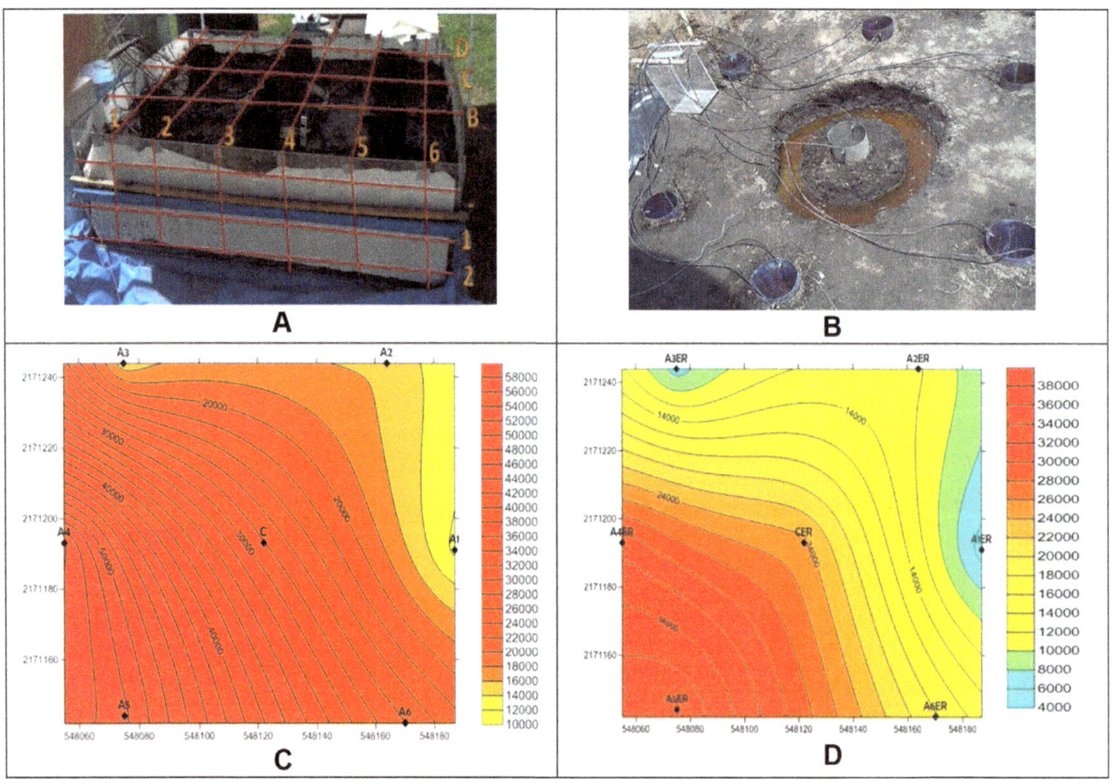

Figure 2. *ER pilot scale ex situ (A) and in situ (B). Middle fraction HC content in the polluted soil before (A) and after (B) its electrochemical treatment in mg HC/kg of dry soil.*

In situ: The amount of F&O was registered in the sampling sections (**Figure 2B**) near the six anodes and cathodes at the beginning (**Figure 2C**) and the end of treatment (**Figure 2D**). **Table 4** shows the removal percentages obtained after a treatment of 7.5 h. In general, a decrease appreciated of pollutant in all sampled points is close to 90%; however, this is not the same in all areas, attributed to soil heterogeneity behavior.

Like in the case of ER *ex situ* treatment efficiency is attributed to transport phenomena occurring during the application of electric field, the use of IrO_2-Ta_2O_5 | Ti anodes, the electrolysis of water, adequate wetting and high clay content in the soil.

Position	Removal after ER in soil (%)			
	Anodes	Center	Center cathode	Cathode
1	55.55	94.63	91.75	12.55
2	21.70	80.16	79.99	
3	52.30	84.48	85.46	
4	44.43	41.03	-18.55	
5	27.65	1.66	75.66	
6	30.11	87.84	21.45	

Table 4. Percentages of HC removal in soil after ER *in situ*.

3.6. Field application of ER

In **Figure 3**, the blue dots ranging from one to five represent the locations of the sampling points located on the orange lines labeled with B, D and F.

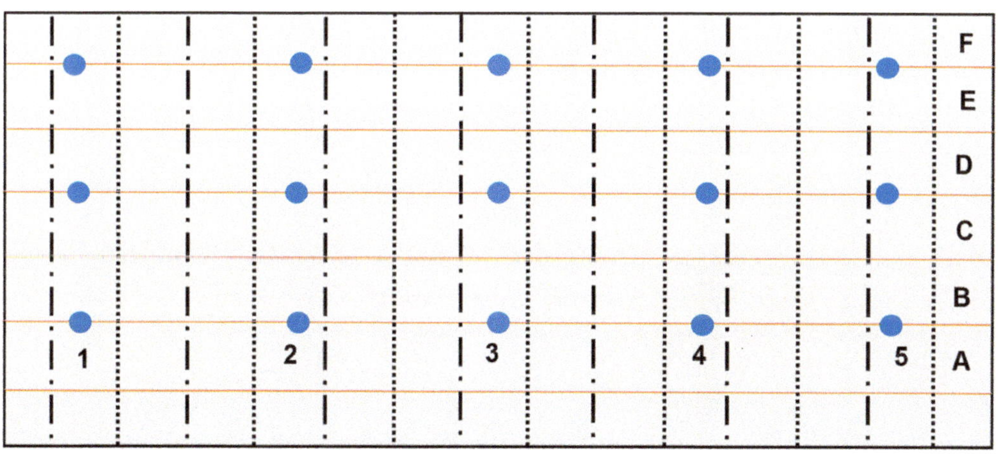

Figure 3. *Representation of the experimental setup process ER.*

Figure 4 shows that for the sampled points, the initial values of the middle fraction HC content (MFHC, **Figure 4A**) determined from the sampling points were higher than 50,000 mg/kg. The electrochemical treatment decreased these values by 74% with average values of 12,000 mg/kg (**Figure 3B**).

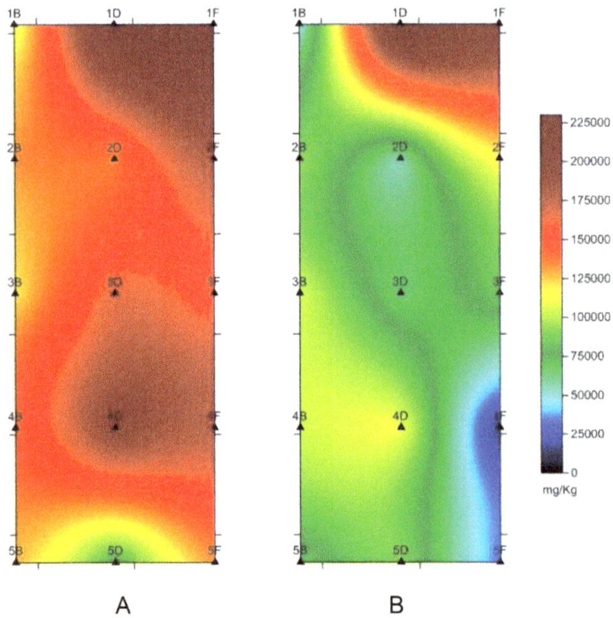

Figure 4. *Middle fraction HC content in the polluted soil before (A) and after its electrochemical treatment (B) in the sampled points shown in* **Figure 3**.

In the case of heavy fraction HC (HFHC), the results are presented in **Figure 5**. The contamination content decreased in all points, except 1D and 5D where the slight increase can be possibly due to the sub-products of the electrokinetic process. The removal rates of the remaining 13 variables are ranging from 11% (1F) to 94% (4F) which demonstrated the feasibility of the field application. It was observed that applying the technique the organic compounds can be removed due to the action of the electric field with the effect of the involved transport processes (electro-migration, electro-osmosis, and electrophoresis), to water electrolysis, the applied electrode configuration and the current.

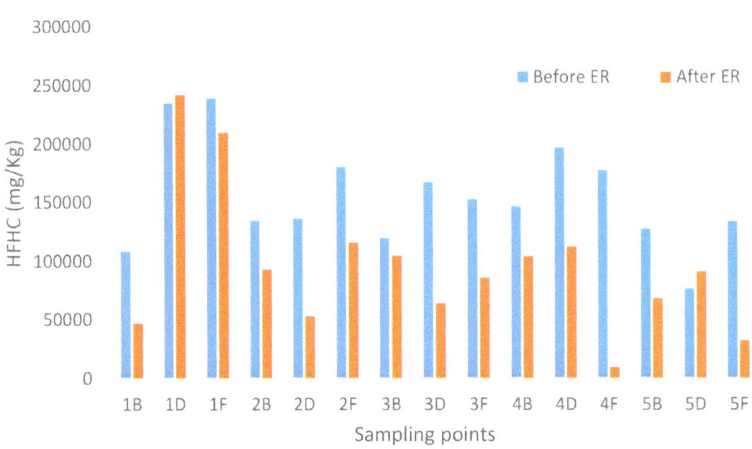

Figure 5. Heavy fraction hydrocarbon content (C_{28}–C_{40}) in different points sampled before and after the process of ER.

The analysis of section B for 16 kinds of poly aromatic hydrocarbons (PAHs) showed that five of them were present in greater abundance. The behavior of these compounds before (blue bars) and after (pink bars) the treatment are presented in **Figure 6**. As it is expected the content of PAHs were various throughout the site; the removal percentages are varying according to the type of compound and the site characteristics: for example, pyrene removal varies 29–90%, the Phenanthrene' removal range is between 18 and 81% and for Benzo (a) pyrene it is 33 and 61%.

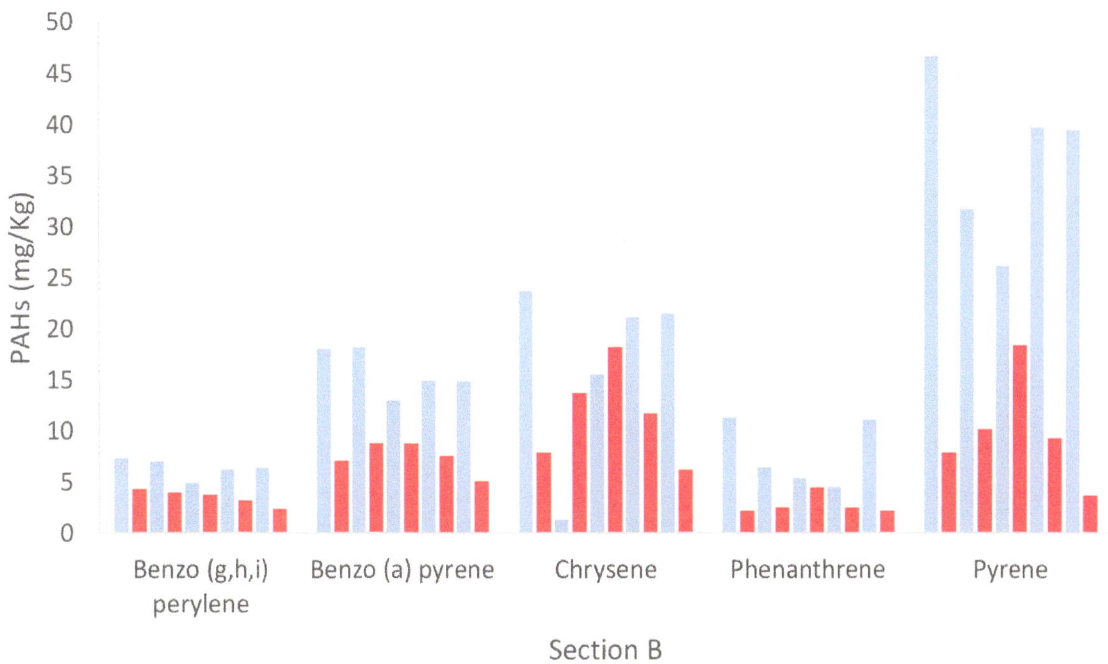

Figure 6. Behavior of PAH content in section B before (blue bar) and after (pink bar) the electrochemical treatment.

As an additional tool to follow the distribution of the contaminant in the soil DC resistivity measurements were taken with the aim of appreciating a decrease in HC, a diminution in resistivity values reflects a decrease of HC content [24]. The purpose of applying geoelectric measure in the contaminated site is to find a fast, economic, non-invasive method that could provide a reliable image on the distribution of soil contamination.

The DC resistivity value depends on several geological factors such as the texture class, the minerals present, the moisture content, porosity, these properties change when the soil is exposed to some type of contamination, in this case by organic compounds, which causes an increase in soil resistivity [24–27].

Behavior analysis of apparent resistivity was performed using the Wenner-Alfa array consisting of an array of four electrodes and can be used in moderate depths, and is relatively sensitive to vertical changes under the subsurface to the center of the array, but little sensitive to horizontal changes [28, 29].

In **Figure 7**, the distribution of the measured apparent resistivity values can be observed at the test site before (left) and after (right) the treatment. It is remarkable that before the remediation process there were two zones where the apparent resistivity was higher than 20 ohmm (marked with white). After the process of ER, the resistivity values decrease to 2–4 ohmm at the same points, which is associated with a decrease in the amount of HC and increase of salts, as the sub-products of ER. This can be validated with the results for middle and heavy fraction HC, in the cases of points 2D (removal rates of 61% HFHC and 71% for MFHC), 2F (removal rates HFHC: 35% and 64% for MFHC) 5F (removal rates: 75% for HFHC and 84% of MFHC).

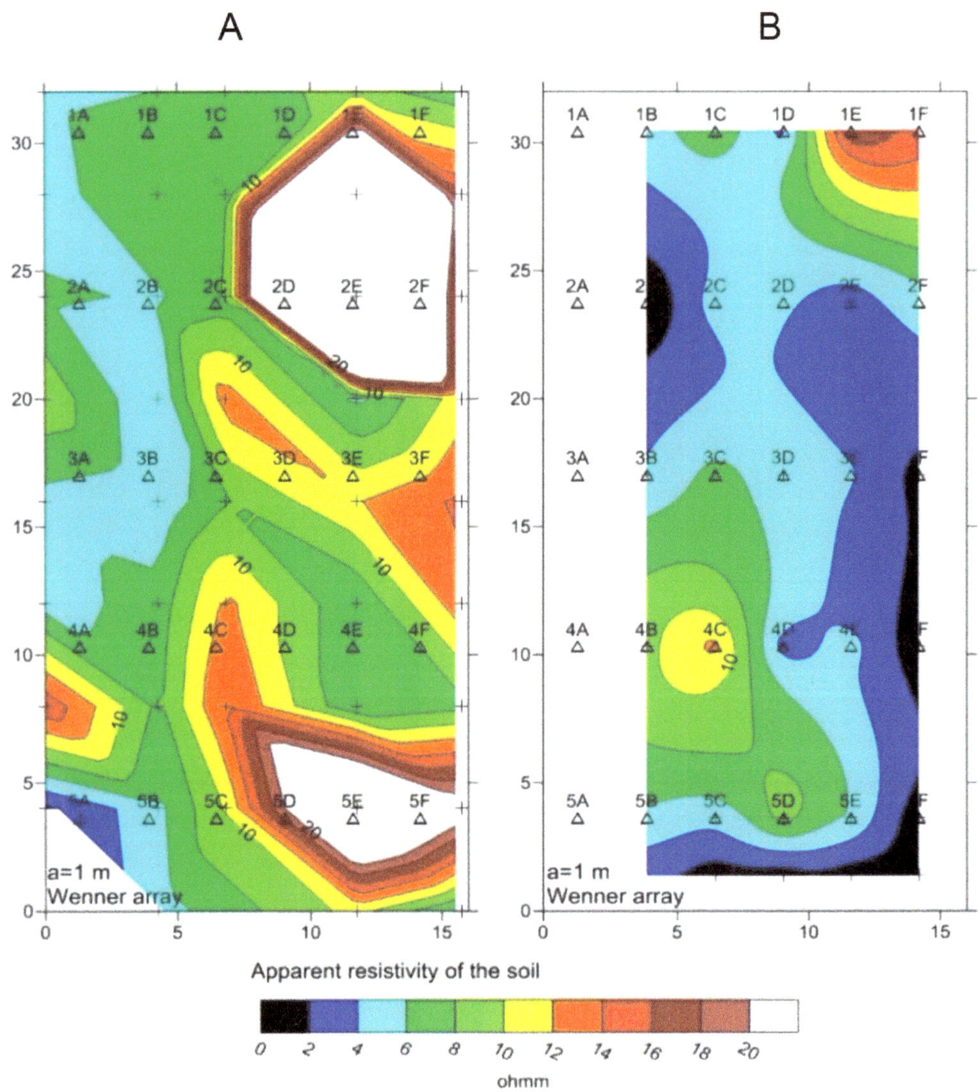

Figure 7. Apparent resistivity behavior before (A) and after (B) the electrokinetic treatment.

According to the obtained results, DC resistivity survey method can be used as an effective tool for monitoring the process of HC removal in soils. However the readings taken do not represent a value of HC concentration, it is an indirect measure of the reduction of these pollutants in the subsurface with respect to an initial value.

4. Conclusions

The success of electrochemical treatment is attributed to several factors: the choice of supporting electrolyte, type of electrodes and their configuration, the distance between them; the cell current, all as a whole allowed the removal of HC from the laboratory scale, pilot and field were appreciable. Removal percentages were ranging from 20 to 90% attributed to soil heterogeneity which does not allow the results to be reproducible in all sampled points, due to geochemical, geophysical and physicochemical factors that occur during the application of electric field such as changes in pH, desorption and/or solubility of the contaminants and redox processes.

The successful field implementation of ER technique makes the ER process an attractive option among the remediation technologies dealing with environmental problems caused by contamination of organic and inorganic compounds. Furthermore due to soil heterogeneity, the ER technique should be used in conjunction with other techniques completing the whole soil rehabilitation process.

Acknowledgements

The authors wish to acknowledge the Consejo Nacional de Ciencia y Tecnología de los Estados Unidos Mexicanos (CONACyT) for its financial support to develop this project and to Magyar Földtani és Geofizikai Intézet for financial support under project TÉT_12_MX-1-2013-0010. M. Pérez-Corona is grateful to CONACyT for her scholarship to get her Ph. D. in Electrochemistry.

Author details

Maribel Pérez-Corona[1], Zsuzsanna Plank[2] and Erika Bustos[1*]

*Address all correspondence to: ebustos@cideteq.mx

1 Centro de Investigación y Desarrollo Tecnológico en Electroquímica, S. C. Parque Tecnológico, Querétaro, Sanfandila, Pedro Escobedo, Querétaro, Mexico

2 Geological and Geophysical Institute of Hungary, Budapest, Hungary

References

[1] Reddy, K.R. and C. Cameselle, *Electrochemical Remediation Technologies for Polluted Soils, Sediments and Groundwater*. 2009, USA: A John Wiley & Sons, Inc.

[2] Saichek, R.E. and K.R. Reddy, *Electrokinetically enhanced remediation of hydrophobic organic compounds in soils: a review, in Critical Reviews in Environmental Science and Technology*. 2005, Taylor & Francis: USA. p. 115-192.

[3] De la Rosa Pérez, D., M. Teutli Leon, and M. Ramirez Islas, *Electro-remediacion de Suelos Contaminados, una Revisión Técnica para su Aplicación en Campo.* Revista Internacional de Contaminación Ambiental, 2007. 23.

[4] Prasad, M., K. Sajwan, and R. Naidu, *Trace Elements in the Environment, Biogeochemistry, Biotechnology and Bioremediation*, 2006, Taylor & Francis: Florida.

[5] Turer, D. and A. Genc, *Assessing Effect of Electrode Configuration on the Efficiency of Electrokinetic Remediation by Sequential Extraction Analysis.* 2004, Elsevier, p. 167-174: Amsterdam.

[6] Martínez Gilbón, A.J., *Electroremedación de suelos contaminados con Hidrocarburos.* 2001, Universidad Autonoma Metropolitana: México, D.F. p. 47.

[7] Hamed, J., Y.B. Acar, and R.J. Gale, *Pb(II) removal from kaolinite by electrokinetics.* Journal of Geotechnical Engineering, 1991. 117: p. 241-271.

[8] Environmental Protection Agency, U.S., *Resource Guide for Electrokinetics Laboratory and Field Processes Applicable to Radioactive and Hazardous Mixed Wastes in Soil and Groundwater from 1992 to 1997.* 1997, EPA, USA.

[9] Dorronsoro Fernández, C. *Descontaminación de Suelos.* Available from: edafologia.ugr.es.

[10] Barrón, F. and I. Labastida, *Determinación experimental de la influencia de dos diferentes electrolitos sobre las variables que influyen directamente en el proceso de electrorremediación de suelos contaminados con hidrocarburos en una celda estandarizada a nivel laboratorio.* 2005, Universidad Autónoma Metropolitana Unidad Azcapotzalco: México, D.F.

[11] Ramírez, D.V., *Caracterización y Posterior Remediación Electro-cinética de Suelo Tipo Vertisol Pélico Contaminado con Hidrocarburo Procedente de una Zona Industrial,* in *Electroquímica.* 2010, CIDETEQ: Quéretaro. p. 181.

[12] López, G.I., *Estudio comparativo de tres diferentes métodos de remediación de suelo tipo Gleysol contaminado con hidrocarburo.* 2011, CIDETEQ: Querétaro. p. 155.

[13] Pérez-Corona, M., *Electro-remediación de suelo contaminado con hidrocarburo in situ.* 2012, CIDETEQ S.C.: Querétaro, México. p. 198.

[14] Ochoa, B., et al., *Electrokinetic treatment of polluted soil at pilot level coupled to an advanced oxidation process of its wastewater.* Physics and Chemistry of the Earth, 2016. 91: p. 68-76.

[15] Romero-Lara, O., *Estudio Termodinámico y Cinetico de la Romoción Electroquímica de Hidrocarburos en Suelo Tipo Gleysol Empleando Diferentes Soportes de Electrodos*, in *Ingeniería Ambiental.* 2011, Universidad de Guanajuato: Guanajuato, México. p. 68.

[16] Gueneau, J.P., *Production and Study of a Ti/TiO_2/Noble Metal Anode*, in *Departamento de Ciencias de los Materiales y Electroquímica.* 2003, Universidad Libre de Brucelas: Bruselas.

[17] Sivapullaiah, P. and B.N. Prakash, *Electroosmotic flow behavior of metal contaminated expansive soil.* Hazardous Materials, 2007. 143: p. 682-689.

[18] Vázquez, M.V., et al., *Integral analysis of the process of electro-remediation of andisols polluted by heavy metals.* Portugaliae Electrochimica Acta, 2004. 22: p. 387-398.

[19] Tran, L.H., et al., *Electrolytic oxidation of polynuclear aromatic hydrocarbons from creosote solution using TI/IRO_2 and TI/SNO_2 circular mesh electrodes.* Environmental Engineering, 2009. 135(10): p. 1051-1062.

[20] Randy, A.H., et al., *Comparison of moisture management methods for the bioremediation of hydrocarbon contaminated soil.* African Journal of Biotechnology, 2011. 10(3): p. 394-404.

[21] Alba, G.I., M.C. Cuevas, and E. Bustos, *Comparing the electroremediation of Gleysol soil contaminated with hydrocarbon with Triton X-114, washing and bioremediation with solid cultures employing agroindustrial residues.* International Journal of Electrochemical Science, 2013. 8: p. 4735-4746.

[22] Pérez-Corona, M., et al., *Electroremediation of hydrocarbons-contaminated vertisol type soil employing different electrode configurations.* Sustainable Environment Research, 2013. 23(4): p. 285-288.

[23] Pérez-Corona, M., et al., *Comparison of different arrangements of electrodes during the electrokinetic treatment of polluted soil with hydrocarbons and its final application in situ.* In *Recent Research Development in Electrochemistry.* 2013, Transworld Research Network. p. 59-80: Kerala.

[24] Pérez-Corona, M., et al., *The cone penetration test and 2D Imaging Resistivity as tools to simulate the distribution of hydrocarbons in soil.* Physics and Chemistry of the Earth, 2016. 91: p. 87-92.

[25] Stewart, M. and L. North, *A borehole geophysical method for detection and quantification of dense, non aqueous phase liquids (DNAPL) in saturated soils.* Applied Geophysics, 2006. 60: p. 87-99.

[26] Fukue, M., et al., *Use of resistivity cone for detecting contaminated soils layers.* Engineering Geology, 2001. 60: p. 361-369.

[27] Sudha, K., et al., *Soil characterization using electrical resistivity tomography and geotechnical investigations.* Applied Geophysics, 2009. 67: p. 74-79.

[28] Loke, M.H., *Tutorial: 2-D and 3-D electrical imaging surveys.* 2001.

[29] Loke, M.H., et al., *Recent developments in the direct-current geoelectrical imaging method.* Applied Geophysics, 2013. 95: p. 135-156.

Permissions

All chapters in this book were first published in SC, by InTech Open; hereby published with permission under the Creative Commons Attribution License or equivalent. Every chapter published in this book has been scrutinized by our experts. Their significance has been extensively debated. The topics covered herein carry significant findings which will fuel the growth of the discipline. They may even be implemented as practical applications or may be referred to as a beginning point for another development.

The contributors of this book come from diverse backgrounds, making this book a truly international effort. This book will bring forth new frontiers with its revolutionizing research information and detailed analysis of the nascent developments around the world.

We would like to thank all the contributing authors for lending their expertise to make the book truly unique. They have played a crucial role in the development of this book. Without their invaluable contributions this book wouldn't have been possible. They have made vital efforts to compile up to date information on the varied aspects of this subject to make this book a valuable addition to the collection of many professionals and students.

This book was conceptualized with the vision of imparting up-to-date information and advanced data in this field. To ensure the same, a matchless editorial board was set up. Every individual on the board went through rigorous rounds of assessment to prove their worth. After which they invested a large part of their time researching and compiling the most relevant data for our readers.

The editorial board has been involved in producing this book since its inception. They have spent rigorous hours researching and exploring the diverse topics which have resulted in the successful publishing of this book. They have passed on their knowledge of decades through this book. To expedite this challenging task, the publisher supported the team at every step. A small team of assistant editors was also appointed to further simplify the editing procedure and attain best results for the readers.

Apart from the editorial board, the designing team has also invested a significant amount of their time in understanding the subject and creating the most relevant covers. They scrutinized every image to scout for the most suitable representation of the subject and create an appropriate cover for the book.

The publishing team has been an ardent support to the editorial, designing and production team. Their endless efforts to recruit the best for this project, has resulted in the accomplishment of this book. They are a veteran in the field of academics and their pool of knowledge is as vast as their experience in printing. Their expertise and guidance has proved useful at every step. Their uncompromising quality standards have made this book an exceptional effort. Their encouragement from time to time has been an inspiration for everyone.

The publisher and the editorial board hope that this book will prove to be a valuable piece of knowledge for researchers, students, practitioners and scholars across the globe.

List of Contributors

Wlodzimierz Breś
Department of Plant Nutrition, Poznań University of Life Sciences, Poznań, Poland

Barbara Politycka
Department of Plant Physiology, Poznań University of Life Sciences, Poznań, Poland

Noureddine Bouaïcha and Sylvain Corbel
Ecology, Systematic and Evolution, University of Paris-Sud, CNRS, AgroParisTech, University Paris-Saclay, Orsay, France

Daniel Sacristán and Ester Carbó
Land-Use Planning Department, Centro de Investigaciones Sobre Desertificación-CIDE (CSIC-University of Valencia-Generalitat Valenciana), Moncada, Valencia, Spain

Ahmed Al-Busaidi and Mushtaque Ahmed
Department of Soils, Water and Agricultural Engineering, College of Agricultural and Marine Sciences, Sultan Qaboos University, Al-Khoud, Muscat, Oman

Michael Aide, Indi Braden and Sven Svenson
Department of Agriculture, Southeast Missouri State University, Cape Girardeau, Missouri, USA

Anna Gałązka and Jarosław Grządziel
Department of Agricultural Microbiology, Institute of Soil Science and Plant Cultivation – State Research Institute, Pulawy, Poland

Patrick Meire and Eric de Deckere
Department of Biology, University of Antwerp, Antwerp, Belgium

Ali Erfani Agah
Department of Biology, University of Antwerp, Antwerp, Belgium
Department of Irrigation and Drainage, Shahrood University of Technology, Shahrood, Iran

Yoshiharu Mitoma
Department of Environmental Sciences, Faculty of Life and Environmental Sciences, Prefectural University of Hiroshima, Shobara City, Hiroshima, Japan

Alina M. Simion and Cristian Simion
Department of Organic Chemistry, Faculty of Applied Chemistry and Material Sciences, Politehnica University of Bucharest, Bucharest, Romania

Shiguo Xu, Yanzhao Fu and Qi Wang
Institute of Water and Environment Research, Dalian University of Technology, Dalian, China

Yi Xu
Institute of Water and Environment Research, Dalian University of Technology, Dalian, China
Shenyang Academy of Environmental Sciences, Shenyang, China

Ivana Smičiklas and Marija Šljivić-Ivanović
Radiation and Environmental Protection Department, Vinča Institute of Nuclear Sciences, University of Belgrade, Belgrade, Serbia

Wilberth Chan Cupul
Biological Control and Applied Mycology Laboratory, Faculty of Biological and Agricultural Sciences, University of Colima, Mexico

Refugio Rodríguez Vázquez
Xenobiotics Laboratory, Center for Research and Advanced Studies of the National Polytechnic Institute, Mexico City, Mexico

Roger Saint-Fort
Department of Environmental Science, Faculty of Science and Technology, Mount Royal University, Calgary, AB, Canada

Maribel Pérez - Corona and Erika Bustos
Centro de Investigación y Desarrollo Tecnológico en Electroquímica, S. C. Parque Tecnológico, Querétaro, Sanfandila, Pedro Escobedo, Querétaro, Mexico

Zsuzsanna Plank
Geological and Geophysical Institute of Hungary, Budapest, Hungary

Index

A

Agricultural Soils, 13, 25, 30, 42-45, 47-48, 52-53, 55-57, 59-61, 103, 136, 197

Algal Communities, 21

Allelochemicals, 1, 8-10, 16-17

Allelopathy, 1-2, 8, 16-18

Anthropogenic Activities, 21, 102, 104, 173

B

Bacterial Diversity, 102, 104, 107-108, 110, 114, 118

Bacterial Growth, 25, 84

Barley, 61, 71, 122, 124, 126-131

Bioaccumulation, 20, 28-29, 31, 38-39, 54-56, 60

Bioreactors, 83, 85-86, 89, 92, 96, 100-101, 112, 211, 224, 227-228

Bioremediation, 19, 102-106, 115-117, 137, 139, 143-145, 147, 152-153, 187, 196, 207, 209-212, 221-223, 225-228, 230-231, 233, 246-247

Biosolids, 60, 63-66, 70, 74, 81

C

Clay Content, 24, 42, 52-56, 181, 183-184, 233, 235, 238, 241

Cmc, 211, 219-220, 222, 225, 227-228

Contamination, 1, 5, 13, 20-21, 23, 25, 42-46, 48, 58, 78, 80, 84, 102, 113-114, 147-149, 154, 182, 187-191, 201, 207, 212-214, 233, 237, 243, 245

Controlled Drainage, 83, 85-86, 89, 101

Crude Oil, 103, 105-106, 117, 145, 153, 230

Cu Toxicity, 43

Cyanobacterial Toxins, 20, 23, 31-33, 36-38

Cyanotoxins, 20-30, 32, 34, 36, 39

D

Denitrification, 83, 85-87, 89-90, 92-97, 99-101

Dgge, 102, 109-112, 114, 118-121

Dioxins, 136, 140-141, 145, 149

Drainage Technologies, 84-85, 89

E

Environmental Change, 155

Extraction, 4, 7, 35, 41, 88-89, 103, 107, 113, 115, 121, 138, 140-142, 145, 149-150, 172, 174, 177-180, 182-184, 189-192, 200, 236-237, 246

F

Fertigation, 1, 5-7, 15

Fertilizer, 2, 6, 13, 20, 23, 63-68, 70, 74, 80, 84, 98-100, 123, 176

Fungal Enzyme Extract, 196, 206

H

Heavy Metals, 1-3, 13, 43-45, 57-60, 62-65, 72-74, 77, 80-81, 136-146, 148-154, 174, 186, 191-192, 195, 247

Human Health, 13, 20, 32, 36, 43-44, 64, 79, 103, 137, 146-147, 188, 197, 226

Hydrocarbons, 102-103, 105, 107, 115-117, 138-140, 143, 148-149, 151-152, 197, 201, 208-209, 212, 228, 230, 232-233, 235, 237, 240, 243, 247

Hydrodynamic Conditions, 156

Hydrophobic, 211-212, 216-223, 227-228, 230, 246

Hydrus-1d, 122-123, 125, 127-129, 131, 133-134

I

Immobilization, 138, 145-146, 154, 172, 182, 184-185, 187, 192-195, 210

Invertebrates, 26

Irrigation, 5, 15-16, 20-22, 24-29, 38, 40, 57, 60, 65, 68, 73, 83-84, 86-89, 92, 101, 122-135, 159, 171, 203, 207

K

Kala Compost, 63-66, 68-70, 74, 76, 78-79

L

Laccase, 196-199, 201-203, 206-210

Land Reclamation, 155-161, 164-170

Living Organisms, 104, 143, 175-176, 182

M

Magnesium, 2, 4, 73, 89, 94, 156, 160

Mechanism, 24, 27, 29, 137-138, 155-156, 162, 165, 168-169, 177, 180, 185, 194, 197, 201, 204, 223, 228, 234

Microcystins, 20-30, 32, 34-41
Micronutrients, 5, 43, 59, 63-64, 176
Mineral Fertilizer, 2
Mining, 13, 174, 176, 188
Mitigation Measures, 155-157, 169
Mobility, 35, 52-53, 55-56, 110, 172, 174, 176-183, 186, 189-191, 197, 233
Moisture Retention, 64
Molecular Methods, 102, 104, 107-108, 114, 117
Mulch, 3, 11-12, 14, 18-19

N
Napl, 211-216, 218-221, 228-229
Ngs, 102, 112-113
Nitrate, 7, 15, 17, 83-86, 89-101, 124, 126-127, 133, 139, 148, 160

O
Overfertilization, 1

P
Pahs, 102-107, 114-117, 120, 138, 141-143, 145, 150, 209, 243
Pesticides, 10-11, 16, 18, 103, 136, 139, 143-144, 152, 197, 201, 207-209
Phosphate Leaching, 122, 130-131
Phosphorous, 80, 84, 122, 125, 128, 221
Phytotoxicity, 12, 17, 20, 27, 30-31, 38
Plant Growth, 9, 11, 63, 65, 68, 73, 104-105, 107, 195
Polluted Soil, 106, 137, 141-142, 145-146, 149-151, 186, 232-233, 240, 242, 247

R
Radioactive Pollutants, 172, 174, 182, 184, 186
Reactive Materials, 172
Replantation, 1, 9-10
Resistivity, 138, 232, 237, 243-244, 247-248

S
Simultaneous Decontamination, 136, 138, 143, 146, 154
Soil, 1-26, 29, 31, 35-40, 42-56, 58-74, 77-82, 84-85, 87-95, 97-99, 101-110, 112, 114-118, 120-166, 168-215, 219-223, 226-243, 245-248
Soil Contamination, 5, 13, 20, 23, 43, 80, 102, 136, 147, 149, 174-176, 185, 237, 243
Soil Improver, 2-3
Soil Matrix, 87, 136, 138, 162-163, 176-178, 182, 212, 221-222
Soil Organic Matter, 17-18, 49, 56, 64, 74, 78, 84, 88-89, 177, 181, 196, 198, 220
Soil Pollution, 1, 11, 15, 23, 58, 60, 137, 150, 152-153
Soil Quality Standards, 42-45, 55-56
Soil Remediation, 141, 150-151, 172, 180, 182, 184, 187, 195, 228-229
Soil Salinization, 155-156, 166, 169-170
Soil Sickness, 1, 8-10, 17-18
Strain Composition, 21
Substrates, 1-2, 5-7, 17, 100, 203
Surfactant, 138-140, 142, 148-149, 211-212, 219-231, 236, 238

T
Trace Elements, 13, 15, 57-58, 64, 67, 102, 149, 175, 179, 185, 190, 246
Treated Wastewater, 63, 124, 134

V
Vadose Zone, 147, 211-215, 228
Volcanic Eruptions, 103

W
Wastewater, 14, 57, 63-64, 81, 122-128, 130, 132-135, 176, 195, 207, 247
Water Quality, 31, 83, 98-100, 122-123, 133, 171